# Current and Future Developments in Nanomaterials and Carbon Nanotubes

*(Volume 1)*

*(Introduction to Carbon Nanomaterials)*

Edited by

**Jagriti Narang**

*Amity Institute of Nanotechnology, Amity University, Noida, UP, India*

&

**Chandra Shekhar Pundir**

*Department of Biochemistry, M.D. University, Rohtak, Haryana, India*

**Current and Future Developments in Nanomaterials and Carbon Nanotubes**

*Volume # 1*

*Introduction to Carbon Nanomaterials*

Editors: Jagriti Narang & Chandra Shekhar Pundir

ISSN (Online): 2589-2207

ISSN (Print): 2589-2193

ISBN (Online): 978-1-68108-595-1

ISBN (Print): 978-1-68108-596-8

## General:

1. Any dispute or claim arising out of or in connection with this License Agreement or the Work (including non-contractual disputes or claims) will be governed by and construed in accordance with the laws of the U.A.E. as applied in the Emirate of Dubai. Each party agrees that the courts of the Emirate of Dubai shall have exclusive jurisdiction to settle any dispute or claim arising out of or in connection with this License Agreement or the Work (including non-contractual disputes or claims).
2. Your rights under this License Agreement will automatically terminate without notice and without the need for a court order if at any point you breach any terms of this License Agreement. In no event will any delay or failure by Bentham Science Publishers in enforcing your compliance with this License Agreement constitute a waiver of any of its rights.
3. You acknowledge that you have read this License Agreement, and agree to be bound by its terms and conditions. To the extent that any other terms and conditions presented on any website of Bentham Science Publishers conflict with, or are inconsistent with, the terms and conditions set out in this License Agreement, you acknowledge that the terms and conditions set out in this License Agreement shall prevail.

**Bentham Science Publishers Ltd.**
Executive Suite Y - 2
PO Box 7917, Saif Zone
Sharjah, U.A.E.
Email: subscriptions@benthamscience.org

BENTHAM SCIENCE

# CONTENTS

# FOREWORD

*"Everything we see around us is made of atoms, the tiny elemental building blocks of matter. From stone, to copper, to bronze, iron, steel, and now silicon, the major technological ages of humankind have been defined by what these atoms can do in huge aggregates, trillions upon trillions of atoms at a time, molded, shaped, and refined as macroscopic objects. By then nanotechnology will have given us specially engineered drugs which are nanoscale cancerseeking missiles, a molecular technology that specifically targets just the mutant cancer cells in the human body, and leaves everything else blissfully alone."*

Richard Smalley spoke these words on 1999 June 22$^{th}$ in a seminar on nanotechnology at Rice University, in which he showed the potential of this new field of research. The term nanotechnology was first introduced during the famous lecture entitled "There's Plenty of Room at the Bottom," by the physicist Richard Feynman held in 1959 in which he considered the possibility to manipulate individual atoms through an organic synthesis more effectively than the methods used up to that point.

So, nanotechnology was discovered recently in the world of science but in nature there are lot of nano-products: volcanoes for example produces nanoparticles, geko's feet have nanostructures that allow them to climb walls as do the leaves of lotus which are highly water repellent.

Matter changes properties gradually, in other words at the nanoscale and this is how we can create new materials and new technological devices. Without realising it, the master glassmaker of the medieval time used these properties. They could stain red glass windows using gold which is yellow. Today is a different story: we can now control the matter at the nanoscale which is why we call them nanotechnologies. We can make carbon nanotubes that are a million times smaller than the head of a pin and hundred times stronger than steel. Actually, nanotechnology is already a part of our life. It is used to make lots of different things: sun creams, hard disks, medicines and even socks that don't smell.

In the world of nanometric carbon the story begins in 1985, when, in his laboratory at Rice University in Houston, Rick Smalley was working on an experiment on carbon aggregates, small particles made by a handful of atoms. He had surely many thoughts on his mind, but he certainly had no idea that he was going to make one of the most fascinating discoveries of the twentieth century; a discovery that ten years later earned him the Nobel Prize for Chemistry. A discovery that would open the way for what we now call nanotechnology.

It was the fullerene, aggregate made of 60 carbon atoms with a weird shape, halfway between the flexible macroscopic world and the rigid quantum world.

Knowledge advances laboriously, bit by bit, like a mountain stream making its way among the rocks, so its beauty and contribution to humanity can be appreciated only on the long term. But sometimes the stream leaps and forms a beautiful waterfall, and we stop, look, take pictures. Sometimes a scientific discovery can be elegant, beautiful.

During the synthesis of fullerenes in 1991, Iijima, observed the presence of secondary products. These derivatives, the carbon nano tubes (CNT), are considered allotropic forms of carbon and have found significant use in the field of materials thanks to their mechanical and electronic properties but also in the biological field for their ability to easily cross the cell membrane in accordance with a mechanism not yet perfectly identified and act as nano-syringe for the introduction of therapeutic molecules, drugs, proteins, genes and antigens within cells.

> *"Even in our vaunted microelectronics of 1999, in our highest-tech silicon computer chip the smallest feature is a mountain compared to the size of a single atom. The resultant technology of our 20th century is fantastic, but it pales when compared to what will be possible when we learn to build things at the ultimate level of control, one atom at a time. My own research these days is focused on carbon nanotubes an outgrowth of the research that led to the Nobel Prize a few years ago. These nanotubes are incredible. They are expected to produce fibers 100 times stronger than steel at only 1/6th the weight almost certainly the strongest fibers that will ever be made out of anything strong enough, even, to build an elevator to space. In addition they will conduct electricity better than copper, and transmit heat better than diamond."*

Again, prophetic words by Richard Smaley.

Carbon nanotubes are nanometric particles capable of combining with various materials giving molecularly doped materials with enhanced physical, mechanical and conductive properties. The fact that nanotubes solely consist of carbon atoms bonded together to form a rigid structure gives the material excellent physical and mechanical properties: the bond between carbon atoms is the most strong link and, consequently, a suitable structure containing this kind of bonds acquires a very high resistance to deformations. A synthetic fiber formed from carbon nanotubes would be the most resistant ever made: it was calculated that a nanotube would have a specific tensile strength at least 100 times greater than steel. It 'should also be noted that the nanotubes also show a considerable flexibility, being able to be folded up to about 90 ° without breakage or damage.

Depending on the type of nanotube (chiral, zigzag, armchair) its electronic structure gives particular conductive properties that can also be transferred to the material that will accommodate these nanoparticles inserted in an appropriate manner.

Today it is possible to insert these reinforcing nanostructures within various polymeric materials, allowing a remarkable improvement of the mechanical properties. The introduction of carbon nanotubes into epoxy resins allows us to make composite materials " doped " from the higher mechanical characteristics while maintaining the physical ones, *e.g.* lightness

The main obstacle linked to the CNT is their insolubility and for any application is required a chemical functionalization with the aim of making them soluble in common organic solvents. In addition to the achievement of a soluble product, it is important to also use nanotubes with a high degree of purity to limit the toxicity, incompatibility with the intrinsic biological systems to impurities.

The following chapters represent the state of the art on CNT, starting from their synthesis functionalisation and purification to their physico-chemical properties as well as electrical, optical, mechanical and thermal properties. Applications are also extensively presented and critically discussed in the fields of sensing, biosensing and biomedicine.

**Roberto Pilloton**
Consiglio Nazionale delle Ricerche
Istituto sull'Inquinamento Atmosferico
Via Salaria Km 29,3 - 00015, Rome
Italy

# PREFACE

The proposed book focuses on the new technologies that are to introduce carbon nanotubes and the methods used to prepare carbon nanotubes are also discussed in detail. This field is very progressing at exponential rate, with a wide variety of research articles and book chapters appearing in the literature every year. This book can be a great resource for budding nanotechnology researchers, undergraduates and post-graduates students who are interested to peruse science, and a career in carbon nanomaterials research. It is one of the most investigative material in the history of nanoscience and mainly responsible for current nanotechnology boom. This book also emphasizes the mechanism proposed for the development of carbon nanotubes. This book is also focused on the unique electrical, optical, mechanical, thermal and vibrational properties of carbon nanotubes. This book also discusses the property changes observed in CNTs due to their structural differences as in single walled and multiwalled CNTs. This book also discusses the methods for synthesis of such unique nano-composite structures. The various aspects of how and where these structures can be applied with enhanced properties are also discussed with evidences from published literatures. This book highlights the significance of CNTs as carrier for delivering wide variety of payloads including drugs, small organic molecules, oligonucleotides, proteins, siRNA, vaccines and nutrients. This book also emphasizes on effects of CNT on cell viability, proliferation, reactive oxygen species (ROS) producing ability, genotoxicity, extra cellular matrix remodelling/tissue remodelling and mutagenicity. Finally it is mentioned about the toxicological considerations of CNT and discusses different ways of minimizing the toxicity.

**Jagriti Narang**
Amity Institute of Nanotechnology
Amity University, Noida, UP
India

# List of Contributors


| | |
|---|---|
| **Ashish Mathur** | Amity Institute of Nanotechnology, Amity University, Noida UP, India |
| **Bhawana Sharma** | Department of Biochemistry, Maharshi Dayanand University, Rohtak, Haryana, India |
| **C.S. Pundir** | Department of Biochemistry, M.D. University, Rohtak, Haryana, India |
| **Jagjiwan Mittal** | Amity Institute of Nanotechnology, Amity University, Noida UP, India |
| **Jagriti Narang** | Amity Institute of Nanotechnology, Amity University, Noida UP, India |
| **Mohammad Faiyaz Anwar** | National Post-Doctoral Fellow, Renal Pathology Lab, Department of Pathology, AIIMS, New Delhi, India |
| **Neelam Yadav** | Centre for Biotechnology, Maharshi Dayanand University Rohtak, Rohtak, Haryana, India |
| **Rajesh Dabur** | Department of Biochemistry, Maharshi Dayanand University, Rohtak, Haryana, India |
| **Ranu Nayak** | Amity Institute of Nanotechnology, Amity University, Noida UP, India |
| **Shikha Wadhwa** | Amity Institute of Nanotechnology, Amity University, Noida UP, India |
| **Sruti Chattopadhyay** | Indian Institute of Technology Delhi (IITD), Haus Khas, New Delhi, India |
| **Suman Lata** | Department of Bio & Nano Technology, Guru Jambheshwar University of Science and Technology, Hisar, Haryana, India |
| **Susanta Sinha Roy** | Department of Physics, Shiv Nadar University, Greater Noida UP, India |
| **Swati Jain** | Amity Institute of Nanotechnology, Amity University, Noida UP, India |
| **Tulika Dahiya** | Shoolni University, Solan, HP, India |

CHAPTER 1

# Say Hello to Carbon Nanotubes

**Ashish Mathur**[1,*], **Shikha Wadhwa**[1] and **Susanta Sinha Roy**[2]

[1] *Amity Institute of Nanotechnology, Sec 125, Amity University, Noida UP, India*

[2] *Department of Physics, Shiv Nadar University, Greater Noida UP, India*

**Abstract:** In this chapter, an attempt is made to introduce carbon nanotubes and the science used to investigate them. This field is progressing at an exponential rate, with a wide variety of research articles and book chapters appearing in the literature every year. Research in this direction is now meeting the industry standards and some promising devices are ready to enter into the market in near future. This chapter can be a great resource for anyone new to carbon nanotube research. It can also introduce the experienced researcher to subjects outside his or her area of study. This chapter can be useful to the undergraduates and post-graduate students who are interested to pursue science and a career in carbon nanomaterials research.

**Keywords:** Application of carbon nanotubes, Carbon nanotubes, Functionalization of carbon nanotubes, Structure of carbon nanotubes.

## 1.1. INTRODUCTION

## 1.2. BRIEF IDEA ABOUT CNT

The discovery of Carbon Nanotubes (CNTs) by Ijima in 1991 evoked a great deal of interest among scientific community right from academia to industry. This new class of material exhibits unique nanoscale properties useful in every field ranging from condensed matter physics to chemistry [1]. With exceptional thermal (thermal stability upto 2800$^0$C in vacuum), electric (about 1000 times higher electric-current-carrying capacity than copper wires) and mechanical properties (about 100 times stronger than steel). CNTs find application in electronic devices such as flat panel displays, field-effect and single electron transistors, rectifying diodes and to improve mechanical properties of composites [2, 3]

* **Corresponding author Ashish Mathur:** Amity Institute of Nanotechnology, Amity University, Noida UP, India; Tel: 09711202697; E-mail: amathur@amity.edu

**Jagriti Narang & Chandra Shekhar Pundir (Eds.)**

## 1.3. STRUCTURE OF CARBON NANOTUBES

One or more graphene sheets roll up seamlessly to form nanotubular structures called carbon nanotubes (single walled (SWCNT) or multi walled (MWCNT)) (Fig. **1**). MWCNT contains few to few tens of concentric graphene cylinders. Arrangement of carbon hexagons in a concentric manner along the tube length with tube ends capped by fullerene like pentagon structures form elongated nanotubes. The carbon atoms are $sp^2$ hybridized to form graphite sheet structure which rolls to form tubules unlike diamond where all carbon atoms are $sp^3$ bonded [4, 5].

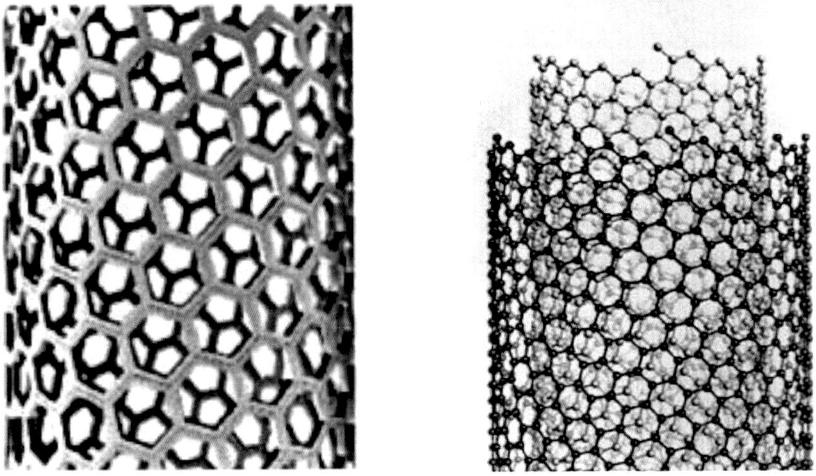

**Fig. (1).** SWCNT and MWCNT [5, 6].

The atomic arrangement upon rolling up of graphene sheets, length and diameter of nanotubes, and nanostructure determines the properties of nanotubes [2]. Briefly, the atomic arrangement in a single walled CNT is of three types: (A) arm chair (B) zig-zag (C) chiral or helical. These three types differ from each other in the sheet direction about which graphene sheet is rolled up to form tubules [7]. The atomic arrangement in carbon nanotubes is shown in Fig. (**2**).

The atomic configurations have implications on material properties. CNTs can be metallic or semi-conducting depending upon the atomic arrangement in a single tube [2]. The coupling between tubes in MWCNTs is weak, thereby causing similar electronic properties of perfect MWCNTs and SWCNTs [7].

A new class of CNTs is double walled carbon nanotubes (DWCNTs) that offer versatility to CNTs for new applications. In a recent study, it was shown that DWCNTs possess merits of both SWCNTs and

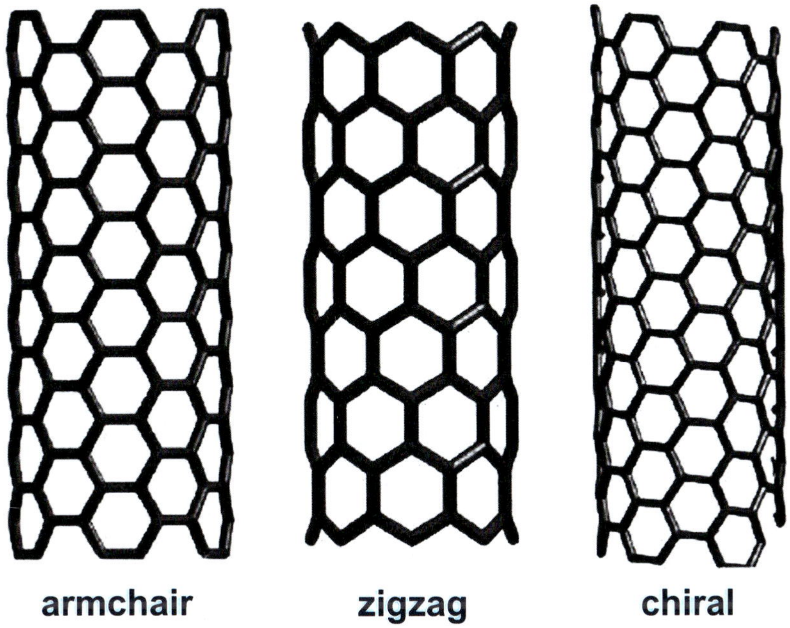


**Fig. (2).** Schematic of arm chair, zig-zag and chiral or helical SWCNT structures [8].

MWCNTs *i.e.* electron emission of DWCNTs has threshold voltage as low as that of SWCNTs and the lifetime (long term emission stability) of DWCNTs is as exceptional as that of MWCNTs [9].

## 1.4. PROPERTIES OF CARBON NANOTUBES

### 1.4.1. Electrical Properties

Electrical transport in nanotubes has attracted significant attention due to its key role in electronic devices. CNTs are virtually perfect 1D conductors. Various mesoscopic phenomena such as single-electron charging, resonant tunneling through discrete energy levels and proximity-induced superconductivity have been observed at low temperatures [3].

MWCNTs have complex structure where every carbon shell has different electronic character, chirality and shell-shell interactions. However, electrical transport is dominated by outer-shell conduction at low bias and temperatures and when MWCNTs are side-bonded to metallic electrodes.

The coupling between the cylinders is weak in MWCNTs, thereby causing electronic properties of perfect MWCNTs to be comparable to those of perfect SWCNTs. The electronic transport in metallic SWCNTs and MWCNTs occur ballistically due to one-dimensional electronic structure [7].

### 1.4.2. Mechanical Properties

Carbon-carbon $sp^2$ bonding offers stiffness and axial strength to the nanotube structure. MWCNTs were found to possess an estimated Young's modulus of 1.26 TPa. Buckling, plastic deformation or fracture of the nanotube results from large displacements, thereby establishing its strength. For length of 1 mm, nanotubes were found to buckle elastically at large deflection angles of $\sim 10^0$. Nanotubes can be distorted without damage for small displacements. The elastic energy stored by the material before failure gives an estimate of toughness. The toughness estimated was 100 keV for a 30 nm diameter nanotube which is about an order of magnitude larger than the strain energy stored in SiC nanorods. The property of nanotubes to elastically sustain loads at large deflection angles facilitate storage or absorbtion of substantial energy [3].

### 1.4.3. Thermal Properties

Carbon nanotubes are thermally stable up to $2800^0$C in vacuum and possess thermal conductivity about twice as high as diamond [2]. The temperature dependence of the thermal conductivity as shown in Fig. (**3**) demonstrates different behaviour than bulk measurements and becomes similar to the latter with increase in MWCNT diameter [3].

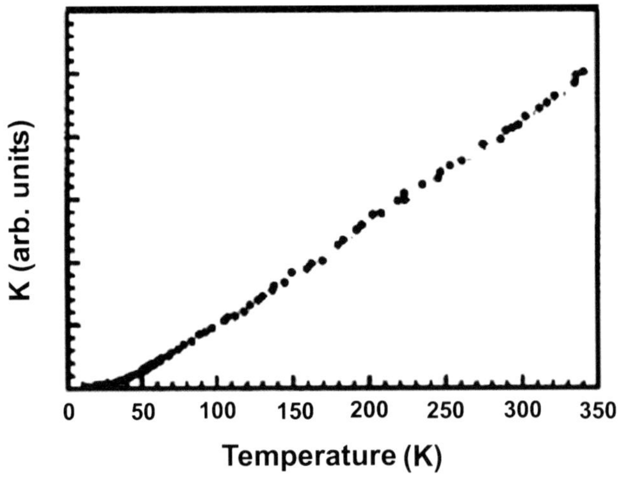


**Fig. (3).** Temperature dependence of the thermal conductivity of bundles of SWNTs [10].

Table **1** summarises the electrical, mechanical and thermal properties of carbon materials.

**Table 1. Electrical, Mechanical and thermal properties of carbon materials.**

| S. No | Material | Band gap (eV) | Young's Modulus (Y) (TPa) | Thermal conductivity (W/m K) | References |
|---|---|---|---|---|---|
| 1. | Diamond | 5.45 | ~ 1 | 2000 | [11] |
| 2. | Graphite | 0.04 | 1 | 3000 | [12] |
| 3. | SWCNT | 1.8-0.18 (semiconducting) | 0.32–1.47 | 200 | [10, 13, 14] |
| 5. | MWCNT | 0.2-2.4 | 0.27–1.26 | >3000 | [4, 7, 14] |

## 1.5. APPLICATIONS OF CNTS

Numerous potential applications of carbon nanotubes including nanocomposites, energy storage and energy conversion devices, water decontamination, sensors, flat panel displays and field emission displays, separation membranes, hydrogen storage, nanometer-sized semiconductor devices, probes and drug-delivery systems have been realised in research laboratories [7, 15, 16]. Ample published reports and research papers exist detailing applications of CNTs. Many CNT applications have already been employed commercially and are now realised in products (Table **2**). However, there are few but noteworthy limitations which restrict the transformation of technologies from bench-top testing to commercial product. These are listed below [7, 17]:

1. Major concentrations of impurities consisting of carbon-coated metal catalyst and other forms of carbon formed during synthesis contaminate the nanotubes.
2. Acid treatment is generally used to remove these impurities which in turn introduce other impurities. This can lead to imperfections in the structure and degradation of nanotube length adding further to nanotube cost.
3. For electronic devices, semiconducting nanotubes are required. However, common synthetic routes lead to a mixture of semiconducting and metallic nanotubes. This raises cost further as metallic SWCNTs needs to be selectively destroyed by electrical heating leaving only semiconducting nanotubes.
4. A single synthetic route to large quantities of SWCNTs of one type is yet unknown.
5. Hydrophobic nature of CNTs limits its application which necessitates structural modification.
6. Numerous reports suggest that the exposure of CNTs and its modified forms pose several health and environmental issues.

**Table 2. Commercial applications of CNTs and modified forms.**

| Nanomaterial | Properties | Commercial Applications | Industry/ Company | References |
|---|---|---|---|---|
| SWCNTs | Ballistic electronic conduction | Electronic devices such as field-effect transistors (FET) and interconnects | IBM, Intel, Unidym | [16, 18] |
| CNTs | Ballistic electronic conduction; high thermal conductivity | CNT ink-electronic devices including displays, solar panels and sensors | Linde electronics (Portugese); Unidym caron for electronics (American); Nanomic Inc.; Samsung (Japan); Nano-Proprietary Inc. (USA) | [18] |
| CNTs | High tensile strength and strong resistance to breakage | Flywheels, bridges | CNano Technology Ltd. (USA) | [20] |
| CNTs | Resistive to high temperatures | Fire resistive products | CNano Technology Ltd. (USA) | [20] |
| CNTs | Less brittle and stronger | Brushes for electric motors | CNano Technology Ltd. (USA) | [20] |
| CNTs | Conductive fuel | Automotive industry | CNano Technology Ltd. (USA) | [21] |
| CNT composites | High electrical and thermal conductivity | CNT yarn, tape and sheet | Nanocomp technologies | [22] |
| CNT composites | High thermal conductivity, excellent mechanical properties | conductive plastics, high-performance fibres, and advanced composite bulk-structural materials | Carbon Nanotechnologies, Inc. (CNI) | [23] |
| CNTs | The mechanical robustness, low buckling force of nanotubes, improved resolution | Scanning probe tips | Seiko Instruments (Chiba City, Japan) | [18] |
| CNTs | extremely small, sensitive, and low in power consumption | Leak detection sensors | Nanomix | [18] |
| MWCNT blended plastics | Improved mechanical properties | Sporting goods | Zyvex Corp. (USA); Mistui Corporation, Toray) | [16] |

*(Table 2) contd.....*

| Nanomaterial | Properties | Commercial Applications | Industry/ Company | References |
|---|---|---|---|---|
| SWCNT | Large field amplification factor | Light emitters | IBM (USA) | [24] |
| CNT composites | Ultra high strength, low weight nanomaterial | Aerospace application | Zyvex Corp.; NASA (USA) | [24] |
| SWCNT/MWCNT | Enhancement in the performance | Absorber coatings for thermal detectors | National Institue of Standards and Technology (USA) | [24] |
| CNTs | High electrical conductivity | Li ion batteries | CNano Technology Ltd. (USA) | [25] |

## 1.6. DEVELOPMENTS IN CARBON NANOTUBE TECHNOLOGY

One of the recent works indicated the discovery of an unusual actuator effect of CNTs *i.e.* a change in the number of electrons on a tube leads to increase in their lengths. This property was found to be useful in manufacturing artificial muscles [18].

Another discovery in the application of CNTs includes a novel new X-ray machine developed by the University of North Carolina at Chapel Hill. A high temperature is not required in this machine to generate the high-energy electrons for producing X-rays. A thin layer of CNTs is used to operate at room temperature rather than the standard metal filaments which are heated inside a vacuum chamber. The use of CNTs maximise the lifetime of the new devices. Portable X-ray machines to be used in ambulances, airport security, and customs operations could be developed as CNT based devices are small and operate at room temperature. Marketing of X-ray machines based on CNTs within next two years is foreseen [18].

Modification of CNTs can be achieved by functionalisation allowing attachment of specific functional groups to their sidewalls or ends through covalent or non-covalent bonding. Functionalisation improves water solubility and biocompatibility of CNTs and selective binding to bio-targets to broaden its application in other domains. Appropriate functionalisation renders dispersivity enhancing surface contact with biological adsorbates [15, 19].

CNT/ polymer composites offer unique mechanical, surface and multi-functional properties. CNTs form strong interactions with the polymer matrix due to nano-scale structure and enormous interfacial area. However, defect sites are generated in composites as CNTs can easily agglomerate, bundle together and entangle. This limits the efficiency of CNTs on polymer matrices [12].

Enhanced dispersion and alignment of CNTs in a polymer matrix is required to enhance mechanical, electric, thermal, electrochemical, optical and super-hydrophobic properties of CNT/ polymer composites. However, this remains a real challenge. Therefore, work is in progress in the direction of improving dispersion and alignment [12].

Filled nanotubes (SiC, SiO, BN, C), metal decorated CNTs, metal filled nanotubes, and fullerene incorporated nanotubes have been developed for different applications [4 and references therein]. Table **2** summarises applications of CNTs, its modifications and commercial products.

## 1.7. TYPES OF CNTS

### 1.7.1. Single Walled CNTs (SWCNTs)

Single-walled carbon nanotube (SWCNT) is made up of a single graphite sheet rolled up seamlessly into a cylindrical tube [7]. The honeycomb-ring structure of SWCNT walls has been revealed by scanning tunnelling microscopy (Fig. **4a**). Fig. (**4b**) shows honeycomb lattice for a SWCNT with a C–C spacing of 0.14 ± 0.02 nm.

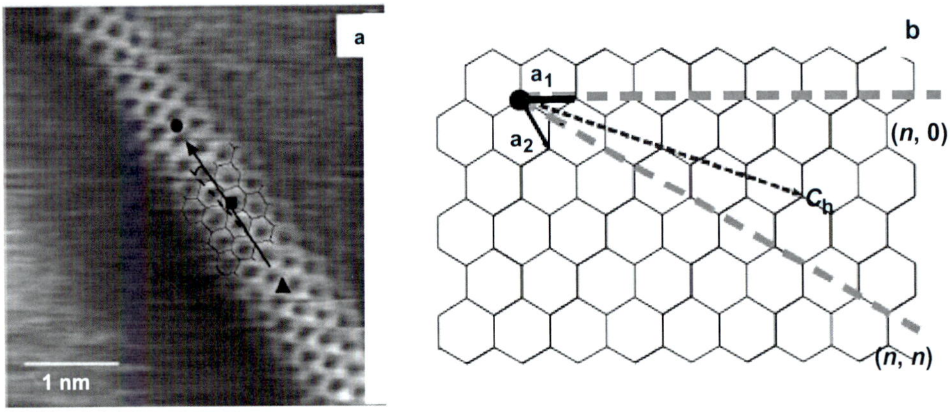


**Fig. (4).** (a) Honeycomb-ring structure of SWCNT as revealed by STM (b) schematic of 2-D grapheme sheet showing lattice vectors $a_1$ and $a_2$ and roll up vector $c_h = na_1 + ma_2$ [26].

The atomic structure of SWCNT has implications on its electronic properties. SWCNTs can be metallic or semi-conducting depending upon the rolling axis of graphene sheet to form nanotube [26]. Due to 1D structure, electronic transport in metallic SWCNTs occur ballistically over long nanotube lengths, enabling them to carry high currents without heating [7]. Semi-conducting nanotubes have bandgaps that scale inversely with diameter, ranging from approximately 1.8 eV for very small diameter tubes to 0.18 eV for the widest possible stable SWNT (refer to Table **3**) [27].

**Table 3. Atomic configuration and related electronic properties.**

| S.No | Atomic configuration | Roll up vectors | Electronic property | References |
|---|---|---|---|---|
| 1. | Zig-zag $(n,0)$ | $n/3$ = integer | Metallic | [7, 26] |
| | | $n/3 \neq$ integer | Semi-conductor | |
| 2. | Chiral $(n,m)$ | $(2n+m)/3$= integer | Metallic | [7, 26] |
| | | $(2n+m)/3 \neq$integer | Semi-conductor | |
| 3. | Armchair $(n,n)$; $C_h$ rotates $30^0$ relative to $(n,m)$ | $n = m$ | Metallic | [7, 26] |

Phonons propagate easily along the nanotube: Thermal conductivity is dominated by phonons at all temperatures [10]. Aligned bundles of SWNTs show a thermal conductivity of > 200W/mK at room temperature [28]. Superconductivity has also been observed, but only at low temperatures [7].

### 1.7.2. Double Walled CNTs (DWCNTs)

Double-walled carbon nanotubes (DWCNTs) are made up of two coaxial single-walled carbon nanotubes (Fig. **5**). A unique form of CNTs, benefits from both SWCNT and MWCNT characteristics, *i.e.* displaying the flexibility of SWCNTs and the electrical and thermal stability of MWCNTs. The most interesting feature of DWCNTs is the possibility of functionalising the outer wall. This allows connections with the external environment, while the mechanical and electronic properties of the inner tube are retained [29, 30].

The applications such as field-emission displays (FEDs), supertough fibres and field-effect transistors takes advantage of extraordinary physical and field emission properties of DWCNTs. The most significant application of DWCNTs is as electron emitters, because they have the advantages of both SWCNT and MWCNT emitters, that is, low threshold voltage and high durability [30].

The fact that the DWNT is unique, is always confirmed by electron diffraction. In all cases, the current-voltage (I~V) curves are linear between 0-5 V, giving us assurance that contacts are Ohmic [32].

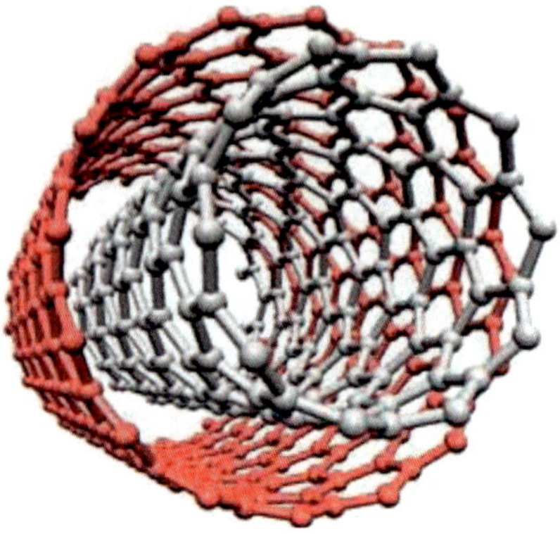

**Fig. (5).** DWCNT explicitly showing a tube within a tube [31].

### 1.7.3. Multi Walled CNTs (MWCNTs)

MWCNTs, first discovered in 1991 by Iijima [6], are composed of few to few tens of concentric cylinders placed around a common central hollow, with the interlayer spacing close to that of graphite (0.34 nm). The diameter varies from 0.4 - few nm and their outer diameter ranges from 2 - 30 nm [4].

The surface of MWCNTs can be manipulated in various ways by functionalisation which offers plethora of applications. The electrical conductivity (as conductive as copper), mechanical properties (15-20 times stronger and 5 times lighter than steel), and thermal conductivity (same as that of diamond; > 5 times that of copper) makes this nanomaterial particularly interesting for industrial applications (Fig. **6**) [33].

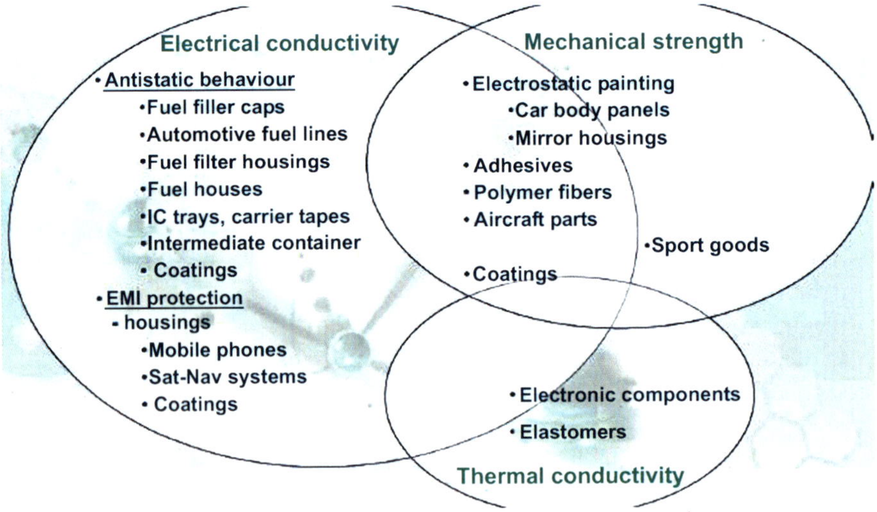


**Fig. (6).** Properties of MWCNTs leading to number of applications [33].

## 1.8. LIMITATIONS OF CNT TECHNOLOGY

Over the last few years, there have been tremendous achievements in CNT research for different applications. However, more efforts are needed for its commercialisation. Major setbacks in the realisation of CNTs based commercial products are listed below [34]:

1. Homogeneity of the device – Perfectly aligned CNTs with high density, same chirality, diameters and lengths assists homogeneity in the device functioning.
2. Specificity of the device – Non - specific binding of analytes other than those of interest creates non-specificity and complexity in real time measurements.
3. Control of morphological features – Device fabrication and its realisation into a commercial product demands highly controlled morphological features. Therefore, controlled and reproducible synthesis and modification methods are required to control diameter, lengths and chirality of nanotubes.
4. Cost issues – synthesis of pure and perfectly aligned CNTs, further purification methods, functionalisation to control other chemical features and integration into a device contribute to raising the overall cost of device fabrications.
5. Potential risks associated with CNTs – Being a new class of material with unique properties in nanoscale poses unknown toxicological hazards. Moreover, conducting such studies is bottlenecked by the small size and tendency of CNTs to aggregate and/or agglomerate in suspensions.

## 1.9. CURRENT DEVELOPMENTS

Carbon nanotubes (CNTs) have already found potential applications in many industries right from defence to electronics to environmental remediation (Table **4**). CNTs possess many desirable mechanical and chemical properties, which have literally overtaken many of the existing advanced materials. CNTs also possess excellent electronic properties which can potentially result in a quantum leap in the electronics industry.

**Table 4. Summary of application areas of CNTs/modified CNTs with their key attributes.**

| S.No | Nanomaterial/ composite | Key attributes | Application | References |
|---|---|---|---|---|
| 1. | SWCNT-lipid bilayer | Enhanced adsorption of bio-molecules at the surface, high sensitivity; biocompatability | Biosensing | [35] |
| 2. | CNT-single stranded DNA | Enhanced adsorption of bio-molecules at the surface, high sensitivity; biocompatibility | Biosensing | [19] |

*(Table 4) contd.....*

| S.No | Nanomaterial/ composite | Key attributes | Application | References |
|---|---|---|---|---|
| 2. | CNT- fluorescein isothiocyanate | - | Drug delivery | [36] |
| 4. | CNT- streptavidin. | Passive diffusion across cell membrane or endocytosis | Drug delivery | [36] |
| 5. | PLPEG-FA/SWNT | The ability of CNT to adsorb near-infrared Irradiation | Cancer treatment | [36] |
| 6. | MWCNTs | Schottky behaviour; High sensitivity in the presence of hydrogen. | Gas sensing | [37] |
| 7. | CNTs | Ballistic electronic conduction; nanodimensions | FETs and interconnects | [16] |
| 8. | CNTs | High electrical conductivity; good chemical stability | Supercapacitors, Li-ion batteries, solar cells and fuel cells | [16] |
| 9. | Nitrogen-doped CNTs and nanofibres | High electrical conductivity; good chemical stability | Lithium-ion batteries (LIBs) | [16] |
| 10. | CNTs | At given temperature, the amount of gas adsorbed or desorbed as a function of pressure. | Hydrogen storage | [38] |

CNTs and their modified forms have been researched over the recent years, especially in the field of bio and chemical sensing. This is due to the size similarity of nanotubes with the analytes such as biospecies enabling strong interactions between them. CNT based biosensors prove to be an ideal one due to the $sp^2$ hybridization of CNTs and the exceptional electronic properties of the nanotubes, amalgamated with their specificity for the immobilized system. Since the electrical properties change on exposure of CNTs with gas molecules such as $O_2$, $NH_3$, $O_3$, *etc.*, CNTs can be used as gas sensors. Enzyme based bio-affinity electrical sensing of proteins and DNA can also be amplified remarkably by using CNTs.

However, the lack of control in synthesis of specific chirality, diameter and length of CNTs limit the commercialization of these potential applications as these key parameters influence the device performance [19, 34].

## 1.10. PROCESSING OF CNTS

### 1.10.1. Current Status of Development and Commercialisation

CNT technology has reached far in terms of developments in synthesis techniques, modification methods and commercial applications, yet there are many challenges which restrict further advancement in CNT technology. One of the challenges is to obtain one type of high purity CNTs for application in electronic devices. Recent development shows that there are methods to separate high-purity SWNT powders according to chirality such as density gradient centrifugation in combination with selective surfactant wrapping or by gel chromatography [39 - 42].

Nanoscopic "peapods," have been produced recently, consisting of fullerenes nested within nanotubes. These are similar to peas in a pod. The uniqueness about this structure is that the location of fullerenes along the tube determines the electronic properties. This implies that the electronic properties are tunable and therefore could have extensive implications on the fabrication of single-molecule based devices [18].

Hierarchical fibre composites called "fuzzy" fibres have been synthesised by growing aligned CNT forests onto glass, SiC, alumina, and carbon fibres. Practical applications may include lightning-strike protection, de-icing, and structural health monitoring for aircraft [43].

CNT yarns and laminated sheets formed by direct CVD or forest spinning or drawing methods may find their potential market for high-end uses in comparison to carbon fibre in near future, especially in the areas where weight-sensitive applications require combined electrical and mechanical functionality. Recent scientific reports demonstrated that yarns composed of high-quality few-walled CNTs have successfully reached a stiffness of 357 GPa and a strength of 8.8 GPa. However, this is only for a gauge length which is comparable to the millimeter-long CNTs within the yarn [43, 44].

Further, forest-drawn CNT sheets have been coated with functional powder which has offered weavable, braidable, and sewable yarns containing up to 95 wt % powder. These interesting materials find their application as superconducting wires, in battery and fuel cell electrodes, and self-cleaning textiles [43, 45].

## 1.10.2. Current Market for CNT Synthesis

There are various industries which are involved in the manufacturing of CNT for commercial purposes as shown in Table. **5**.

**Table 5. A list of Industries/companies involved in CNT synthesis on large scale.**

| CNT type | Synthesis technique | Industries/companies |
|---|---|---|
| MWCNT | CVD | Hyperion Catalysis International, Inc. (Cambridge, MA), |

*(Table 5) contd.....*

| CNT type | Synthesis technique | Industries/companies |
|---|---|---|
| MWCNT | - | Mitsui Corp. (Japan) |
| SWCNT | High-pressure carbon monoxide (HiPco) flow method | Carbon Nanotechnologies, Inc. (CNI) (US) |
| SWCNT/MWCNT | carbon decomposition and catalyzed chemical vapor deposition technique | NanoDevices, Inc. (Santa Barbara, CA), Samsung (Tokyo, Japan), Nanomix, Seiko Instruments (Chiba city, Japan) |
| MWCNT | CCVD | Cnano Technology (USA) |
| MWCNT | CCVD | Nanocyl, Belgium |
| MWCNT | CCVD | Arkema Inc. (France) |
| MWCNT | CCVD | Showa Denko K.K (Japan) |
| MWCNT | CCVD | Bayer MaterialScience AG (Germany) |

### 1.10.3. Challenges and Future Prospects

All synthesis methods produce CNTs with number of impurities, the nature and quantity of which depends upon the technique used. The impurities include carbonaceous particles such as nanocrystalline graphite, fullerenes, amorphous carbon and metals (typically Fe, Co, Mo or Ni) incorporated during synthesis as catalysts. These impurities cause impediments in achieving desired properties of CNTs and in comprehensive characterisation and applications. Therefore, to develop efficient and simple purification methods remains a fundamental challenge in CNT technology [31]. Further, the cost of high-purity samples of SWNTs is about $750 per gram as reported by Ray Baughman of the University of Texas, Dallas [18].

Another challenge in the processing of nanotubes, despite the emergence of large-scale flow methods such as HiPco, is the strong tendency for nanotubes to agglomerate and their inability to maintain long-term order.

Sorting nanotubes by electrical type creates another big challenge for fuel-cell applications. This may also have an implication on longer-term uses such as biological materials and embedded membranes [18].

Large scale industrial production of high purity CNTs still suffers a major setback and requires further development.

With further development in this area, above stated issues can be addressed and CNTs can have a special place in the market with its mark in almost all potential applications.

## 1.11. APPLICATION OF CNTS IN LIFE SCIENCES AND HEALTH

Nanotubes offer benefits over spherical nanoparticles for some applications. Large inner volumes of nanotubes can be filled with various chemicals and biomolecules with the size ranging from tiny molecules to large proteins. Distinct inner and outer surfaces of certain types of nanotubes can be differentially functionalised with specific drugs internally and to evade immunogenic response externally. Further, drug loading is made simple due to the open-mouthed structure of nanotubes [46]. These properties of nanotubes give way to plethora of biomedical applications discussed here in detail.

### 1.11.1. CNTs in Diagnostics

#### *1.11.1.1. Probing and Imaging*

Small and uniform sizes, high conductivity, high mechanical strength and flexibility of carbon nanotubes may prove to be essential in their use as nanoprobes. Nanoprobes find their application in diverse areas, such as high resolution imaging, nano-lithography, nanoelectrodes, drug delivery, sensors and field emitters. MWNT tips are conducting, with the advantage of being slender and the possibility to image features (such as very small, deep surface cracks). Such features are not possible to probe using larger, blunter etched Si or metal tips. Imaging of biological molecules such as DNA and amyloid-b-protofibrils (related to Alzheimer's disease), with higher resolution is also possible using nanotube tips, compared to conventional methods. Further, the tips are resistant to break down upon contact with the substrate due to the high elasticity of carbon nanotubes [47].

Imaging applications within live cells and tissues are made possible due to excellent optical properties of nanotubes. SWCNTs photoluminescence show exceptional photostability which allows longer excitation time as opposed to organic fluorophores or quantum dots. Also, visibly opaque tissue displays highly attenuated absorption, autofluorescence, and scattering characteristics in the range of 700–1400 nm. The fluorescence profiles of many semiconducting nanotubes overlap with this range, providing their observation in whole blood and thick tissue. Nanotube fluorescence was also used to image SWCNT in tissue sections and measure their concentration in blood [16]. SWNTs have been used as near-infrared fluorescent tags allowing selective probing of cell surface receptors and cell imaging.

#### *1.11.1.2. Biosensing*

The performance of biosensing devices has been found to improve substantially by using CNT-based electrochemical transducers. Alignment of nanotubes is vital as aligned CNT "forests" allow efficient electron transfer between the electrode and the redox centers of enzymes. Huge benefits from the augmented response of the biocatalytic-reaction product have been observed in case of bioaffinity devices utilizing enzyme tags [48].

The integration of carbon nanotubes for biosensing application has great potential, in particular, the high surface area of SWCNTs (~1600 $m^2$ $g^{-1}$) along with retention of electrical conductivity to attain high biomolecule densities is of huge interest. Further, the electrical conductivity is amplified by the ballistic transport of charge carriers through SWCNTs, with high current densities (109 A $cm^{-2}$) in the presence of oxygen. It is not very difficult to disperse and assemble CNTs, screen-print, and potentially inkjet-print them to fabricate miniaturised device configurations with controlled transparency. This is further augmented by the possibility that they can be functionalised covalently and/or non-covalently. These properties allow CNTs to be a vital component of universal sensor platforms, offering amalgamation of exceptional optical and electronic properties with bio-recognition and enzyme catalysis to augment functionality [49].

Recently, a biosensor incorporating CNTs coated with a thin layer of protein-recognising polymer has been produced which can use electrochemical signals to detect minuscule amounts of proteins. This allows the detection of a range of illnesses using a crucial new diagnostic tool [50].

The two types of electrochemical biosensors that incorporate the use of CNT electrodes are [48]:

1. Amperometric (oxidase or dehydrogenase) enzyme electrodes accelerated oxidation of NADH or hydrogen peroxide using CNT molecular wires to achieve efficient electron transfer to enzyme redox centres.
2. Bioaffinity devices CNT support platforms are used for improved detection of the product of the enzyme label or of the target guanine.

Table. **6** shows different CNT based bio-sensing platforms for various disease diagnosis.

**Table 6. The use of CNTs in biosensors for diagnosis of different diseases/disorders.**

| Material | Disease/disorder detected | Type of detection/biosensor | References |
|---|---|---|---|
| CNTs | Lyme disease | Bioaffinity sensor | [51] |
| SWCNTs | Cancer biomarker | Immunosensors | [52, 48] |
| CNTs | Prostate cancer biomarker | Immunosensor | [53 |

*(Table 6) contd.....*

| Material | Disease/disorder detected | Type of detection/biosensor | References |
|---|---|---|---|
| CNTs | Central nervous system disorder; neurodegenerative diseases or brain tumours | Electrode coated with CNTs | [54] |
| SWCNT | Prostate cancer biomarker | Immunosensor | [55] |
| CNTs | Oral cancer biomarker | Immunosensor | [56] |

### *1.11.1.3. Point of Care Devices*

The successful treatment of cancer demands highly sensitive techniques for measuring cancer diagnosis markers at ultra-low levels during very early stages. This requires the methods to facilitate an early detection and sufficient selection of the treatment of such diseases and thereby leading to increased patient survival rates. At present, diagnostic tests used are not sensitive enough and detect protein levels at advanced stages of the disease. Point of care (POC) devices which are smaller, faster, and cheaper are highly desired. POC devices will largely improve the monitoring of the disease progress and patient therapy.

POC cancer diagnostics using SWCNT based devices have been found to be quite useful. Densely packed "forests" of single-walled carbon nanotubes (SWCNTs) with capture antibodies attached to the nanotube ends on a solid surface were used in an immunosensor. Cancer biomarkers bound to their respective antibodies on the SWCNTs surface upon addition of serum from prostate cancer patients to the immunosensor. Biomarkers were then detected electrochemically [56].

Large number of alkaline phosphatase enzymes and secondary antibodies per CNT can be carried on MWCNTs to achieve high detection limit for proteins in buffer. The enzyme reaction product α-naphthol was found to preconcentrate by the use of carbon nanotubes by adsorptive accumulation. There are many Pont of care devices such as:

***Lab-on-a-Chip***

Another dimension of microfluidic/lab-on-a-chip technology is nanotechnology on a chip. It enables biological tests to measure the presence/activity of chosen substances making the process quicker, more sensitive, and flexible when nanoparticles are used as tags or labels. Lab-on- a- chip incorporating nanotechnology is a new concept for total chemical analysis systems. The technology involves micro-total analysis systems that are different from simple biosensors as they conduct an overall analysis [57].

Semiconductors and metals form vital components of electrical detection systems. Carbon nanotubes and semiconductor nanowires, for instance, are being studied widely as sensor components. Fluid systems integrated with mechanical devices are also being explored. Furthermore, porous media can be formed in fluid channels for sample concentration and filtration [58 and references therein].

### 1.11.2. CNTs in Therapeutics

#### *1.11.2.1. Drug Delivery*

The constant advancement in the new delivery vehicles has effectively improved therapeutic effect of drugs. Conventional drug delivery systems include viral vectors, liposomes, cationic lipids, polymers, and nanoparticles. Some safety concerns have been associated with the use of viral vectors, while nonviral suffer from poor penetration of some therapeutic agents into cells. Carbon nanotubes have received great attention as potential drug delivery vehicles due to the following reasons [16, 60]:

1. CNTs are readily internalized by cells.
2. They have shown to exhibit low cytotoxicity over the period of few days, after suitable surface functionalisation.
3. Higher surface area to volume ratio of nanotubes as compared to nanospheres, offers potential to be functionalised with a wide variety of functional agents.
4. They can hold higher loadings of therapeutic agents.
5. More than one functional group can be introduced on the same nanotube, offering attachment of multiple therapeutic agents such as targeting molecules, contrast agents, drugs, or reporter molecules.
6. Functionalised CNTs are non-immunogenic.
7. Suitably functionalised CNTs are highly dispersible in aqueous solutions.
8. They are biocompatible.
9. Stable supramolecular assemblies with nucleic acids can also be formed, paving a way to varied applications such as gene therapy, genetic vaccination and immunopotentiation enhancement.
10. Biologically active moieties can be attached to CNTs, followed by their delivery into the cell cytoplasm or nucleus.

Treatment of cancer and other such complex diseases which require site specific therapeutic activity, can be treated using CNTs. The ability of nanotubes to carry multiple therapeutic moieties or other functional molecules facilitate probing, imaging and targeting at only specific sites. Thus, multimodal options are available for the treatment of such diseases using CNTs / functionalised CNTs [61].

Nanotube spearing using nickel-embedded nanotubes has been discovered recently. The technique involves the delivery of macromolecules immobilized on the nanotubes by penetrating cell membranes. The process is driven by magnetic field strength, which along with other factors such as nanotube speed and duration of spearing can be varied to affect penetration efficiency. Numerous benefits of this approach include decreased cytotoxicity, increased control and more efficient transduction method than standard transfection reagents [46]. In one example, CNT internalised into those cells caused selective cancer cell killing by hyperthermia due to high thermal conductivity of CNTs. In another example, CNT-based vaccine delivery system has been developed which *in vivo* has brought out strong antipeptide - antibody responses in mice. Moreover, CNTs do not produce immunogenic effect as no antibodies were found to be produced against the CNT backbone alone. All these examples suggest that CNTs play an important role in the construction of novel and efficient vaccines [62].

#### *1.11.2.2. Nucleic Acid Delivery*

CNTs are able to form stable supramolecular assemblies with nucleic acids, thus leading a way to variety of applications including gene therapy, genetic vaccination and immunopotentiation development. For nucleic acid delivery to cells, numerous cationic systems are under investigation, for ex. ammonium functionalised CNTs have been tested to form stable complexes with nucleic acid *via* electrostatic interactions. Gene transfer and expression are the main aims of such complexes as it is quite difficult for plasmid DNA alone to penetrate into cells and reach their nucleus. Both single- and multi-walled cationic CNT can form stable complexes with plasmid DNA pCMV-Bgal. After complex formation, apparent effect of functionalised CNTs on the expression of b-galactosidase was observed in gene transfer experiments. Five to six times higher levels of gene expression were obtained as compared to DNA alone. Covalent functionalisation of external walls of nanotubes with polyethyleneimine (PEI) enhanced the efficiency of DNA transfer PEI [60].

#### *1.11.2.3. Protein Transport*

The sidewalls of acid-oxidized single-walled carbon nanotubes can adsorb variety of proteins by non-specific binding to form non-covalent protein-nanotube conjugates. CNTs transport proteins inside various mammalian cells *via* the endocytosis pathway. The internalised protein-nanotube conjugates can penetrate into the cell cytoplasm to perform specific biological functions, after being released from endosome. The proteins that have been transported inside the cell using CNTs include streptavidin (SA), protein A (SpA), bovine serum albumin (BSA) and cytochrome *c* (cyt-*c*). Thus far, it has been reported that relatively pure, well solubilised short carbon nanotubes are nontoxic after internalisation into mammalian cells. Apparently, no adverse effects to cell viability have been observed [63]. SWNTs have been demonstrated as nonviral molecular transporters for the delivery of short interfering RNA (siRNA) into human T cells and primary cells. The ability of CNTs to deliver and RNA interference efficiency are much higher that those of various existing non-viral transfection agents. CNTs were proposed to be used as standard molecular transporters for several types of biologically significant cells, from cancer cells to T cells and primary cells, with additional advantage of improved silencing effects over traditional liposome-based non-viral agents [64].

#### *1.11.2.4. CNTs in Tissue Engineering*

Bone tissue regeneration requires a scaffold that is steadily replaced by growing bone tissue. Such scaffold should possess following properties:

1. Similar mechanical properties as that of bone tissue to be replaced.
2. Excellent biocompatibility with surrounding tissue.
3. Large porosity and pore size.
4. High pore interconnectivity.
5. Biodegradability.

No existing artificial scaffold has met all these requirements yet, which opens a way to the development of novel biomaterials and scaffold fabrication techniques [65]. Properties such as mechanical strength of tissue scaffolding and conductivity can be reinforced by dissolving a tiny proportion of CNTs into a polymer, where CNTs are used as additives. These nanocomposites offer benefits of being injectable, thermally crosslinkable, and cytocompatible *in vitro*. Further, reinforcing effects of SWCNTs can be augmented by their dispersion into polymer, which in turn is improved by surface functionalisation [66].

CNT-based scaffolds have demonstrated positive impacts on cell proliferation and differentiation, directing bone formation. Cell tracking and labeling, sensing cellular behavior, augmenting cellular behavior, and enhancing tissue matrices are other applications of CNTs [65, 66].

#### *1.11.2.5. CNTs Application as Implants*

Carbon nanotubes have been shown to be used as a culture substrate for neural cells. Further, biomodified SWCNTs have been demonstrated to be used as a biocompatible platform for possible neuroprosthetic implants due to their unique properties. This has been validated by the *in vitro* study in which the behaviour, viability and differentiation of NG108-15 neuroblastoma cells developed in the presence of biomodified CNTs was observed over time. More importantly, the biocompatibility of these layer-by-layer (LBL) made SWCNTs was demonstrated.

The important phenotype characteristics such as neutrite outgrowth of the cells grown on this film were retained. Later, it was evidenced that LBL SWCNT films can be utilized to stimulate the neurophysiological activity of NG108-15 cells [64]. More effective deposition of calcium-containing minerals take place in the presence of carbon nanotubes, fibres and nanocomposites of ceramics on the implants. Adsorption and conformation of proteins that mediate specific osteoblast adhesion (such as fibronectin and vitronectin) were found to be improved on nanomaterials such as 3D nanofibrous scaffolds. Nanophase materials were estimated to draw more proteins to their surface as compared with traditional materials, due to their transformed surface energetics, facilitated by a higher exposed surface area and changed electron distribution. The bioactivity of the selected proteins on the nanophase material is altered owing to greater wettability and the presence of surface features close to the size of the proteins. Hence, nanostructured materials may aid bonding between an implant and surrounding bone by enhancing bone cell interactions, which in turn will improve orthopedic implant efficiency [57 and references therein. Sheets of SWNTs have been used to construct electromechanical actuators by applying small voltages (a few volts), to strips of polymer laminated nanotube sheets suspended in an electrolyte. In response, the sheets are bent to large strains, mimicking the actuator mechanism of natural muscles. As a result of this interesting behaviour of nanotube sheets, several applications, such as nanotube-based micro-cantilevers for medical catheter applications and as novel substitutes, especially at higher temperatures, for ferroelectrics emerges [47].

### 1.11.3. Functionalisation of CNTs and Biomedical Applications

Nanotubes also facilitate selective modification through the attachment of functional groups chemically; hence find application as molecular probes, demonstrating applications in chemistry and biology [47]. Carbon nanotubes possess highly hydrophobic surfaces and are insoluble in aqueous solutions. Functionalisation or surface modification renders solubilisation, biocompatibility and low toxic effects of CNTs. Surface functionalisation of carbon nanotubes take place *via* covalent or non-covalent bonding. Covalent functionalisation involves chemical bond formation with nanotube sidewalls *via* chemical reactions while non-covalent functionalisation makes use of hydrophobic interaction with CNT, leading to dissolution of CNTs in aqueous solution [67].

#### *1.11.3.1. Covalent Functionalisation of Carbon Nanotubes*

Some of the examples of covalent functionalisation include oxidation with oxidizing agents such as nitric acid, attaching hydrophilic polymers such as poly(ethylene glycol) (PEG) to oxidized CNTs, yielding CNT-polymer conjugates, cycloadditions by photochemical reaction of CNTs with azides or carbyne. Oxidation introduces carboxyl groups at the ends of tubes and at the defects on the sidewalls. After oxidation, CNTs become soluble in water, however, they tend to aggregate in the presence of salts. This makes them unsuitable for direct use for biological applications. Cylcoaddition reaction occurs on the aromatic sidewalls unlike oxidation reaction taking place at the ends. Introduction of amino-terminated PEG can be used for further conjugation of biological molecules such as peptides or drugs. The intrinsic physical properties of CNTs including photoluminescence and Raman scattering are often destroyed after covalent functionalisation as the nanotube structure is disrupted. This reduces the potential of CNTs for optical applications [67]

### *1.11.3.2. Non-covalent Functionalisation of Carbon Nanotubes*

Coating CNTs with amphiphilic surfactant molecules or polymers lead to non-covalent functionalisation of CNTs. The chemical structure of CNTs is not disrupted, however sonication employed in the functionalisation process causes shortening of length. Unlike covalent functionalisation, the physical properties of CNTs are effectively preserved and aqueous solutions of CNTs are obtained by using non-covalent approach. This enables CNTs promising for multiple biomedical applications including optical applications. Proteins can be immobilized on SWCNTs functionalized by an amine-reactive pyrene derivative. Single-stranded DNA molecules have been widely used to solubilize SWNTs, however, recently it was suggested that DNA functionalisation of SWNTs might not be stable in biological environments containing nucleases. Several amphiphiles have been non-covalently attached to CNTs to render solubility in aqueous solutions. Other traditional surfactants studies include sodium dodecyl sulfate (SDS) and Triton X-100. Amphiphiles requiring relatively high critical micelle concentrations (CMC) to solubilise CNTs are not stable unless excess of surfactant molecules is added in the solution. Such large amounts of surfactants may disrupt biological structures/molecules and are therefore not suitable for their use in biological environments. For biological applications, an ideal coating molecule to be used for non-covalent functionlisation of CNTs should possess following features:

1. It should be biocompatible and nontoxic.
2. It should be stable enough to oppose detachment from the CNT surface in biological solutions.
3. The CMC values of the amphiphilic coating molecules should be very low in order to remove excess coating molecules from the CNT suspension without causing the coating removal from the surface of CNTs.
4. The coating molecules should possess various functional groups which can

bioconjugate with antibodies or other molecules to produce several functional CNT conjugates for diverse biological applications [67].

SWCNTs can be functionalised with biomolecules such as polyethyleneglycol and are then conjugated to antibodies. This SWCNT-antibody conjugate then selectively recognises positive breast cancer cells with small nonspecific binding to other cells such as negative T-cells and Herceptin. Selective SWCNT-antibody binding to cells is imaged by detecting the intrinsic NIR photoluminescence of nanotubes. Hence, SWNTs serve as novel NIR fluorophors and offers specific and sensitive biological detections and imaging. Further, it has been shown that surface functionalisation of nanotubes strongly affects the interactions between carbon nanotubes and living cells [68]. Table **7** enlists therapeutic effect of some of the functionalised CNTs (*f*-CNT).

**Table 7. Therapeutic effect of functionalised CNTs.**

| ***f*-CNT** | **Therapeutic effect** | **References** |
|---|---|---|
| PEGylated SWCNTs | Chemotherapy | [69] |
| Au-SWCNT | Viral disease treatment | [70] |
| DNA-MWCNT or siRNA-MWCNT | Central nervous system disorder/brain tumor treatment | [61, 66] |
| Methotrexate-CNTs | Cancer treatment | [61] |
| B and T cell peptide epitopes-CNTs | Vaccination | [61] |
| Amphotericin B-CNTs | Fungal disease treatment | [61] |
| DNA-CNT | Gene delivery | [61] |
| Phospholipid-PEG-SWCNT-DNA | Gene delivery | [71] |
| Phospholipid-PEG-SWCNT-siRNA | Gene delivery | [71] |
| Ammonium-CNT | Nucleic acid delivery | [60] |
| Polyethyleneimine-MWCNT | Nucleic acid delivery | [60] |

## 1.12. CHALLENGES AND CURRENT DEVELOPMENTS

The use of CNTs in biological systems is impaired by inherent cytotoxicity of CNTs. To explore full potential of CNTs, new strategies will be required to control interaction between CNTs and live cells and to limit their toxicity [72]. Most studies on the application of CNTs in biomedical sciences are based on *in-vitro* experiments and assessments. Therefore, the *in-vivo* fates and potential health risks associated with the exposure of CNTs is still lacking due to the inadequate data and inconsistency in the results published. This issue needs to be addressed by fully studying the nature of CNTs, before its implementation in biological systems without any concern about potential health risks involved [66].

Bio-distribution, translocation and elimination of CNTs after systemic administration have shown advancement recently. The recent development possibly forms a stepping stone towards safe biomedical applications of CNTs. Various published results also demonstrate that appropriate chemical and material modifications of CNTs can reduce toxicity significantly and improve biocompatibility [66].

## 1.13. FUTURE PROSPECTS

The development of a noninvasive approach to release nanotubes from endosomes into cells would greatly advance the application of CNTs in biological systems. A controlled and non-destructive transport of nanotubes inside the cells with specificity to a target area would widen the therapeutic potential for CNTs. The efficacy would be improved further by more efficient mode of unloading the cargo. Further, intensive research to determine the long-term effects of functionalised and/or conjugated nanotubes in-vivo is needed. The issues related to toxicity, interaction with biologic components like blood and tissue, metabolism of CNTs, and the possibility of in-vivo nanotube excretion after the therapeutic agent has been delivered have raised concerns regarding safe use of CNTs for biological applications. These issues must be addressed before administrating CNT based devices, drugs or scaffolds in the biological system [16, 67].

## 1.14. APPLICATIONS IN ENVIRONMENT SECTOR

Carbon nanotubes possess extraordinary properties such as small pore size, structural uniformity, antimicrobial nature, high thermal conductivities, and excellent mechanical properties. These properties have been exploited in numerous applications including environmental applications. Contribution of CNTs in environmental applications has been in the following areas [73]:

1. Sorbents
2. High-flux membranes
3. Depth filters
4. Antimicrobial agents
5. Environmental sensors
6. Renewable energy technologies
7. Pollution prevention strategies

### 1.14.1. CNTs as Sorbents

Adsorption behaviour of sorbates is affected by hydrophobicity and capillarity of carbon nanotubes and orientation of sorbates in microporous carbons. Adsorption by non-functionalised nanomaterials is dominated by physisorption [74]. High equilibrium rates, adsorption capacity, low sensitivity to pH range, have been reported by adsorption studies [73]. Table **8** summarises the properties of carbon based nanomaterials for their application as effective sorbents.

**Table 8. Summary of the advantages of carbon based nanomaterials over conventional sorbents [73].**

| Limitations of traditional sorbents | Advantages of carbon based nanomaterials |
|---|---|
| Limited density of surface active sites | high surface area to volume ratio and high adsorption capacity |
| High activation energy of sorptive bonds | effectiveness over a broadpHrange |
| Slow kinetics and non-equilibrium of sorption in heterogeneous system | rapid equilibrium rates |
| Low mass transfer rate to the sorbent surface | manipulatable surface chemistry |
| Low porosity environments causing limited transport due to large dimensions of traditional sorbents | High porosity and controlled pore size Distribution |

Hydrophobic, dispersion and weak dipolar interactions of organic contaminants determine sorption energies in case of nanomaterial surface similar to conventional systems. Most common examples of environmental applications of nanosorbents include the removal of hazardous organic chemicals such as trihalomethanes, polycyclic aromatic hydrocarbons, or naphthalene, [73 and references therein].

The physical and chemical properties of carbon based nanomaterials can be altered by bottom-up synthesis and selective and specific functionalisation to attain sorption characteristics that suits particular application. For instance, hydrophilic functional groups such as –OH, and –COOH attached to CNTs enhance sorption of polar and low molecular weight compounds in contrast to conventional sorbents. Adsorption of inorganic contaminant to carbonaceous nanomaterials is marked by specific complexation reactions. The key determinant in case of inorganic sorption is the surface functional group density, rather than total surface area. Lastly, oxyanion metal contaminants are adsorbed by ternary surface complex reaction involving cationic bridging of the metal on the negatively charged surface of carbon nanomaterial [75]. Carbon based nanomaterials can also be employed as scaffold with high surface area for the sorption of oxides or macromolecules. Nanoscaffolds customized to adsorb complex ions, metals, and radionuclides can be synthesised by covalent chemistry routes. CNTs decorated with ceria ($CeO_2$) nanoparticles as scaffolding agent have been synthesised for the removal of arsenate and chromium [75, 76].

### 1.14.2. Aligned CNTs as High-Flux Membranes

Novel membranes comprising of CNTs may cut down the energy and cost of desalination considerably. Wide functionalisation pathways for CNTs offer incorporation of tunable properties for the development of membranes to enhance steric selection with chemical and/or charge-based selectivity. Pore diameter, surface hydrophobicity, and nanotube tip functionalisation for channel gating can be altered for membrane application. High-flux aligned CNT membranes have demonstrated potential for advanced water treatment applications. Flow rates of ~ 4-5 orders of magnitude higher than conventional fluid flow have been achieved by aligned CNT membranes [77]. Hydrophilic functional groups incorporated by either oxidation or chemical modification of the CNT tips lead to further improvement in the binding energy between water and the nanotubes, augmenting membrane flux [78].

### 1.14.3. CNTs for Composite Filters

Small and uniform pore sizes of nanotubes, facilitating separation of small molecules, offer advantage over traditional filter materials such as activated and porous carbons used for water purification. The flow rate is not reduced despite such small pore sizes, owing to the distinctive nature of the flow of solvents through the nanotube cavities. Furthermore, separation can be made selective by chemical modification of nanotubes surface (Fig. 7) [16]. Mechanical stability of CNT based membranes can be enhanced by forming nanocomposites. The latter technology can also be used to disrupt polymer packing of the active layer in conventional reverse osmosis membranes. MWNTs composites have been found to improve mechanical strength and durability in addition to enhancing water permeability [73]. Porous aggregate structures of SWCNTs with narrow diameter, high surface area, and thermal resistance are also applicable in regenerative hybrid ceramic filters. MWCNT and SWCNT filters have shown to be effective in removing bacterial and viral pathogens from water [79].

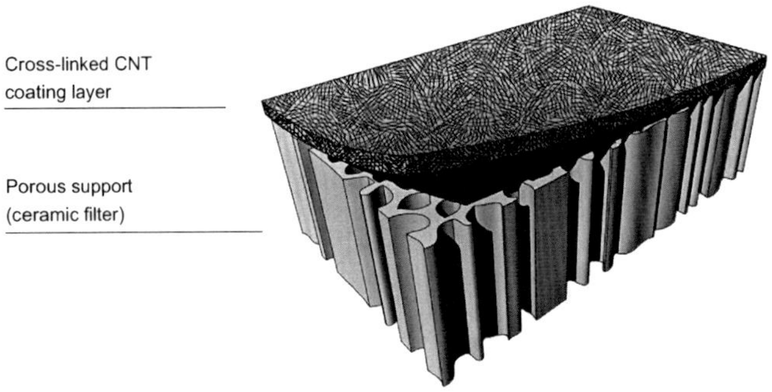


**Fig. (7).** Schematic of CNT hybrid filter [73].

Furthermore, the oxidized MWCNTs have also been used in cigarette filters to adsorb nicotine and tar [80].

### 1.14.4. CNTs as Antimicrobial Agents

To act as antimicrobial agent, CNTs must be in direct contact with target microorganisms [81]. Due to hydrophobic nature of CNTs, suspension of CNTs in water is particularly complicated; therefore adequate CNT-microbe contact is not achieved for disinfection. Hence, antibacterial activity of CNTs could be realised by coating the reactor surface with CNTs that is in contact with the pathogen laden water [82]. Non-aligned single or multi-walled nanotubes, that are easy to separate, could also be used in a packed column/bed filter. CNTs can be applied to surfaces such as water filtration membranes to prevent biofilm formation and the subsequent biofouling [82]. Deliberate perturbation of cell membranes is a potent method for pathogen control. Bacterial colonization and biofilm development in drinking water systems, medical implant devices, and other submerged surfaces are some of the challenging issues which may be addressed with novel antimicrobial surface coatings containing CNTs, thus offering smart engineering solutions to such problems.

### 1.14.5. CNTs for Environmental Sensing

Chemical, biological, thermal, optical, stress, strain, pressure, and flow sensors are made possible due to unique electrical conductivity, chemical stability, high surface area, mechanical stiffness, and possibility of functionalisation of CNTs to augment conventional carbon electrode sensor platforms. Generation of large electric fields at low voltages is assisted by the sharp tips of nanotubes. This property of CNT enables fabrication of portable, battery operated, and small-scale sensors. A major breakthrough has been the emergence of CNT nanowire sensors. Analyte composition or concentration is determined by the change in the conductance of CNT due to the adsorption of charge species to the surface of the CNT. Large SWCNTs with fully exposed surface area exhibits fast response time, low detection limits, and high sensitivity. Health and safety monitoring, infrastructure management, chemical and material efficiency in manufacturing, and cost-effective regulatory platforms are some other environmental applications of CNTs. Biosensor research also includes monitoring environmental microbial ecology and detecting microbial pathogens [73].

### 1.14.6. CNTs for Renewable Energy

CNT anodes may have the potential to enhance the sensitivity and spatial resolution of radiation detectors used in nuclear power plants [83]. CNTs being stronger and lighter may find application in wind harvesting devices as they exhibit structural improvements. Optical properties of CNTs have been exploited in organic-inorganic hybrid photovoltaic cells and dye-sensitised solar cells (DSSC). For anode replacement, thin films of SWNTs offer low cost, transparent, highly conductive, and physically flexible. One of the challenges is the separation of semiconducting and metallic SWNTs for optimisation for specific application with significantly high efficiencies.

### 1.14.7. Pollution Prevention Using CNTs

Pollution prevention refers to:

1. "Source Reduction" and the drop in use of raw materials, energy, water, or other resources to diminish or eliminate creation and/or accumulation of waste.
2. Use of less toxic and renewable reagents and processing materials, where possible,
3. Fabrication of more environmentally benign manufactured products.

The application of CNTs in computer displays may lower the environmental impacts by reducing toxic heavy metals. This, in turn, would result in dramatic reduction in material and energy requirements, without compromising the expected improved performance for consumer needs. Recent developments include the use of CNTs in the field emission displays (FEDs) in display technology which are now being commercially available [84].

### 1.14.8. Challenges in the Area

CNT application as anode replacement with low cost, transparency, high conductivity, and physical flexibililty has been achieved to a significant degree of success; however, optimisation is restricted by the complexity involved in the separation of semiconducting and metallic SWNTs, required for a specific application. It is well known that nanoscale inorganics is impaired by high charge recombination [85]. CNTs may serve as conductive scaffolds which are expected to facilitate charge migration to the electrode, thereby reducing charge recombination effects [86].

Some of the advantages of substitution of anode material with CNTs are as follows:

1. Improved charge transfer.
2. Optimisation of triple phase contact layer between carbon electrode, noble metal catalyst and the proton conductor.
3. Corrosion resistance.
4. Excellent gas diffusivity.

5. Appropriate path for water removal.

CNT based nanosorbents are effective in reducing organic contaminants, residual taste and odour but their application in waste water treatment is handicapped by their cost and possible toxicity concerns. These issues have direct impact on further research in the area [73].

### 1.14.9. Current Developments

Modelling and prediction of water quality and quantity have been planned by the Water and Environmental Research Systems Network (WATERS Network), by forming an integrated sensor network. The environmental observatories will be linked and cache of data for modelling will be made accessible. Accurate and flexible sensors will be required to perform this task at large scale and with accurate forecasting [73 and references therein]. Employing SWNTs as a scaffolding agent resulted in decrease in the rates of platinum recrystallization. This caused lengthening of the lifespan of the fuel cell [87]. Freestanding fuel cell electrodes with improved conductivities and simplicity of fabrication methods have been made possible by exploiting structural and mechanical properties of CNTs [73].

### 1.14.10. Future Prospects

The current status of CNT development for environmental application demands further research in the following areas:

#### *1.14.10.1. Sensing*

Rapid, specific and sensitive detection of microbes using nanobased sensors would definitely influence domestic security, drinking water systems, and subsurface monitoring of biodegradation.

#### *1.14.10.2. Membrane Filters*

Though a vast research on nanotube based filtration membranes is being conducted worldwide, their translation into commercial products and their development still appear to be far from reality [16].

#### *1.14.10.3. Antibacterial Agents*

There is a need for the development of hybrid SWCNT ceramic filters for water treatment application based on the bacterial and viral pathogen removal from water. The aim is to achieve filters with resistance to high temperatures and significant durability [73]. CNTs as scaffolding agents holds a big promise in future for antimicrobial Ag nanoparticles [88] or semiconducting photocatalysts such as $TiO_2$ [73].

## 1.15. APPLICATION IN FOOD SECTOR

A great deal of advancement in the application of nanomaterials has been made till date. However, the use of carbon nanotubes in the packaging of food products has not been massive. There have been some efforts in the area and carbon nanotubes have been incorporated into polymer structures (liquids, solutions, melts, gels, amorphous and crystalline matrices) to increase their mechanical properties such as tensile strength and elasticity and moisture barrier properties [89]. Fig. (**8**) demonstrates the applications of nanotechnology in food sector.

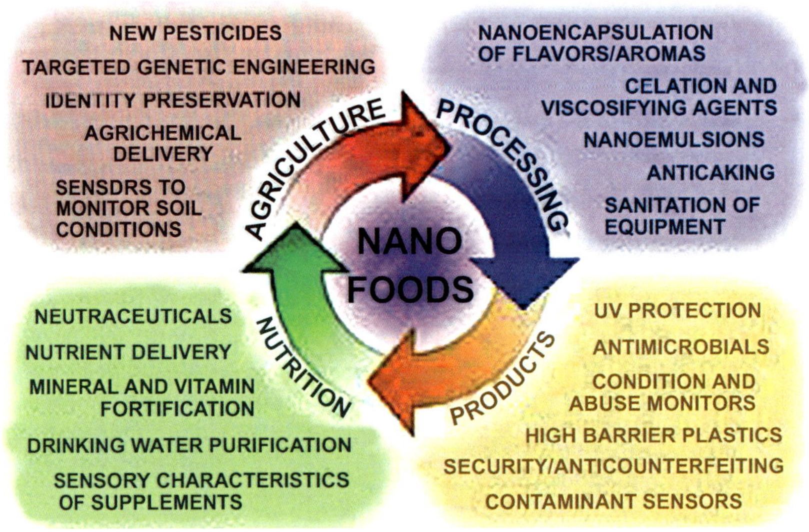


**Fig. (8).** Nanotechnology applications in food sector [90].

The aim of 'precision farming' is to maximise output (*i.e.* crop production) while minimising input (*i.e.* fertilisers, pesticides, herbicides, *etc*) by monitoring ecological variables and applying specific action. It integrates the use of global satellite positioning systems, computer and remote sensing devices to gauge extremely localised ecological conditions to determine crop growth or identifying the location of problem precisely. Variants such as soil conditions, plant development, seeding, fertilizer, chemical and water use can be modified by using centralised data. The use of such advanced technique lowers production costs and improves production potentially [91]. Precision farming also minimises environmental pollution by reducing agricultural waste. Nanosensors and monitoring systems are foreseen to have a large impact on precision farming methodologies. Following are the benefits offered by nanosensors in food technology [91]:

1. Trapping and measurement of individual proteins or other small molecules can be made possible by nanosensors utilising carbon nanotubes.
2. Specific functionalisation of nanosurfaces enables activation of an electrical/chemical signal induced by contaminant such as bacteria or specific enzymatic reaction.

### 1.15.1. CNTs as Fillers in Nanocomposites for Food Packaging

Packaging of food is necessary to protect the food from dirt/dust, oxygen, light, pathogenic microorganisms, moisture, and other destructive or harmful substances when food is not consumed instantly after its manufacture. Further, the packaging must also be:

1. Safe
2. Inert
3. Not expensive to produce
4. Lightweight
5. Easily disposable or reusable
6. Able to withstand extreme condition during processing or filling
7. Impervious to a host of environmental storage and transport conditions
8. Resistant to physical abuse.

No polymer in its pure form displays all the required mechanical and barrier properties for every possible food packaging application, modifications to attain complex multilayer films or polymer blends are being carried out and utilized more often [90]. Nanoreinforcements have been found to be effective in improving moisture barrier of polymer films. Fillers must be less permeable, have excellent dispersion in the matrix and a high aspect ratio in order to enhance the barrier properties of the polymers [90 and references therein]. The dispersion properties of CNTs can be improved by suitable functionalisation, for example, acid treatment of CNTs introduce carboxylic functional groups on CNT surface, thereby increasing its dispersion in hydrophilic solvents. Research indicated 200% improvement in water vapour transmission rate by using CNT nanocomposite with polymer matrix than pure polymers. Increased tensile strength/modulus, toughness and thermal stability of polymers such as PVOH, polypropylene has been observed by addition of CNTs as fillers, even in low concentrations (~ 0.1 wt%) [92 and references therein].

### 1.15.2. CNTs as Sensors

Spoiled or inedible food products can be easily recognized by consumers due to specific odours, colours or other sensory characteristics. Therefore, a method of detection is required to detect the presence of gasses, aromas, chemical contaminants and pathogens, or response to changes in environmental conditions. Chemical and electro-optical properties of nanoscale particles may be able to serve as potential solution to this issue. This would not only ensure quality control but would also augment food safety and lessen the incidences of food-borne illnesses. Consumers will be able to get fresh and flavoursome food products. Electrochemical based detection by nano-sensors offers useful approaches for food environment. These operate by specific binding of selective antibodies to a conductive nanomaterial such as carbon nanotube followed by monitoring changes in conductivity of the material as soon as the analyte binds to the antibodies. For example, when Microcystin-LR (MCLR) (a toxin produced by cyanobacteria), binds to the surface of anti-MCLR-coated single-walled carbon nanotubes, conductivity changes that occur due to this specific binding are detectable down to MCLR concentrations ~ 0.6 nM in drinking water, which is within the limits set by the World Health Organization (WHO) [90].

Other electrochemical systems based on CNTs include detection of Cholera-toxin [90]:

1. Food colorants in soft drinks and in ketchup or chilli powder
2. The concentration dependent intensity changes of the colorant-specific oxidation peaks.
3. Antioxidant, flavour compound and vitamin content in vanilla beans and apples

Many advantages of detection using CNT based sensors over conventional methods such as HPLC include:

1. Rapid and high-throughput
2. Simple and cost effective
3. Low power requirements
4. Recycling is easier
5. No need of exogenous molecules or labels.

Other uses of CNTs and its composites in food technology are summarised in Table **9**:

**Table 9. Uses of CNTs and its composites in food technology.**

| Material | Application in food sector | Key attributes | References |
|---|---|---|---|
| CNTs | Food packaging | Anti-microbial effect | [90] |
| CNTs | Refrigeration of food | Amplified heat transfer coefficients | [93] |
| CNTs | Nanosensors for crop management; food production and processing; soil quality and plant health monitoring | Specific and sensitive detection | [94] |
| CNTs | Nanoherbicide | capability of killing underground weed seeds, specific binding properties | [95] |
| CNTs | radio frequency identification tags (RIFD tags) | Storage of electric charge | [96] |
| CNT-metal composite | Electronic packaging | improved mechanical, physical, electrical and thermal properties | [97] |
| CNT Epoxy Composites | Electrically conductive adhesive | Enhanced conductivity, elasticity, strength, and durability. | [97] |
| CNT- lead free solder composite | Solder Joints | Improvement in shear strength | [98] |

### 1.15.3. Regulation on the Safe Use of 'Nano'

As recommended by the U.K.RS/RAE2004, chemicals including nanoparticles or nanotubes must be considered as new substances. Following are the regulations set by the U.K.RS/RAE2004:

1. Full safety assessment of the composition of nanoparticles must be carried out before they are declared 'safe' to be used in the food products.
2. The use of nanoparticle materials in the ingredients lists of food products must be indicated.
3. The discharge of nanomaterials into the environment must be avoided as far as possible [99].

Precise safety standards and testing procedures are needed for the introduction of new nanotechnology according to The European Union regulations for food and food packaging [100].

### 1.15.4. Challenges and Future Prospects

Nanocomposites help to protect the food against environmental factors passively and also incorporate properties to the packaging material. Therefore, the stability of foods is enhanced, or at least their eventual inadequation to be consumed is indicated. However, many safety concerns about nanomaterials have already been raised. The size range of nanomaterials is such that they may penetrate into cells and finally remain in the system. There is an inadequacy in scientific data related to the migration of most types of nanoparticles (NPs) from the packaging material into food, as well as their fate inside the biological system. This demands accurate information on the effects of nanoparticles, especially nanotubes, to human health following chronic exposure [92]. A challenge is to ensure product safety by filling various data gaps which exist in the research related to the safe use of CNT nanomaterials within the food sector. To fill these gaps, information regarding the following is required [90]:

1. Nanomaterial migration through polymer films or the long-term stability of polymeric films.
2. The interaction of nanomaterial biomolecules and cellular components.
3. The interrelationships between various physical properties of nanoparticles, especially nanotubes (size, shape, surface charge, *etc.*).
4. Toxicity or pharmacokinetic properties of CNTs.
5. Suitable and steady methods to identify, characterise and quantify nanomaterials in complex food matrices.
6. Toxicity of nanomaterials following oral routes of exposure.
7. Biodegradability of nanomaterials.

A big challenge is the public acceptance of food products incorporating nanomaterials. Research indicates that consumers remain wary about the use of nanotechnology or nanomaterials in food products. There are also concerns about ethical and moral issues among them in addition to health and environmental risks associated with the consumption of 'nanofoods'. These include labelling and the right to choose, development of human abilities, engineering of living matter, privacy, equitability of socioeconomic distribution and genuineness of the food supply [90]. All the aforementioned issues related to successful and safe execution of these applications will require continuous dialogue between the scientists, companies and the consumers. The advantages of nanotechnology will surely play a key role in supplying healthier foods safely if these issues are resolved [90]. Further advancement in the food industry can be achieved using nanotechnology in the following areas [101]:

1. Improved security of production, processing and shipping of food products – using sensors for pathogen and contaminant detection.
2. Maintenance of historical environmental records of a particular product and tracking of individual shipments - using electronic devices.
3. Sensing, localization, reporting, and remote control of food products – using 'smart' systems that can enhance efficiency and security of food processing and

transportation.
4. Delivery systems that carry, protect, and deliver functional food ingredients to their specific site of action.

Detection of microorganisms, toxic proteins, and degraded products in food and beverages are some of the applications for which development of a multiwalled CNT-based biosensor is underway [92]. R & D focuses mainly on the following [102]:

1. Altering and optimising the physical appearance of food including colour, texture, flavour, and consistency.
2. Releasing flavours and nutrients in a controlled manner.
3. Improving the absorption of nutrients.
4. Expanding the shelf-life of foods and enhancing stability.
5. Eliminating unwanted molecules from foods (*e.g.* through nanofiltration).
6. Enhancing traceability and safety of food products by using nanosensors.
7. Developing functional foods, and facilitating consumers to alter food depending on their own nutritional requirement or tastes.
8. Developing new and improved food packaging materials enabled with nanosensors which will be able to identify food spoilage organisms and activate a colour change reflecting that the shelf life is ending/ended.
9. Nanoscale-sensing devices are also under development to enable the food/food ingredients to be traced back to the source of origin.
10. Development of 'Electronic Tongue' technology, made up of sensor arrays to indicate condition of the food.

## 1.16. APPLICATION IN ELECTRONICS AND CONSUMER GOODS

Currently, carbon nanotubes account for a 28% market share of overall nanomaterials demand. Asia-Pacific leads, followed by North America and the European Union in terms of production capacity [103].

Since the last decade there are substantial advancements in nanotechnology which offers a wide range of industrial and consumer products worldwide. Out of many nanoparticles, carbon nano-composites are currently being used in various applications, from basic consumer goods to critical national defence technologies. Some concerns have been raised regarding the potential for exposure and health risk of nano-composites containing CNTs due to comparisons to other high aspect ratio fibres. CNT applications can be divided into three major categories on the basis of their technological development: i) present applications, ii) near-term applications (will be in the marketplace in the next ten years), and iii) long-term applications, (will be in the market after ten-year time period) [104].

### 1.16.1. Carbon Nanotubes in Field Emission and Lighting Applications

One potential application for CNTs that would require large quantities of material is the field emission display (Fig. **9**), which has been widely talked about as the first major application of CNTs; and prototypes have been demonstrated by large companies such as Samsung and Motorola. However, the product has not been commercialized yet, for reasons that are unknown.

**Fig. (9).** Proof of concept CNT Lamp [105].

CNTs have been considered as preferred field emitters due to their low threshold voltage, good emission stability and long emitter lifetime [106 - 109]. These characteristics make them useful in the fabrication of cathode-ray lighting elements and flat panel displays. Also, Samsung released the fabrication of a 4.5 inches diode type flat panel display using arc-based SWCNTs (Fig. **10**) [110]. Although SWCNTs were demonstrated to have excellent emitting performance [110], they simply degraded at high emission current [111]. In this sense, double walled nanotubes (DWCNTs) and/ or thin MWCNTs have been found to be the best field emitting materials. A field emission display (FED) using CNTs as the field emission source has been developed by Samsung Display Innovation (SDI), a division of Samsung Corporation. The Samsung CNT-FED is micro-fabricated on display glass and features a triode structure, where a bias voltage is applied between the bottom (cathode, ITO) and middle (gate) electrodes, and the emission is focused by the top electrode. Each pixel contains many CNT emitters. The CNT emitters are applied to the device by screen-printing of a custom-made CNT paste, or by direct growth of CNTs by CVD. In the former case, commercially available CNTs are mixed with many ingredients, and the paste is wiped through a screen mask that is placed over the device. Many types of CNTs (*e.g.*, SWCNTs, MWCNTs, DWCNTs with various diameters and lengths), obtained from many

**Fig. (10).** Field-emission display (5 inches in diagonal) based on CNTs (Samsung SDI).

suppliers, have been evaluated. It is believed that SWCNTs generally exhibit a higher field emission density, yet lower lifetime, while MWCNTs exhibit lower emission density, yet longer lifetime. Therefore, a careful trade-off must be made to optimise device performance. Currently, uniform-length, high-quality, small-diameter MWCNTs are preferred (MWCNTs are also easier to disperse than SWCNTs); a price of approximately US$10/gram would be suitable for commercial production of the CNT FED. So far, SDI has made prototype CNT FED displays in the 5–15 inches range. A lifetime of 30,000 hours is sought, and the SDI team believes this can be achieved by its technology. Also, as the display size increases, maintaining good vacuum sealing is an increasing challenge, and inadequate vacuum decreases the lifetime as the emitters decay more quickly. There is much other work in emitter applications of CNTs; however, the Samsung display is by far the most serious commercial effort. Iljin Nanotech (Seoul, Korea) mentioned development efforts using CNTs for field emission light sources and reports international patent filings on the topic. The group of Prof. Shoushan Fan at Tsinghua University (Tsinghua-Foxconn Nanotechnology Research Center, Beijing, China) is testing prototype field emission lamps using CNT-coated wires sealed inside phosphor lined tubes. These are sought as ambient light sources to replace fluorescent tubes; however, the power consumption must be reduced significantly for practical applications. Field emission from the ends of CNT yarns is also being studied [112]. Thermionic CNT yarn emitters are sought as a long-life replacement to $LaB_6$ filaments that are commonly used in electron microscopes [113]. In general, CNTs are sought for field emission because of the high current density due to their high aspect ratio; however, many other nanomaterials (*e.g.*, small-diameter nanowires, diamond nano-crystals, nanofabricated metal tips) exhibit structure-based and surface-based field emission and will be an important part of competing technologies.

The application of CNTs as X-ray emission sources has been led by Otto Zhou's group at the University of North Carolina, and is being commercialized by Xintex, Inc. (co-founded by Prof. Zhou) [114]. In a CNT-based X-ray source, CNTs are placed in a triode configuration as a field electron emitter; the emitted electrons are incident upon a target (typically Mo) which, in turn, emits X-rays and is aimed toward the sample [115]. Overall, imaging systems using CNT X-ray emitters offer promise over current technology due to their small size, fast response time, and capability for multiplexing and array imaging. Similar to field emission displays, CNT X-ray emitters can be arrayed as "pixels," and imaging is facilitated by the rapid response time of the emitters along with frequency division multiplexing using a single flat detector [116]. CNTs can be integrated in X-ray emitters by direct growth on the emitter substrate, or by post-growth deposition, and therefore, face similar manufacturing challenges as field emission displays. In a prototype device, purified SWCNTs were electrophoretically deposited in a tangled morphology on a metal substrate, which was then incorporated in a sealed emitter tube [117]. A CNT emitter tube is stable for over 100 hours of operation (36 million cycles), with a constant voltage pulse of 10 μs width at 100 Hz. The emitter current exceeds 1 $A^0$ [118].

### 1.16.2. Carbon Nanotubes in Electronics

It should be pointed out that the integration of nanotubes into electronic devices still is a long-term application with the prospect of commercial products beyond the ten-year time frame. In order to fulfil Moore's law, various semiconductor companies like Intel, IBM, AMD, UMC and TSMC are commercially manufacturing silicon-based transistors at the 65-nm technology node. Intel intends to commercialize transistors at the 45-nm technology node by the end of 2007 or early 2008 [119]. By an evolutionary process, a conservative estimate for the smallest conceivable silicon transistor at this point is 16 nm, expected to hit markets by 2018.

### 1.16.3. CNTs as Interconnects

Due to their metallic characteristics nanotubes can also function as interconnects. As interconnect feature sizes shrink, copper resistivity increases due to surface and grain-boundary scatterings and also due to surface roughness [120]. In contrast, carbon nanotubes exhibit a ballistic flow of electrons with electron mean-free paths of several micrometres, and are capable of conducting very large current densities [121]. Carbon nanotubes have been shown to withstand current densities up to 1010 $A^0/cm^2$, exceeding copper by a factor of 1000. A bundle of closely packed parallel metallic carbon nanotubes, or a large defect-free MWCNT are proposed to be used for the next generation of interconnects. A bundle of 400 SWCNTs can be as narrow as 20 nm in diameter since making a bundle and connecting the tubes in parallel serves two main purposes. Firstly, connecting in parallel helps in overcoming the problem of high value of contact resistance of a single nanotube and secondly bundles ensure a lower capacitance comparable to that of a single nanotube, as opposed to the high value for a monolayer of nanotubes. Moreover, with respect to resistance, nanotubes are favourable in high aspect ratio structures like vias, where also the highest current densities are expected [122]. In addition to the on-chip (short as well as *via*) interconnects, due to the mechanical flexibility and compliance of nanotube bundles, they are also being considered for back-end (die to package, in packaging) flexible interconnects and these may be realized in a time frame faster than the integration of on-chip nanotube-based structures. Another important achievement is the development of an electromechanical memory device by Nanterro Inc., a start-up company focused on bringing on the first major electronic application of nanotubes. Owing to their mechanical resilience and electrical conductivity, CNTs can perform as sensitive nanoelectromechanical devices such as non-volatile memory. Using a crossbar layered architecture, Nanterro is developing a bistable memory device that can be actuated electrostatically [123].

### 1.16.4. Transparent Conductors

Work at the University of Florida, led by Prof. Andrew Rinzler, is developing CNT films as conductive transparent electrodes for applications that include organic light-emitting diodes, organic solar cells, and supercapacitors. One of the features claimed for these transparent films is the high surface area of the nanotubes for coupling electronically to the active organic medium. Eikos Corporation is also developing CNT films as transparent conductors, particularly for displays; the films can be deposited by roll-to-roll and conventional printing techniques [124]. One of the very first commercial successes for MWNTs was to make use of their electrical properties for electro-static discharge (ESD) applications; the nanotubes prevent charge build-up on the surface of the insulating plastic. Today, large (tonnage) quantities of multiwall nanotubes are manufactured by the company Hyperion Catalysis International and supplied as premixed material in thermoplastics to the General Electric (GE) Company for use in their engineering plastics that are employed in automobile body parts. This remains as one of the biggest bulk application of nanotubes today. The applications also include high-end microelectronics products where static electric discharge from packaging components can destroy electronic chip components. In addition to the ESD application, electromagnetic interference (EMI) shielding applications have also been pursued [125] for nanotubes at high loadings, utilizing the extremely high conductivity of nanotubes. Carbon fibres have been used in this area and unless nanotubes offer significant advantages compared to the traditional fillers, it is not clear if nanotube-based plastics would outperform conventional materials in EMI shielding.

## 1.17. CARBON NANOTUBES IN MISCELLANEOUS APPLICATIONS

One of the applications of nanotubes that is in the market today is that of nanotube-based probe tips. The nanotube-based tips provide better resolution for imaging and have better durability because of their elasticity, unlike conventional silicon tips that suffer from frequent tip crashes. Today, nanotube-based (mainly good-quality multiwall) AFM probe tips can be bought from scanning-probe instrument manufacturers [126]. Another interesting application that has been touted for nanotubes is their use as membrane filters. Several groups have published papers in recent years suggesting that membranes made from aligned nanotube arrays (both SWNTs and MWNTs) could be used for selective transport of species [127 - 129]. Activated and porous carbons have been traditionally used as filters (*e.g.*, water purification) but the advantage of nanotubes is that the pore sizes are very uniform and small, so that separation of really small molecules becomes possible. Although the pore size is small, the flow rate is not reduced due to the unique nature of the flow of solvents through the nanotube cavities. Another advantage could be that the nanotube surfaces may be chemically modified to make the separation very selective. Although several groups are working to develop nanotube-based membranes for filtration, molecular as well as gas separation, these are still far from commercial products and their development could take more than a decade. In recent years, work has progressed in getting stable dispersions of nanotubes in various solvents (nanotube inks) [130], which can be directly spun or printed onto various substrates. Easily printable (*e.g.*, inkjet printing) circuits and films are extremely useful for a range of applications ranging from flexible electronics to antennas. Nanotubes have also been spun into fibres using various techniques ranging from extrusion of nanotube slurries to drawing from aligned arrays. High weight percentage dispersions of SWCNTs have been spun directly or using polymer binders to continuous fibres reaching macroscopic lengths [131, 132]. The fibres can be also assembled into sheets, membranes and other products. Several applications have been demonstrated for such nanotube fibres and their products, for example as high-brightness incandescent bulbs [133], reinforcements in polymers, shielding elements and in ballistic protection. Direct growth of aligned nanotubes can be used to make distinct products, such as brushes [134]. The tiny brushes with nanotube bristles have been demonstrated for use in a variety of applications including cleaning of microscale spaces, conducting contact brushes, and probe arrays. Another possibility is the use of aligned nanotube arrays as dry adhesive tape; one recent study shows that properly engineered nanotube arrays could have adhesive strengths even rivalling gecko feet [135]. Below is the list of consumer products based on CNTs (Table **10**).

**Table 10. Enlists various common consumer products being researched or available as commercial products.**

| Application | Product | Material | References/Company/Industry |
|---|---|---|---|
| Sports goods | snow skis, baseball bats, and bicycle frames, golf shafts, tennis racquets, baseball bats | CNT-polymer composite | [136] |
| Electronic goods | Semiconductor chips; flat panel displays | CNTs | [137]; Intel, IBM, AMD |
| Energy applications | Fuel cell and solar cell | CNTs | [137] |
| Packaging | Food packaging, sensors in packaging | CNT composites | [97] |
| Sensors | Biosensors, chemical sensors | CNTs | [138]; Nano-Proprietary Inc. |
| X-Ray systems | - | CNTs | [138]; ZincRay Systems |
| Appliances | Refrigerator, washing machines, air/water filteration and purification | CNTs | [139, 140] |
| Miscellaneous | Anti-fog, self cleaning, adhesives, Nanoinks | CNTs | [16, 130] |

## 1.18. CNTS FOR ENERGY APPLICATIONS

### 1.18.1. Application of CNTs in Fuel Cell Electrodes and Catalyst Supports

Fuel cells are electrochemical devices that convert chemical energy directly into electrical energy. They combine hydrogen and oxygen to generate electricity along with water and heat as its by-product. The simplest form of fuel cells contain a single fuel cell consisting of anode and a cathode, with an electrolyte between the electrodes. Hydrogen reacts with a catalyst, generating a proton and an electron at the anode. The electron travels through a circuit generating a current, whilst a proton passes through the electrolyte. Oxygen forms water once it reacts with the proton and electron and liberates heat at the cathode. A single cell generates about 0.7 V, which can power a single light bulb. The output efficiency can be increased anywhere from several watts to multiple megawatts when cells are stacked in series. Emissions from fuel cells are considerably lower than those from the cleanest fuel combustion processes as these fuel cells involve an electrochemical process rather than combustion. Other advantages include durability and high efficiency [1, 140].

On the basis of the electrolyte employed, fuel cells can be classified as [2, 141]:

1. Polymer Electrolyte Membrane Fuel Cells (PEMFCs)
2. Direct Methanol Fuel Cells (DMFCs)
3. Alkaline Fuel Cells (AFCs)
4. Molten Carbonate Fuel Cells (MCFCs)
5. Phosphoric Acid Fuel Cells (PAFCs)
6. Solid Oxide Fuel Cells (SOFCs)

There is growing exploration of CNTs as permeable electrodes in fuel cells (where the catalyst is in contact with the proton/electron charge transfer media and is electrically connected to the support electrode), and more generally as support materials for nanostructured catalysts (Fig. **11**). SWCNT and MWCNT soot have been widely investigated as supports in the dispersion and stabilisation of Pt or Pt-metal alloy electro-catalysts for PEM fuel cells. Typically, MWCNT have higher conduction than SWCNT due to the larger outer shell which is considered to be conductive relevant, whilst the latter usually provide greater surface area in preparation of supported Pt catalysts. For instance, investigations into support/catalyst interactions have shown increased catalytic activity in both SWNT and MWNT systems [2, 141].

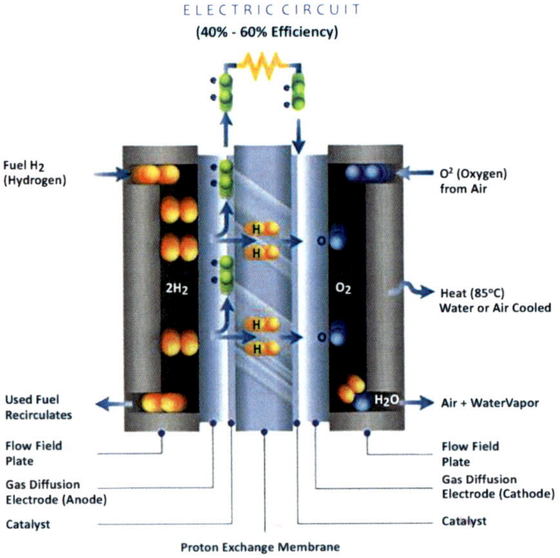


**Fig. (11).** Fuel cell set up [1, 140].

The main focus for PEMFC research and development is to produce a commercially viable product. In order to reach this goal, several cost and durability requirements have been set. In recent publication [7, 142], Ballard Power has outlined some of the key challenges required to meet the increased durability of 5000 h stack life and the required steps to lower the cost of PEMFCs. Understanding of what factors affect performance losses and material degradation is essential in finding adequate solutions. It is suggested that advanced tools for performance diagnostics, ex-situ structural characterization and accelerated durability testing are key areas of improvement for providing a strategy to improve fuel cell performance and durability. Reducing performance losses and degradation directly translate into cost reduction as well because price per kW and over the lifetime also decreases [9, 143]. For automotive applications, since the PEMFC technology is considered as the replacement for the ICE technology, the cost target has been set to be $30/kW [7, 8, 142, 144]. The cost targets, created by the U.S program for the development of fuel cells [8, 144] for the year 2010 and 2015 are $45 for 30 kW and 80 kW, respectively for the net system. The actual 'commercial' cost of a PEMFC stack is difficult to calculate because appropriate consideration of the volume production must be taken. Nonetheless, a study by Isa-bar on (2001) reports the breakdown of cost of each component. The cost of the membrane electrode assembly (MEA) is estimated to be more than 50% of the total PEMFC component costs. The polymer electrolyte membrane and precious metal cost contributes to more than 80% of the MEA cost. The study by GM group also indicates that the cost of PEM and platinum add up to greater than 70% of the total PEMFC cost.

### 1.18.2. CNTs in Li-Ion Batteries

Rechargeable Li-ion battery (Fig. **12A**) with high energy density, high working potential and long cycling life has been considered as one of the most promising power sources of portable systems and electric vehicles.

The lithium-ion battery is currently one of the most important and widely used rechargeable batteries. Its electrodes are based on intercalation materials between which lithium ions are transferred through an electrolytic medium during charging and discharging. Carbon materials and lithium containing lamellar compounds of $Li_xMO_2$ (M = Mn, Co, Ni, *etc*, $x \approx 1$) usually serve as the anode and cathode, respectively, in commercial Li-ion batteries. The redox reaction can be expressed as:

$$x\mathrm{Li} + \mathrm{C} \Longleftrightarrow \mathrm{Li}_x\mathrm{C}.$$

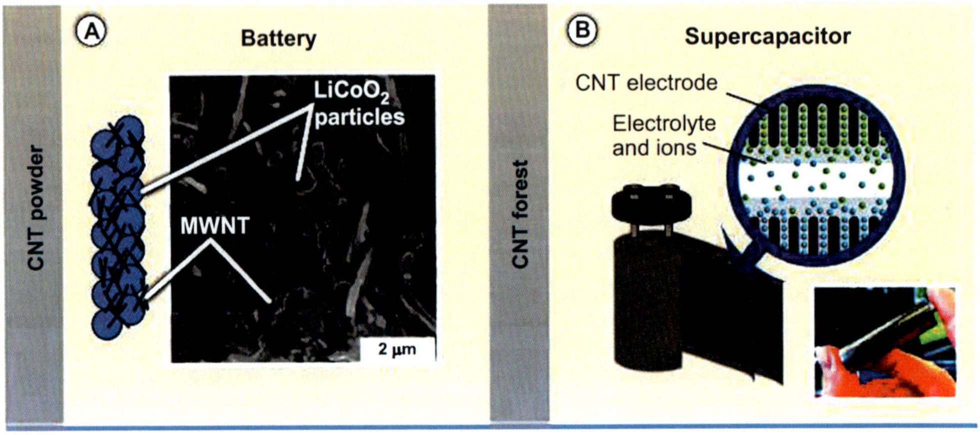


**Fig. (12).** Energy-related applications of CNTs (A) Mixture of MWCNTs and active powder for battery electrode [67, 145]. (B) Concept for supercapacitors based on CNT forests. [Source: FastCap Systems Corporation].

When the insertion of $Li^+$ into C occurs, the cell is discharged; when de-insertion proceeds, the cell is charged. CNTs are considered to be excellent candidates for high performance electrode material in Li-ion batteries owing to their unique structure and superior properties. The abundant nanosized pore structures of CNTs may provide hosting sites for storing large amount of Li+ with sound dynamics; good chemical stability and high mechanical strength and elastic modulus could make it possible for CNT electrodes to have a long cycle life; high electrical conductivity originating from their good structural completeness may improve the power density of CNT-based electrodes; and a short diffusion distance for the lithium ions may bring about a high discharging rate (Fig. **12B**). Therefore, application of CNTs for Li-ion batteries has drawn much interest in scientific community [43]. Lithium ion batteries incorporated with CNTs appear to be the major commercial success for MWCNTs. The main constituent of the anode in Li-ion batteries is meso-carbon microbeads that become intercalated with Li ions as the batteries are charged. Dimensional changes occurring through repeated cycling, decrease electrical connectivity between micro-beads, degrading performance and limiting cycle life. The addition of ~10% MWCNTs to the anode keeps the micro-beads in electrical contact by bridging the nanotubes from bead to bead. In addition to increased cycle life, overall performance is enhanced by improvement of the overall conductivity of the anode. While this represents a significant commercialization success, Li ion batteries constitute a small fraction of the overall battery market [43]. For Li-ion batteries, in addition to a MWCNT anode MWCNTs coated with $LiCoO_2$ as the cathode has also been used. In these applications, smaller diameter nanotubes were typically found to have better performance, presumably due to their increased accessible surface area per unit mass. A key finding is that nanotubes mixed with other carbons (*e.g.* graphite micropowders or acetylene black) show enhanced capacities over either material alone. With the advantage of technical feasibility and cost effectiveness, the above electrode additive route may accelerate the practical applications of CNTs/CNFs in lithium-ion batteries [43]. According to a recent report, the market predictions for lithium-ion battery was $113.1 M in 2008 and is predicted to be $381.9 M by 2015, at an annual growth rate of 25.3% (2009 to 2015) [146]. CNT composites for Li-ion battery electrodes were one of the first nanotube applications marketed by Showa Denko, which has a capacity of producing 500 tons/year CNTs. They have collaboration with various other companies producing CNT based products. A brief summary of some leading research companies in this field is shown in Table **11**.

**Table 11. A comparison between CNT modified batteries by various leading companies across the globe.**

| Researching companies | Novelties in batteries |
|---|---|
| Toshiba | Toshiba found that the battery losts only one percent of its capacity after conducting tests on a new battery by discharging and fully recharging it thousand times at 77 degrees. |
| A123Systems | A123 Li-Ion battery reaches to its “high capacity” within five minutes. The temperature range is -30 to +70 $^0$C. These batteries offer 4 times higher thermal conductivity than conventional Lithium-Ion cylindrical cells. They A patented nanophosphate technology is employed in this product. |
| Altair | This is a nanobattery with a one-minute recharge. Nano-structured lithium titanate spinel oxide (LTO) is employed in the technology. |
| Next Alternative Inc. | A CNT lead/lead-acid battery will deliver 380 miles (as compared to 50-100 miles range covered by a typical lead acid battery) distance between charges. This battery could also be recharged with 10 minutes (as compared to a typical lead-acid battery having a recharge time between 4 and 10 hours). |

## 1.18.3. CNTs in Supercapacitors

The term ‘supercapacitors’ or ‘ultracapacitors’ originates from their remarkably high energy density compared with conventional capacitors (usually 1–2 orders of magnitude higher) (Fig. **12B**). Projections of future development using novel electrode materials indicate that a much higher specific capacitance and power density are likely to be achieved. There are high expectations regarding the use of CNTs in their application as the electrodes of supercapacitors, owing to their advantages of good chemical stability, unique pore structure and low electrical resistivity. Following are the advantages of supercapacitors over the conventional capacitor and battery [1, 147]:

1. Specific capacitance in several Farads to several hundred Farads (105 -106 times ordinary capacitors).
2. Virtually unlimited cycle life - can be cycled millions of time.
3. Low impedance – enhances load handling when put in paralleled with a battery.
4. Rapid charging – discharging, supercapacitors charge in seconds.
5. High power density.
6. Simple charging methods. No special charging or voltage detection circuits required like batteries.

Supercapacitors provide high power density and intermediate energy density (>1000 W/kg, >1Wh/kg). When combined with a fuel cell, a system with simultaneously high power and energy densities is produced (>1000 W/kg, >100 Wh/kg). Supercapacitors enable applications requiring ultrahigh power delivery rate, such as directed energy weapons, rapidly accelerating vehicles, and high-pulse-power communications systems. They could also be combined with batteries to provide a high power drain capability that is not detrimental to battery life. Supercapacitors, while possessing lower energy storage capacity than batteries, are able to deliver their energy at far greater rates, making them important for numerous applications, for example, in future electric vehicles, where high rates of acceleration are intermittently needed. Chemically modified MWCNTs can be incorporated as an additive material to enhance the electrical conductivity of phenolic resin based activated carbon electrode that is most widely used in supercapacitor industries as electrode active material. CNTs and their composites have shown promising features of high specific capacitance, good stability and sound electrodynamics for serving as the electrodes of supercapacitors. There is a great need for further investigations focusing on the optimization of the surface and pore structure of CNTs, and on improvement of electrode fabrication techniques and the energy density, to enhance the integrated performance of the CNT-based supercapacitors, aiming at important applications in electric/hybrid vehicles and micro-electrochemical systems. Today several companies such as Maxwell Technologies, Siemens Matsushita, HEC, Panasonic, TOKIN and several others are investing in supercapacitor development [148].

### 1.18.4. CNTs in Solar Cells

The current most commonly used method of capturing solar energy is with photovoltaic (solar) cells. At present, solar cells are usually made of silicon crystals or organic polymer dyes. Organic dye cells are cheaper but less efficient than silicon cells [1, 149]. One of the materials being used in new solar cells is carbon, in the form of carbon nanotubes, buckyballs, or other forms. Both structures hold promising futures as materials for more efficient solar cells, though current efficiencies are lower than commercial cells. Further research and experimentation will allow the efficiency of carbon-based cells to increase greatly, possibly surpassing that of silicon-based cells, allowing them to be utilized in greater extent than our current solar technology [2, 150]. Silicon, in its crystalline form has proven to be the most reliable material for solar cell fabrication. The major issue with crystalline silicon is its poor light absorbing property along with the size and fabrication limitations [5, 151]. Organic dye sensitised solar cells (DSSCs), are the newer type of solar cell. They are designed to be cheaper than silicon cells, with the trade-off of being less efficient. DSSCs consist of a thin transparent semiconductor, usually made of conducting glass that acts as one electrode, with another glass plate coated with platinum as the other (counter) electrode. Light sensitive dye molecules are attached to the semiconductor, and an electrolyte, usually iodine-triiodide (I- / I3-), immerses the setup. Efficiencies have reached up to 12%. However, 12% efficiency seems quite small as compared to the 30% of silicon cells; the cost trade-off makes up largely for the loss in efficiency. A big advantage is that DSSCs continue to operate well in poor sunlight and cloudy conditions [6, 152]. Research into DSSCs is focused on replacing components of cells with carbon nanotubes. The first all-carbon solar cell was fabricated using carbon-based electrodes with an active layer composed of sorted semiconducting single-walled carbon nanotubes and C60 fullerene. Recently, scientists have reported solar-cell devices with power conversion efficiencies (PCEs) of 0.1–1.3%. These solar cells consisted of an all-carbon photoactive layer using semiconducting single-walled carbon nanotubes (SWCNTs) as the light-absorbing component and charge-donating (donor) material, with a fullerene layer as the charge-accepting (acceptor) material [153 - 155].

Different types of CNT solar cells use different types of nanotubes to achieve the properties they want, and the purity of the CNTs used is a very important factor in how well a cell works. There are multiple ways in which carbon nanotubes can be utilized in solar cells (Fig. **13**). The research into DSSCs is focused on replacing key components of organic dye cells with carbon nanotubes will make them cheaper, easier to produce, longer lasting, and more flexible than they currently are (Fig. **14**).

Another type of carbon solar cell involves coating a carbon fibre with poly-silicon to create a hair-like solar cell. The poly-silicon is doped with boron and phosphorus so that the p-type is on the inside and the n-type on the outside. When sunlight hits the cell, electrons are displaced throughout the n-layer and holes throughout the p-layer, creating an electric field. Current efficiencies of these types of cells are still very low, but as with all types of solar cells, it can be improved with research and experimentation. This type of solar cell is very thin and can be very flexible, allowing for it to be used in many different applications [11, 156].

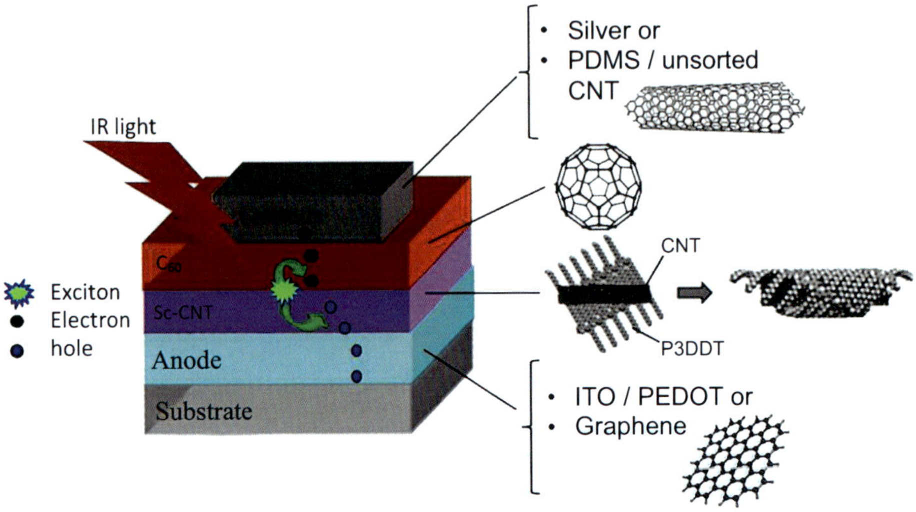


**Fig. (13).** Structure of the carbon-based solar cell, showing the components of each layer and the process of electron-hole pair (exciton) generation and separation when light is absorbed. Devices with both standard electrodes and carbon-based electrodes were fabricated. P3DDT and PEDOT are conductive polymers, while PDMS is a flexible silicone polymer. CNT: Carbon nanotube. ITO: Indium tin oxide. Sc-CNT: Semiconducting CNT. $C_{60}$: A fullerene [43].

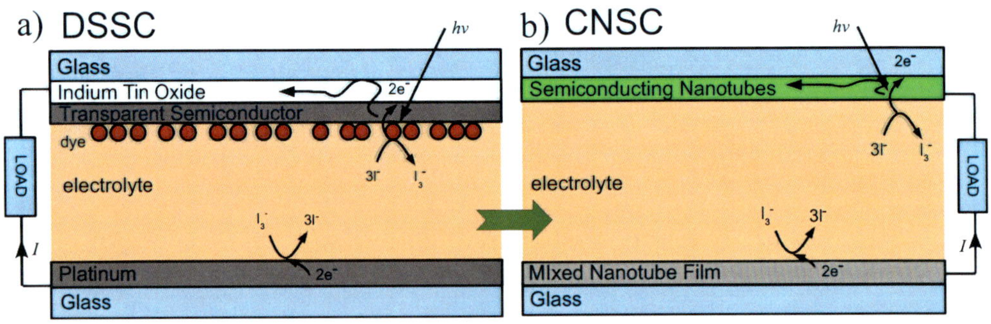


**Fig. (14).** Comparison of organic cells, one using dyes, the other CNTs [6, 152].

### 1.18.5. Challenges

One major detrimental factor in the widespread use of solar panels is their cost. Replacing components of cells with carbon nanotubes is one way in which prices can be reduced, the process of making silicon solar cells is time consuming, expensive, and wasteful, and is a major reason why they are taking so long to catch on. ACSCs can be produced quicker, are made of a cheap and very abundant element, and can be produced with very little material wasted, due to CNTs coming from a solution rather than a solid block like silicon. The ability of ACSCs to absorb light from different wavelengths than silicon cells can allow them to be deployed in tandem and absorb more light than either alone, extracting more energy per unit area [10, 157].

### 1.18.6. Future Prospects

The structure of carbon nanotube solar cells allows for several benefits in its applications that other solar cell technology cannot match up to. The expensive metals needed such as platinum that are used in the structure of other solar solar cells, as well as the layer of rare indium necessary for their function is replaced with carbon nanomaterials. These carbon nanotube solar cells are then implemented with a thin film structure. The film structure of carbon nanotube solar cells creates many benefits for its physical properties. One of the main advantages of the carbon solar cell film is its flexibility. This "carbon nanotube fibre is flexible and will not break after being bent, folded, or even tied many times" [14, 158]. The thin film of carbon nanotube solar cells does not require another physical structure such as a panel in its development for use. Its light weight and flexibility allow carbon nanotube solar cells to have the potential to be employed where thick solar panels cannot. These films could possibly scale entire buildings or be incorporated into clothing all while harnessing solar energy. The integration of solar cells with transparent electrodes has been announced to be undertaken by Unidym, Inc, with Nippon Kayaku. Thin-film photovoltaic technologies promise to be one of the great technologies in the solar market. These are expected to contribute strongly to the industry's growth from $10.6 billion in 2006 to $18.5 billion by 2011 according to Solarbuzz (a leading solar energy research and consulting company) [159].

## 1.19. APPLICATION OF CNT IN AEROSPACE AND DEFENCE

After making its significant mark in sectors such as electrical and optical devices including flat panel displays, sensors, energy storage and biomedical applications, CNTs are now being tested for aerospace application. To be used in the latter, the material must be light, strong and structurally efficient for its use in air vehicles. With extraordinary stiffness and high specific tensile strength, CNTs show good capability for use as reinforcing elements in polymer composites. The amalgamation of carbon nanotubes and polymer matrix results in greatly increased strength and stiffness of the composite product with negligible increases in weight. New materials containing polymers reinforced by CNTs demonstrate excellent mechanical properties combined with low density. The ability of a composite material to resist vibration and flame is also enhanced [160]. Only a small amount of nano-filler (< 5 wt%) is required to obtain a nanocomposite material with significant enhancement in mechanical, thermal, electrical, and other desired properties such as gas impermeability, dimensional stability, and flame retardance. Furthermore, the composite density or light transmission of matrix resins is not compromised in achieving the above stated benefits [161]. CNT based composites provide lighter material with better performance/functionality which helps to reduce weight, and operating cost [162].

### 1.19.1. CNT Technology in Aerospace and Defence

#### *1.19.1.1. Sensors*

Some of aerospace applications where the use of sensor technology emerges are:

1. A requirement of a device with enhanced measurement capability and compact size
2. Less power requirements for electronics
3. A requirement of a portable device for composition measurements of Earth's atmosphere
4. A need for highly miniaturized gas detectors allowing Earth Science Enterprise Plans for *in situ* measurements in order to authenticate satellite observations.
5. Nanosensors ensuring suitable and reliable operation in the microgravity or reduced gravity environment, particularly those operating in multiphase media.

The unique microgravity considerations make this deliverable particularly NASA-specific. Interpretation of sensor data, such as localization of hazards will be enabled by employing Microgravity transport modelling capability. Complexities of multiphase (gas, liquid, solid) behaviour in micro or reduced gravity must be accounted for by sample acquisition for accurate monitoring. It must also require little/no crew time and expendables. External environment monitoring of any hazardous conditions in the extra-vehicular environment can also be one of the applications of recommended sensors. Hazards in this case include reactive chemicals, erosive dust, and radiation [163]. A chemical sensor has also been developed at NASA Ames incorporating nanostructures, such as single walled carbon nanotubes (SWCNTs) and metal oxide nanobelts or nanowires. Various gases these sensors have been exposed to include nitrogen dioxide, acetone, benzene, nitrotoluene, chlorine, and ammonia. These sensors demonstrated high degree of selectivity and sensitivity at room temperature. Large surface area, low surface energy barrier and high thermal and mechanical stability are key attributes of nanostructured chemical sensors leading to higher sensitivity, lower power consumption, portability of the device and a more robust solution as compared to most state-of the-art systems. This makes them attractive particularly for security, defence and space applications, as well as a variety of commercial applications.

Furthermore, wireless capability with the sensor chip of such sensors can be used in mobile networking, fixed-base detection and warning systems for military bases, facilities and battlefields, monitoring filter breakthrough, personnel badge detectors, embedded suit hermit city sensors and other applications [164]. Furthermore, applying mechanical stress to nanotubes modulates their electrical conductivity.

### 1.19.2. Nanotube Composites

A need for new materials arises from the fact that airframes and spacecraft undergo a wide variety of stress loadings in different orientations. Therefore, these materials are expected to improve the endurance of components and the overall efficiency of aerospace vehicles. The other desired characteristics of aerospace materials include corrosion resistance, response to temperature variations, and drag coefficients, to name a few. CNT nanocomposites offer a great potential as they exhibit various excellent properties such as superior strength to weight ratios, stiffness, strength, high-temperature-resistance and electrical and thermal conductivity [165]. Nanotubes in plastic composites have been expected to enhance stiffness, strength and toughness, and electrical and thermal conductivity. These are anticipated to be the largest bulk application of these reinforcers. Nanotubes could act as fillers in diverse multifunctional composites due to their wide range of exceptional properties. For example, by suitably engineering nanotube–polymer interfaces, the strength and damping properties can be improved. Blending nanotubes with plastics provides materials that could serve to shield microelectronics from electromagnetic interference. Similarly, microwave-absorbing capability of nanotubes could be utilized to heat temporary housing structures, offering applications in space exploration [165]. Carbon nanotubes amalgamated with epoxy resin have been shown to improve electrical properties of the composite. These composites have been prepared with the aim to obtain a light, thin and mechanically strong composite material to cover electric circuits against external electromagnetic interference. This is one of the main concerns with aerospace vehicles [97, 166,].

CNT-polymer composites offer key benefits such as:

1. Weight reduction of electrical wire & cable by 50% with aluminum-like electrical conductivity. Estimated power wire reduction is translated into considerable weight reduction for large aircraft.
2. Huge fuel reduction & operating cost reduction for airlines and others.
3. Polymer nanocomposites demonstrate improved thermal properties, delayed melting and dripping, and reduction in the rate of burning (by more than half), leading to their application as fire retardants.

The technology makes use of CNTs which can conduct larger amounts of electric current than copper while with only one-sixth the weight and the cables composed of electrically conductive carbon nanotubes in a polymer. Some examples of fire retardation applications include reduction of fire risk in enclosed spaces in vehicles, submarines, aeroplanes, and ships.

#### *1.19.2.1. Aerospace Fabrics*

CNT based fabrics made with high strength composite materials possess potential applications. Aligned, individualized CNTs can be used to make resin pre-impregnated fabrics and high fibre volume fraction composite materials possessing morphologies similar to traditional carbon fibre materials. These fabrics, owing to their multifunctional properties, may fill the gap in defence, aerospace, automotive and consumer markets [167].

### 1.19.3. Current Status and Challenges

Aerospace composites with desired properties are already being developed and are being used efficiently in aerospace structures. In aerospace structures, these composites are used in:

1. Highly loaded parts with thick gages.
2. High fatigue loads (fuselage and wing structure, *etc*).
3. Areas susceptible to corrosion (fuselage, *etc*).
4. For critical weight reduction (empennage, wings, fuselage, *etc*).

Applications of nanocomposites are currently at their embryonic stage and need further research as the results in the performance of such material are dispersed. There are some key issues related to the large scale production and processing of nanocomposites which restrict its use in major applications including aerospace and defence [168]. These include:

1. Structure control of CNTs
2. Dispersion of CNTs in polymer matrix
3. Interfacial interaction between CNTs with polymer
4. Contacts between individual CNTs

Also, there are some practical challenges as seen by aerospace industry in order to take advantage of CNT nanocomposite technology which include [3]:

1. Industrial scale production is required
2. Repeatability and reliability in measurements are highly emphasized by the aerospace industry
3. Production cost and performance ratio are mandatory

4. Environment, health and safety issues concerning the exposure by technicians during manufacturing and assembly, end users and environment after the disposal

### 1.19.4. Future Prospects

Carbon nanotube technology has shown to be considerably suitable and promising for chemical detection, therefore can be extended to produce gas or liquid chemical sensors that would have benefits such as extremely low power, ultra-miniature size, versatility, and cost reduction. These carbon nanotube sensors would facilitate distributed, wireless gas sensing, providing efficient multi-point measurements, or larger convenience and flexibility. In addition, carbon nanotube chemical sensors are suitable for sensing different analytes and could be configured to form an array to systematically and cost-effectively monitor multiple analytes [164]. In the short term, high-value applications for nanotube composites foresee the biggest markets in the sectors such as aerospace (which needs lightweight, high-strength, high-temperature-resistant composites), which can absorb the added costs associated with CNT composite production and processing. Markets are looking forward to the cost reduction of nanotubes, once it becomes similar to that of carbon fibres (or even to that of the much cheaper reinforcer, carbon black), products such as nanotube-filled rubber tyres are expected to emerge [165]. Fig. (**15**) demonstrates estimated market for nanotechnology in aerospace industry by 2015.

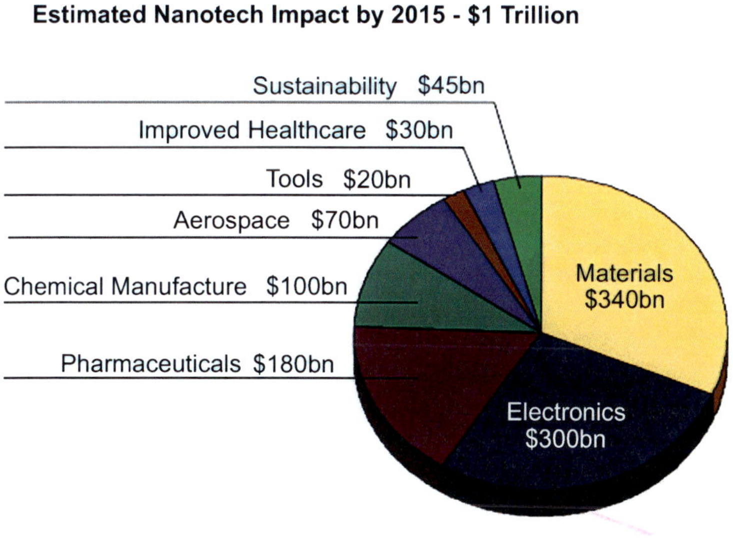


**Fig. (15).** Estimated market for Nanotechnology in Aerospace Industry (Source: U.S. National Science Foundation).

Having said that, most of the aerospace and defence application of CNTs have still been identified as long term application. NASA sees the big market for such nanocomposites in future, which is clear from the following statement:

> *"In futuristic scenario, aircraft could weight as little as half of the conventional aircraft manufactured with today's materials. Such novel materials would be extremely flexible, allowing the wing to reshape instantly and remaining extremely resistant to damage at the same time. In addition, these materials would have "self–healing" functionality. The high strength to weight ratio of nano materials could enable new airplane design that can withstand crashes and protect the passages against injury"*...NASA 2001 [162].

Future prospects of composites include their application in the following areas [169]:

1. Lighter, stronger structural components
2. Active flow control
3. Distributed propulsion
4. Electric propulsion,
5. Advanced fuel cells,
6. High efficiency electric motors

## 1.20. APPLICATION OF CNTS IN AUTOMOTIVE AND TRANSPORTATION

Intriguing mechanical and electrical properties of carbon nanotubes have made many applications possible such as light weight and strong vehicle or aircraft body enabled with *in situ* health monitoring, advanced aircraft or car brake carbon based composite discs for more efficient heat dissipation, strong windscreens with de-icing properties. Properties of CNTs that lead to such wide range of applications have been discussed under section 2 [170]. Less wear, better gliding, thinner coating, less lubricants, longer service intervals and pronounced fuel saving are some of the desired features expected from high mechanical stability [171]. Carbon nanotubes as reinforcing material, as discussed earlier, has shown to enhance properties of the polymer matrix while maintaining its functional integrity (Fig. **16**). Electrical conductivity in polymer matrix can be introduced by adding only a few percent loading of carbon nanotubes. This could lead to a solution to many problems with static electricity that could be a spark of fire within a vehicle [170].

Mechanical strength of nanotubes is reflected in the fact that deformations of nanotubes are exceptionally elastic. In case of a car crash, all the buckles and twists unfold and show no signs of damage at all [173]. This implies that nanotubes can sustain large strains under compression. Geometric parameters of nanotubes, which can be modified, define its flexibility and compliance [16].

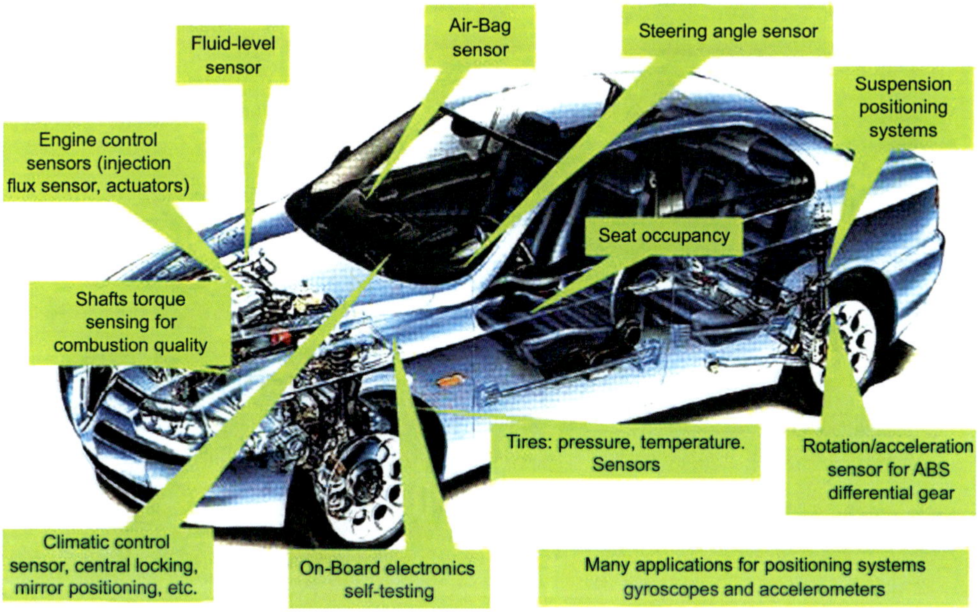


**Fig. (16).** Automotive application of nanotechnology where CNTs may hold a great potential [172].

### 1.20.1. CNT in Sensing

Sensors such as pressure, flow, thermal, gas, and chemical sensors based on CNTs find huge application in automotive industry. To determine when the air filter is dirty, a pressure sensor can be used. It measures the pressure difference at the filter's air intake and on the other side of the diaphragm. Similarly, mass air flow rate, engine speed and air temperature can be calculated by measuring the pressure from intake fold. A pressure sensor to control variable stiffness air springs is required in vehicles equipped with active-suspension systems. Crash sensing is another area where CNTs based sensors can be beneficial. This application requires fast response, within fractions of seconds. Other important areas where CNT based sensors can be effectively employed include vehicle dynamic control (VDC), roll over detection, and antitheft systems. During automobile skidding, VDC systems assist the driver. Lateral sliding of the vehicle can also be detected using CNT based sensors. Vans, trucks, and sports utility vehicles manufacturers are employing rollover detection sensors more often. The tipping over of the vehicle is determined by CNT-based sensors. These days, automakers provide antitheft systems installed in the cars to prevent car theft. Conventional antitheft systems are based on tilt detection systems, implying that the vehicle cannot be parked at higher inclinations. Also, the performance of conventional antitheft systems is affected in extensively varying automotive temperature environment. CNT-based sensors hold a promising future application in such cases. Nanotube sensors can be employed in other areas including ignition control, headway control, transmission control, vehicle navigation, tyre condition, and cabin air quality monitoring [174].

### 1.20.2. CNTs as Hydrogen Storage Material for Transport Application

Structures or materials with excessive storage capacity per unit volume and weight for gases such as hydrogen or methane are highly desired. The objective is to power fuel cells for ultralow-emission vehicles or for electric power generation using these materials. One of the biggest challenges in this area is to store these gases without using excessive pressure or huge tanks. Nanotubes, due to appropriate morphology, offer a solution to this problem by adsorbing hydrogen molecules within the tubes [173]. The adsorption potential inside the nanotubes at 50 K, was found to be about 25% higher than on the flat surface of graphite. The curvature of the surface and the related higher number of carbon atoms interacting with the hydrogen molecule facilitate hydrogen adsorption. Low temperatures further enhance hydrogen adsorption in nanotubes than on a flat surface [175].

### 1.20.3. CNT Nanocomposites

Carbon nanotubes are known to have a tensile strength 20 times, and strength-to weight ratio of 100 times that of steel, therefore could be the ultimate high-strength fibre. There have been extensive examples where nanotubes have been demonstrated to be a potential reinforcement material for enhancing properties of composite materials, polymers, metals and concrete. Being a low density material, nanotubes are very beneficial for cars, aeroplanes and space vehicles [173]. High specific strength, thermal conductivity and low coefficient of thermal expansion make metal matrix/CNT composites (MM/CNT) potentially viable in advanced load bearing applications such as aerospace and automotive structures and in electronic packaging. Lower weight of the material causes large savings in energy [176]. Commercial Al-MWNT (with strength 0.7-1 GPa and density 2.6 $gcm^{-3}$) composites possess strengths equivalent to stainless steel at one-third the density. Apparently Al-MWCNT composites are inexpensive in comparison to other Al alloys such as Al-Li alloys with similar strengths, hence more economical [43]. Metals and polymers reinforced by CNT find their application in break shoes, cylinder liners, piston rings, gears due to enhanced properties such as high strength, wear resistance, good thermal conductivity, low density [176].

### 1.20.4. Challenges

CNT technology faces many challenges which need further research. Some of the challenges are given below:

1. Optimisation of individual nanotubes is a necessity
2. Bonding between nanotubes and the material to be reinforced must be strong enough so that they can actually carry the loads
3. The distribution of the load within the nanotube must be uniform to ensure that the outermost layer does not disintegrate [173].
4. Increase in the viscosity is observed even when low wt% (< 5%) of the nanotubes is added to polymer (*e.g.*, epoxy) as reinforcing material, thereby making processing very difficult.

Besides many noticeable advantages of nanotubes-polymer composites, there is a concern regarding good interface between nanotubes and the polymer matrix and achieving good load transfer during loading from the matrix to the nanotubes. The weak bonding between nanotubes and polymer matrix is due to atomically smooth nature of nanotubes which must be engineered properly.

Secondly, nanotubes perform differently in response to a load, as compared to individual nanotubes. The organisation of nanotubes into aggregates (concentric arrangement within MWNTs and as bundles in SWNTs) causes this non-uniform distribution of load throughout the structure. This problem can be resolved by forming 3D nanocomposite of aligned interlaminar carbon-nanotube arrays directly onto SiC fibre stacks. This 3D composite offer improved multifunctional properties along the thickness direction. The 3D structure allows fastening of adjacent plies of nanotubes arrays in the composite, which offers interlaminar strength and toughness under various loading conditions. Incredible enhancements in the interlaminar fracture toughness, delamination resistance, through-thickness mechanical properties and damping, were noticed without compromising the inplane properties of the composites [16]. Finally, despite very high aspect ratios, nanotubes are not perfect reinforcing structures as compared to continuous fibre reinforcements in composites. The fact that CNTs come under the category of short fibres must be appreciated [16].

### 1.20.5. Current Developments and Commercialisation

Apart from many challenges on the way of CNT technology, it has reached far in terms of processing methods and industrial applications, many of which have now been translated into real products. For instance, carbon nanotubes are added to the some plastic part by general motor cars. The composite allows better surface qualities as they can be electrified during painting [173]. Currently 'Nanocyl', a

Belgian company, has demonstrated several successful attempts to use CNTs in different automotive applications such as automotive fuel system line components (fuel pumps, fuel filter housings, fuel line connectors fuel line clips, fuel reservoirs, fuel filler pockets), thermoplastics, epoxy resins, body parts, e-painting [177]. Massive weight reductions can be achieved by reducing the weight of the equipment upon introduction of CNTs in the parts or replacement of the part by CNT based structure. For example, a structure containing nanotubes, which can receive radio waves and turn them into sound, has been created. Thus, this structure acts as a radio which fits onto a pinhead.

### 1.20.6. Future Prospects

The use of CNTs in automotive application is expected to result in light weight and stronger vehicles, improvement in structural and thermal barrier and flame-retardant properties [178]. Nanotubes may replace carbon fibre as they are relatively lighter and offer higher-performance. Another possibility for a wide range of systems such as computers and instruments is to be built into nanotube structures without adding extra weight. Weight reduction and simplicity of design would be expected from such developments.

Carbon nanotubes are predicted to power future vehicles as well as constructing structural components. It has been shown that nanotube capacitors which can store roughly as much electrical power as a battery with the benefits such as shorter discharge and recharge times can be produced. CNTs have been found to be considerably successful as hydrogen storage material. Carbon nanotubes have emerged as miraculous material with their multifaceted application in almost every commercial field; however, practical engineering based on CNT integration into real products may not be straightforward.

## 1.21. APPLICATION IN CONSTRUCTION INDUSTRY

Concrete is reinforced with other reinforcing materials such as glass, polypropylene, carbon fibres and carbon nanotubes in order to get its mechanical properties improved. The unique properties making carbon nanotubes suitable as reinforcing material are [179].

1. CNTs are stronger than steel, but lightweight
2. CNTs can bear torsion and bending without breaking
3. Young's modulus is 1-5 TPa for SWCNTs as compared to carbon fibres with Young's modulus of about 400 GPa
4. Extraordinary thermal and electrical properties, thermally stable upto $2800^0$C in vacuum and thermal conductivity is about twice that of diamond; electrical conductivity is about 1000 times that of copper.

5. CNTs can behave as conductors or semiconductors depending upon diameter and helicity.
6. CNT are also highly flexible, being capable of bending in circles or forming knots. Like macroscopic tubes, they can buckle or flatten under appropriate loadings [180].

These properties make CNTs a material of choice for reinforcing concrete and other construction materials to achieve multifunctional high performance construction material.

### 1.21.1. CNTs in Building Materials

Concrete and other construction material can be reinforced by using carbon nanotubes. Nano-enhanced products are expected to be broadly applicable in the following areas:

#### *1.21.1.1. Surfaces and Coatings*

New materials containing CNTs are appearing to be advantageous in multifunctional coatings. Various companies are using nanomaterials to amplify vital characteristics of their product: These features include stain-resistance and colour durability, self-cleaning, increased hardness and scratch-resistance, corrosion and UV resistance, antimicrobial properties and enhanced thermal performance. Carbon nanotubes were found to be effective at killing E. coli bacteria as indicated by the research from Yale University. A new, solvent less, eco-friendly antifouling coating for marine applications has been recently developed by Nanocyl (BIOCYL™). MWCNTs incorporated in protective thermal coatings have also been developed to be used potentially in automotive and aerospace applications to reduce insulation weight [181]. Similarly, MWCNT-containing paints resists attachment of algae and barnacles, hence is used to reduce bio-fouling of ship hulls. CNTs prove to be a possible alternative to environmentally hazardous biocide-containing paints. CNTs incorporated in anticorrosion coatings for metals can lead to enhancement in coating stiffness and strength whilst offering an electric pathway for cathodic protection [43].

#### *1.21.1.2. Thermal Performance*

Potential energy efficient benefits to architectures have been attained by altering molecular properties affecting thermal performance [182].

#### *1.21.1.3. Structural Materials*

Carbon nanotubes offer strength and lightness to the structural materials, hence, are expected to be an ideal material to be used in construction materials. However, the development in building materials involving such nanomaterials is still in its infancy and needs further research [182]. Composites of MWCNTs and plastic materials demonstrate improvement in mechanical properties such that considerable weight reduction has been achieved with equal mechanical performances, thereby saving energy and resources [181].

### 1.21.2. Challenges

Some challenges with respect to material properties limit the use of nanotubes in building materials. These include:

1. Uniform dispersion and entanglement of CNTs is difficult to achieve [183, 184]
2. The adhesion between nanotubes and matrix is not proper, due to its graphitic nature causing ‘sliding’. Furthermore, since CNTs are assembled in bundles, additional sliding inside the bundle occurs which prevents nanotubes from exhibiting their unique mechanical properties.

A large number of carbon nanotubes-polymer interfaces are involved in slippage, creating friction and in turn energy dissipation. This leads to mechanical damping caused by viscoelastic behaviour at the interfaces. Such damping is imperative to many commercial applications, for example isolating objects from vibration. This requires good dispersion of nanotubes in the matrices to optimize this effect [165].

A relatively high cost involved with the production and processing of nanomaterial reinforced materials as compared to conventional materials implies that they are less likely to feature in high-volume building materials. Such technology appears to be far off from application as yet. Nonetheless, there is some progress in this area and the 'nano' tag motivates further advancement in building products [182].

### 1.21.3. Current Developments

Recently, a new carbon nanotube fibre similar to textile thread in appearance has been unveiled by the scientists from Rice University in Texas. These fibres possess excellent electrical and thermal conductivities similar to metal wire. An industrially scalable process for making the threadlike fibres has been described by researchers. These fibres outperform high-performance materials that are commercially available in a numerous ways [185].

### 1.21.4. Future Prospects

Futuristic goals include achievement of improved mechanical and shrinkage properties, which would reduce energy consumption during cement production, designing self-powered sensors to observe and record the performance of installations and development of innovative concrete materials.

On the basis of research leading to products intended specifically for the construction industry, CNT materials can be divided into three broad categories which include:

1. CNT composites made with existing construction material
2. CNT ropes for use as structural components
3. CNT heat transfer systems.

As pointed out earlier, CNTs, owing to their exceptionally high strength, toughness and aspect ratios, form an excellent reinforcing material. Potential candidates for CNT matrix materials include polymer, cement and glass. CNTs or nanofibres as reinforcement material may present an advantage that they do not interfere with optical transmittance of glass, making glass-CNT composites of particular interest.

Longer CNTs formed into ropes may possibly provide applications such as suspension bridges, where improved mechanical properties CNTs including tensile strength and elasticity would allow further benefits than existing technology. Carbon nanotubes may provide a material of choice for the construction of structures such as space elevators. This would allow cargo and goods to be shifted from space to the earth and back with minimum energy requirements. Though it remains a big challenge to engineer such structures, CNT seems to be the only material of choice so far as they can bear the massive structural loads required [180].

The thermal conductivity of carbon nanotubes presents other applications. Composite materials with improved thermal resistance may be possible, in view of the fact that an adequate density of CNTs could allow rapid heat conduction away from the contact surface to heat sinks. Futuristic applications include development of insulating materials and heat pipes, taking benefit from the variation in thermal conductivity across the tubes and along their lengths. Another application would be in the heating of buildings by replacing the existing liquid based systems for heating floors [180].

## 1.22. APPLICATION IN TEXTILES

Due to continuous advancement in technology ultra-modern gadgets are required for several day to day applications. In this regard, a new market is emerging for smart clothing. There are a range of opportunities that exist in the textile market to extend the functionality and performance of textiles to meet market demands. These smart nano-textiles will revolutionize the clothes we wear, the furnishings in our homes, and the related materials used in industry. Smart fabrics may be used in a variety of applications ranging from security to healthcare and entertainment. Nanotechnology/nanoparticles enhance the functionalities of textiles keeping the look and feel of the fabric is having a huge impact on the textile industry. The market for textiles using nanoparticles is predicted to reach \$13.6 billion in 2007 and will increase to \$115 billion by 2012. Textiles, being a common interface can be an ideal substrate for integrating sensors to monitor the wearer and the environment. Textiles offer a platform for incorporating sensing, monitoring, and information processing devices. Smart textiles can sense and react to environmental conditions and can be used to detect various physical phenomena like mechanical, thermal, chemical, electrical, or magnetic signals. Therefore, the fundamental components within smart textiles are sensors, actuators, and control electronics. The sensing elements, and control electronics must be integrated into the textile while retaining the usual physical and comfort properties of clothing. A means of smooth integration is required to develop accurate textile sensors. This is why nanoparticles are important to the smart textiles industry, enabling the incorporation of new functionalities at various production stages at the fibre-spinning level, during yarn/fabric formation, or at the finishing stage [186]. Carbon nanotubes, due to their remarkable properties, are one of the important candidate used in textile industries for making smart textiles (Table **12**). In the last decade, a number of different techniques for CNT-based fibre assembly have emerged. Specifically, there are two main types of processes by which CNTs fibres can be assembled (Fig. **17**).

**Table 12. CNT application and properties in textile in comparison to other nanofillers and their properties [188].**

| Nanofillers | Properties/Applications |
|---|---|
| Carbon nanofibres | Enhanced tensile strength, high chemical resistance and electrical conductivity |
| Carbon black nanoparticles | Increase abrasion resistance and toughness, high chemical resistance and electrical conductivity |
| Clay nanoparticles | High electrical, heat, and chemical resistance, Block UV light, anticorrosive, flame retardant. |
| Metal oxide nanoparticles ($TiO_2$, $Al_2O_3$, ZnO, MgO) | Photocatalytic ability, High electrical conductivity, UV absorption, photo-oxidizing capability against chemical and biological species, antimicrobial/self-sterilization |
| Carbon nanotubes | 100 times more tensile strength than that of steel at one-sixth the weight, high electrical conductivity similar to copper, excellent thermal conductivity |

**Fig. (17).** Scanning electron microscopy images of a carbon nanotube yarn twisted during solid-state spinning [187].

The first is solution spinning, where the nanotubes are dispersed in surfactant or solubilised in acid and then processed into a fibre. The second is solid-state spinning, where fibres are drawn from a vertically grown array of nanotubes or directly collected from the aerogel formed in the reaction furnace. Solution spinning is readily scalable but has not yet shown the ability to process long nanotubes. Solid-state spinning has exhibited the best mechanical properties for CNT fibres to date. However, the process scalability and fibre morphology are problematic. Future developments in CNT-based fibres will stem from two areas: improvements in the starting material (*i.e.* longer, defect-free, type-specific CNTs), or improvements in the processing and post-processing of the fibres. Given the wide range of achievements in CNT synthesis and fibre processing of the past decade, the prospect of high-performance CNT-based fibres is quite promising, particularly as researchers continue to apply the effective techniques of rigid-rod polymer processing to SWCNT fibres [186].

### 1.22.1. Sports

Nano-textiles are making an impact on the conventional sportswear industry. Nano-textiles have helped textile industry to produce clothes adapted to wearer's needs, while maintaining their comfort. For example, sensors have been integrated into adidas running shoes to alter its shock-absorbing characteristics according to person's style, body weight, speed and surface friction. Advanced wireless sensor has been incorporated in 'smart' running shoes released by Nike. These 'smart' sensors connect to an iPod/music player containing various playlists as well as tracks distance, time, speed, and calories burned. Another example of improved performances using engineered textiles includes 'nanotech swimsuits' for Olympic swimmers to help reduce drag.

### 1.22.2. Healthcare

Motivation behind smart textiles for healthcare is the need to observe patients for longer periods due to rehabilitation or chronic illness. Conventional clinical visits allow only a brief window on the physical condition of the patient. Wearable devices based on CNTs provide the option to observe physiological signals constantly in a realistic setting. This makes it vital for the future of the healthcare system, given the worldwide aging population. Nano-textiles provide a shift in the focus of healthcare expenses from treatment to prevention and wellness promotion.

## 1.23. HOME FURNISHING

Smart homes can be created using universal sensing in home furnishings. Wearable sensors made up of smart textiles can be used for tele-monitoring elderly or isolated individuals in comfortable surroundings at home.

## 1.24. MILITARY/SECURITY

The security and well being of the people working in extreme conditions needs be increased which requires real-time information technology to intervene. 'SmartShirt' by Sensatex™ has been developed by Georgia Tech Research Corporation for military applications. Smartshirt integrated with optical and conductive fibres, act like a sensor. Otical fibres are used to detect bullet wounds with its precise location. A range of other sensors are used to watch over body's vital signs.

## 1.25. FASHION/LIFESTYLE

High-tech advanced textiles have found their way to fashion industry with the developments in new emerging materials. Fabric with touch pads incorporated into jackets has been developed by Eleksen for additional control of different portable devices. Initially snowboarding jackets were developed for ease of access to the control buttons. A fabric keyboard has also been developed Eleksen for personal digital assistants (PDAs) that can be rolled out, easily stored, and transported (Fig. **18**). Textronics, a producer of electronic textiles, has developed fabrics that can provide warmth, illumination, conduction, and sensing abilities (Fig. **19**). Recently, NuMetrex heart rate monitoring sports bra has been developed which integrates conductive knitted sensors linked wirelessly to a heart rate monitor watch.

**Fig. (18).** Elektex fabric-based keyboard (30 x 11 cm; weight, 65 g) [188].

**Fig. (19).** Lumalive textile garment from Philips features flexible arrays of coloured light-emitting diodes fully integrated into the fabric (Philips) [188].

The extraordinary strength and lightness of the nanotube sheets suggest they could find use in high-end applications such as military aircraft or motor sports. Also, because the conductive sheets are so thin that they're nearly transparent when sandwiched between two sheets of plexiglass, they could be incorporated into automobile windshields as antennas or heating elements, or could find a home in photovoltaic cells or in artificial muscles. Despite a wide range of potential applications in the textile sector, the relatively high cost and uncertain toxic effects of CNTs has so far prevented large-scale use of CNTs in this area. However, a number of applications are at different stages of R&D. Different spinning processes can used to make CNT fibres and yarns from agglomerated CNTs, or CNT fibres can be interwoven with other polymer yarns [189].

In addition, there may be other possible applications of CNTs for example:

1. Where synthetic fibres have been surface coated with a layer of CNT.
2. Where fabrics have been printed on with a CNT-containing material.
3. Where CNTs are integrated into other polymers, through *in situ* polymerisation or dispersal in molten polymers, to produce composite master batches.

Because of the aggregation of CNTs, their uniform dispersion in polymer matrices is generally problematic, and requires chemical modification of the CNT surface. The short-term applications for textiles are expected to be mainly in the form of composites of CNTs with other polymers to improve mechanical or functional properties. It is also expected that first products incorporating CNT materials would be for the high-value sportswear, due to the current high cost of CNTs. It is of note that CNTs are likely to undergo various transformation during manufacturing, use and end-of-life treatment of the products. The health effects of potential exposure to the product-derived CNTs may be different from neat CNTs that are generally used in toxicological studies. Research is therefore needed to characterise CNTs in consumer products to inform the risk assessment process.

### 1.25.1. Challenges

Production of polymers and master batches, fibres, yarns, fabrics and finished products (*e.g.* apparel), use stage involving degradation of the matrix processes, recycling stage and disposal may all lead to nanoparticle generation in the environment. Processes such as melting, mixing, co-extruding, granulating, melting, spinning, weaving, sizing, knitting, bleaching, dyeing, printing, washing, drying/fixing, cutting, sewing, shaping, washing could lead to release of CNT as the nanotubes are more stable and will not be degraded as polymers. UV-radiation, exposure to chemicals, water, or oxidation of polymer molecules causes degradation of polymers. At least 75% of fashion textiles, in the UK are landfilled and not recycled after their use [190]. 5.5 kg textiles per inhabitant in Switzerland are disposed off annually and burned in municipal waste incinerators. Dispersion of CNT in a polymer by various processes such as the dry powder mixing method, ultrasonic treatment, *in situ* polymerisation, or using a twin-screw extruder may lead to some emission of CNT. Garments usually suffer a 10% weight loss during their life cycle [191]. CNT-containing textiles are less likely to wear and tear; however some fibre fragments may come off during use, which ends up in wastewaters, or in the environment as dust. Heat exposure of textiles during ironing or drying cycle may also lead to emission of CNT fibres in the environment. The effect of (UV) light and the physicochemical nature of the

CNTs coming off the textiles are unknown. Garments are worn close to the human body over longer periods of time. Possible exposure of nanofibres from the worn clothes may cause skin irritability. Recycling and disposal (especially at the vicinity) may enhance the risk of human exposure to CNTs if nanotubes are not destroyed upon products decomposition (*e.g.* by shredding or low-temperature incineration). This poses a big challenge for CNT based textiles.

MWCNT and SWCNT have been demonstrated to be burnt off at 740°C completely under oxidative conditions. Therefore, CNT/polymer composites are likely to incinerate completely, while any remaining CNTs may agglomerate with other particles and removed by the filter. Incineration in open fires may not destroy the CNTs, consequently, enhancing the risk *via* inhalation. Modern developments in smart nano-textiles may affect many aspects of our daily lives and produce clothing that is "smart". A major problem in wearable sensors at present is the interconnections, with conventional silicon and metal components being highly incompatible with the soft textile substrate containing carbon nanostructures. Creating a wearable garment integrates textile and fashion design with input from the end users, such as healthcare workers, defence forces, and sports physicians. Market trends suggest great opportunities for nanotechnology within the textile market; given the current pace of development, smart nanotextiles will be an integral part of our lifestyle in coming years.

## 1.26. APPLICATION OF CNTS IN SECURITY

### 1.26.1. Military Applications

Nanotechnology research is being conducted by all branches of the U.S. military including the Defence Advanced Research Projects Agency (DARPA), Office of Naval Research (ONR), Army Research Office (ARO), and Air Force Office of Scientific Research (AFOSR). Composite materials are the main area of research by The Air Force [192]. The Navy Research Laboratory's Institute for Nanoscience is involved in the study of quantum dots for their application in nano-photonics and identifying bio- materials [193]. Various military applications of nanotechnology exist today. The most palpable are in materials science. Carbon nanotubes, diamond films and fibers possess higher strength-to-weight ratios than steel. This property of carbon materials can be explored in lighter and stronger armor and parts for vehicles, equipment, and aircraft. These advanced military equipments would serve to protect soldiers from improvised explosive devices (IEDs) and small-arms fire in a better way. The U.S. Army has shown much interest in nanotechnology for use in soldiers' uniforms, equipment, and armor. A uniform that can provide flexible armor protection for soldiers' limbs by using shear thickening liquids (liquids that can solidify upon applying force) is expected to be created. Such armor would facilitate considerable reduction in the weight that a soldier must carry. Numerous other features of the uniform would include sensors, medical treatment capabilities, communications, and individual environmental control for the soldier and integrated sensing systems to be woven into the garment's fabric [194]. A more precise control of fuel combustion and blast of explosives would be offered by nanotechnology. Smaller and more powerful rockets, propellants, warheads, bombs, and other such explosive devices can be made possible through nanotechnology. Nanotechnology would also facilitate more powerful batteries, fuel cells, photovoltaic panels, and possibly simpler methods of generating electrical power for slower release of energy. Recently, piezoelectric fibers have been developed, which in future may be incorporated in fabrics to generate their own electricity, which would completely eliminate the need for batteries [195]. In electronics, creation of ultra-small computers and sensors could be possibly made using nanotechnology. There are high hope of creating an integrated package that could sense, distinguish, make decisions, record and report information, and offer control input to other devices. Nanophotonic materials find their application in the fiber optic waveguides that can actually strengthen the light beams passing through them. These materials enable them to be used for long-distance, strategic-level communications systems and/or high-power narrow-beam lasers. Other possible applications are optical computing, data storage, and signal processing.

### 1.26.2. Homeland Security Applications

Nanotechnology could play a major role in homeland security applications. Highly sensitive and 'smart' sensors could be created using nanomaterials which are capable of identifying any hazardous materials present in the air. For example, carbon-nanotubes are comparatively inexpensive and consume minimal power, therefore could serve one of the ideal materials to serve the above stated purpose.Emergency responder devices are the other areas where nanotechnology could leave its mark in homeland security. Rescuers are bestowed upon by more flexibility offered by lightweight communications systems which require negligible power and possess a large contact radius. Nanotech robots could play a major role in disarming bombs and saving trapped victims, minimising the risks to rescue workers [196].

## 1.27. CONCLUSION

The properties and characteristics of CNTs are quite promising and there is a huge potential in these structures to be the building block of futuristic devices. These tubular structures have tremendous properties, some of which has already shown great potential in futuristic devices and heavily published in literature. These structures have already proven to serve as safer and more effective alternatives drug delivery methods. In addition to these, many other potential applications have been proposed for carbon nanotubes, including hydrogen storage and energy conversion devices, sensors, field emission displays and radiation sources, nanometer-sized semiconductor devices, atomic force microscopy (AFM) probes and future applications in computers. Overall, recent studies regarding CNTs have shown a very promising glimpse of what lies ahead in the future of nanoscience and nanotechnology.

## CONFLICT OF INTEREST

The authors declare no conflict of interest, financial or otherwise.

## ACKNOWLEDGEMENTS

Thanks to all scientists referenced throughout the chapter whose valuable work has guided the way through to this research work.

## REFERENCES

[1] Iijima S. Carbon nanotubes: past, present, and future. Physica B 2002; 323: 1-5. [http://dx.doi.org/10.1016/S0921-4526(02)00869-4]

[2] Thostenson ET. Ren-Z, Chou TW. Advances in the science and technology of carbon nanotubes and their composites: a review. Compos Sci Technol 2001; 61: 1899-912. [http://dx.doi.org/10.1016/S0266-3538(01)00094-X]

[3] Popov VN. Carbon nanotubes: properties and application. Mater Sci Eng Rep 2004; 43: 61-102. [http://dx.doi.org/10.1016/j.mser.2003.10.001]

[4] Komarov FF, Mironov AM. Carbon Nanotubes: Present and Future. Physics and Chemistry of Solid State 2004; 5(3): 411-29.

[5] Dai L, Patil A, Gong X, *et al.* Aligned nanotubes. ChemPhysChem 2003; 4(11): 1150-69. [http://dx.doi.org/10.1002/cphc.200300770] [PMID: 14652993]

[6] Iijima S. Helical microtubules of graphitic carbon. Nature 1991; 354: 56-8. [http://dx.doi.org/10.1038/354056a0]

[7] Baughman RH, Zakhidov AA, de Heer WA. Carbon nanotubes--the route toward applications. Science 2002; 297(5582): 787-92. [http://dx.doi.org/10.1126/science.1060928] [PMID: 12161643]

[8] Terrones M. Science and Technology of the Twenty-First Century: Synthesis, Properties, and Applications of Carbon Nanotubes. Annu Rev Mater Res 2003; 33: 419-501. [http://dx.doi.org/10.1146/annurev.matsci.33.012802.100255]

[9] Saito Y, Nakahira T, Uemura S. Growth Conditions of Double-Walled Carbon Nanotubes in Arc Discharge. J Phys Chem B 2003; 107: 931-4. [http://dx.doi.org/10.1021/jp021367o]

[10] Hone J, Whitney M, Piskoti C, Zettl A. Thermal conductivity of single-walled carbon nanotubes. Phys Rev B 1999; 59.

[11] Chung DD. Materials for thermal conduction. Appl Therm Eng 2001; 21: 1593-605. [http://dx.doi.org/10.1016/S1359-4311(01)00042-4]

[12] Xie XL, Mai YW, Zhou XP. Dispersion and alignment of carbon nanotubes in polymer matrix: A review. Mater Sci Eng 2005; 49: 89-112. [http://dx.doi.org/10.1016/j.mser.2005.04.002]

[13] Coleman JN, Khan U, Blau WJ, Gun'ko YK. Small but strong: A review of the mechanical properties of carbon nanotube–polymer composites. Carbon 2006; 44: 1624-52. [http://dx.doi.org/10.1016/j.carbon.2006.02.038]

[14] Yu MF, Lourie O, Dyer MJ, Moloni K, Kelly TF, Ruoff RS. Strength and breaking mechanism of multiwalled carbon nanotubes under tensile load. Science 2000; 287(5453): 637-40. [http://dx.doi.org/10.1126/science.287.5453.637] [PMID: 10649994]

[15] Upadhyayula VK, Deng S, Mitchell MC, Smith GB. Application of carbon nanotube technology for removal of contaminants in drinking water: a review. Sci Total Environ 2009; 408(1): 1-13. [http://dx.doi.org/10.1016/j.scitotenv.2009.09.027] [PMID: 19819525]

[16] Endo M, Strano MS, Ajayan PM. Potential Applications of Carbon Nanotubes. Appl Phys (Berl) 2008; 111: 13-62.

[17] Donaldson K, Aitken R, Tran L, *et al.* Carbon nanotubes: a review of their properties in relation to pulmonary toxicology and workplace safety. Toxicol Sci 2006; 92(1): 5-22. [http://dx.doi.org/10.1093/toxsci/kfj130] [PMID: 16484287]

[18] Ouellette J. Building the Nanofuture with Carbon Tubes. American Institute of Physics 2003; pp. 18-21.

[19] Daniel S, Rao TP, Rao KS, *et al.* A review of DNA functionalized/grafted carbon nanotubes and their characterization. Sens Actuators B Chem 2007; 122: 672-82. [http://dx.doi.org/10.1016/j.snb.2006.06.014]

[20] http://www.cnanotech.com/2013/07/03/applications-of-carbon-nanotubes-2.html

[21] http://www.cnanotechnology.com/en/automotive.html

[22] http://www.compositesworld.com/

[23] http://www.unidym.com/technology/cnt_application_electronics.html

[24] Baddour CE, Briens C. Carbon Nanotube Synthesis: A Review. Int J Chem React Eng 2005; 3: 1-22. [http://dx.doi.org/10.2202/1542-6580.1279]

[25] http://www.cnanotechnology.com/en/pdf/CNano_Press_Release_Conductive_Paste_6-8-2010.pdf

[26] Odom TW, Huang JL, Kim P, Lieber CM. Atomic structure and electronic properties of single-walled carbon nanotubes. Nature 1998; 391: 62-4. [http://dx.doi.org/10.1038/34145]

[27] Elliott JA, Sandler JK, Windle AH, Young RJ, Shaffer MS. Collapse of single-wall carbon nanotubes is diameter dependent. Phys Rev Lett 2004; 92(9): 095501. [http://dx.doi.org/10.1103/PhysRevLett.92.095501] [PMID: 15089483]

[28] Hone J, Llaguno MC, Biercuk MJ, *et al.* Thermal properties of carbon nanotubes and nanotube-based materials. Appl Phys, A Mater Sci Process 2002; 74: 339-43. [http://dx.doi.org/10.1007/s003390201277]

[29] Flahaut E, Bacsa R, Peigney A, Laurent C. Gram-scale CCVD synthesis of double-walled carbon nanotubes. Chem Commun (Camb) 2003; (12): 1442-3. [http://dx.doi.org/10.1039/b301514a] [PMID: 12841282]

[30] Yamada T, Namai T, Hata K, *et al.* Size-selective growth of double-walled carbon nanotube forests from engineered iron catalysts. Nat Nanotechnol 2006; 1(2): 131-6. [http://dx.doi.org/10.1038/nnano.2006.95] [PMID: 18654165]

[31] Prasek J, Drbohlavova J, Chomoucka J, *et al.* Methods for carbon nanotubes synthesis—review. J

Mater Chem 2011; 21: 15872-84.
[http://dx.doi.org/10.1039/c1jm12254a]

[32] Kociak M, Suenaga K, Hirahara K, Saito Y, Nakahira T, Iijima S. Linking chiral indices and transport properties of double-walled carbon nanotubes. Phys Rev Lett 2002; 89(15): 155501.
[http://dx.doi.org/10.1103/PhysRevLett.89.155501] [PMID: 12365998]

[33] Applications and Benefits of Multi- Walled Carbon Nanotubes. http://www.cefic.org/Documents/Other/Benefits%20of%20Carbon%20Nanotubes.pdf

[34] Fam DW. Palaniappan A1, Tok AIY, Liedberg B, Moochhala SM. A review on technological aspects influencing commercialization of carbon nanotube sensors. Sens Actuators B Chem 2011; 157: 1-7.
[http://dx.doi.org/10.1016/j.snb.2011.03.040]

[35] Advances LI. Planar Lipid Bilayers and Liposomes. Elsevier 2008.

[36] Klumpp C, Kostarelos K, Prato M, Bianco A. Functionalized carbon nanotubes as emerging nanovectors for the delivery of therapeutics. Biochimica et Biophysica Acta 2006; 1758: 404-12.
[http://dx.doi.org/10.1016/j.bbamem.2005.10.008]

[37] Trojanowicz M. Analytical applications of carbon nanotubes: a review. Trends Analyt Chem 2006; 5(5)

[38] Darkrim FL, Malbrunot P, Tartaglia GP. Review of hydrogen storage by adsorption in carbon nanotubes. Int J Hydrogen Energy 2002; 27: 193-202.
[http://dx.doi.org/10.1016/S0360-3199(01)00103-3]

[39] Shanov V, Yun YH. Synthesis and Characterisation of Carbon Nanotube Materials. Journal of the University of Chemical Technology and Metallurgy 2006; 41(4): 377-90.

[40] Rafique MM, Iqbal J. Production of Carbon Nanotubes by Different Routes— A Review. Journal of Encapsulation and Adsorption Sciences 2011; 1: 29-34.
[http://dx.doi.org/10.4236/jeas.2011.12004]

[41] Khare R, Bose S. Carbon Nanotube Based Composites- A Review. J Miner Mater Charact Eng 2005; 4(1): 31-46.
[http://dx.doi.org/10.4236/jmmce.2005.41004]

[42] Meyyappan M, Delzeit L, Cassell A, Hash D. Carbon nanotube growth by PECVD: a review. Plasma Sources Sci Technol 2003; 12: 205-16.
[http://dx.doi.org/10.1088/0963-0252/12/2/312]

[43] De Volder MF, Tawfick SH, Baughman RH, Hart AJ. Carbon nanotubes: present and future commercial applications. Science 2013; 339(6119): 535-9.
[http://dx.doi.org/10.1126/science.1222453] [PMID: 23372006]

[44] Koziol K, Vilatela J, Moisala A, *et al.* High-performance carbon nanotube fiber. Science 2007; 318(5858): 1892-5.
[http://dx.doi.org/10.1126/science.1147635] [PMID: 18006708]

[45] Lima MD, Fang S, Lepró X, *et al.* Biscrolling nanotube sheets and functional guests into yarns. Science 2011; 331(6013): 51-5.
[http://dx.doi.org/10.1126/science.1195912] [PMID: 21212350]

[46] Goldberg M, Langer R, Jia X. Nanostructured materials for applications in drug delivery and tissue engineering. Journal of Biomaterial Sciences, Polymer Edition 2007; 18(3): 241-68.

[47] Ajayan PM, Zhou OZ. Applications of Carbon Nanotubes. Carbon Nanotubes. Top Appl Phys 2001; 80: 391-425.
[http://dx.doi.org/10.1007/3-540-39947-X_14]

[48] Wang J. Electrochemical biosensors: towards point-of-care cancer diagnostics. Biosens Bioelectron 2006; 21(10): 1887-92.
[http://dx.doi.org/10.1016/j.bios.2005.10.027] [PMID: 16330202]

[49] Kim SN, Rusling JF, Papadimitrakopoulos F. Carbon nanotubes for electronic and electrochemical detection of biomolecules. Adv Mater 2007; 19(20): 3214-28. [http://dx.doi.org/10.1002/adma.200700665] [PMID: 18846263]

[50] Carbon Nanotubes Form Ultrasensitive Biosensor to Detect Proteins. 2010. http://www.sciencedaily.com/releases/2010/06/100627155118.htm

[51] Nanotube sensor detects Lyme disease. 2013. http://physicsworld.com/cws/article/news/2013/jun/13/nanotube-sensor-detects-lyme-disease

[52] Yu X, Munge B, Patel V, *et al.* Carbon nanotube amplification strategies for highly sensitive immunodetection of cancer biomarkers. J Am Chem Soc 2006; 128(34): 11199-205. [http://dx.doi.org/10.1021/ja062117e] [PMID: 16925438]

[53] Mellker RJ, Dennis DM. Application of Biosensors for Diagnosis and Treatment of Disease United States Patent 006974706B1 2005.

[54] Hu H, Ni Y, Montana V, Haddon RC, Parpura V. Chemically functionalized carbon nanotubes as substrates for neuronal growth. Nano Lett 2004; 4(3): 507-11. [http://dx.doi.org/10.1021/nl035193d] [PMID: 21394241]

[55] Rusling JF, Kumar CV, Gutkind JS, Patel V. Measurement of biomarker proteins for point-of-care early detection and monitoring of cancer. Analyst (Lond) 2010; 135(10): 2496-511. [http://dx.doi.org/10.1039/c0an00204f] [PMID: 20614087]

[56] Malhotra R, Patel V, Vaqué JP, Gutkind JS, Rusling JF. Ultrasensitive electrochemical immunosensor for oral cancer biomarker IL-6 using carbon nanotube forest electrodes and multilabel amplification. Anal Chem 2010; 82(8): 3118-23. [http://dx.doi.org/10.1021/ac902802b] [PMID: 20192182]

[57] Sahoo SK, Parveen S, Panda JJ. The present and future of nanotechnology in human health care. Nanomedicine (Lond) 2007; 3(1): 20-31. [http://dx.doi.org/10.1016/j.nano.2006.11.008] [PMID: 17379166]

[58] Craighead H. Future lab-on-a-chip technologies for interrogating individual molecules. Nature 2006; 442(7101): 387-93. [http://dx.doi.org/10.1038/nature05061] [PMID: 16871206]

[59] Wang YQ, Fan L, Zhao WM, Qian L, Feng C, Liu L. Pacemakers and pacemaker electrodes United States Patent20130158643A1 2013.

[60] Bianco A, Kostarelos K, Prato M. Applications of carbon nanotubes in drug delivery. Curr Opin Chem Biol 2005; 9(6): 674-9. [http://dx.doi.org/10.1016/j.cbpa.2005.10.005] [PMID: 16233988]

[61] Prato M, Kostarelos K, Bianco A. Functionalized carbon nanotubes in drug design and discovery. Acc Chem Res 2008; 41(1): 60-8. [http://dx.doi.org/10.1021/ar700089b] [PMID: 17867649]

[62] Lacerda L, Bianco A, Prato M, Kostarelos K. Carbon nanotubes as nanomedicines: from toxicology to pharmacology. Adv Drug Deliv Rev 2006; 58(14): 1460-70. [http://dx.doi.org/10.1016/j.addr.2006.09.015] [PMID: 17113677]

[63] Kam NWS, Dai H. Carbon Nanotubes as Intracellular Protein Transporters: Generality and Biological Functionality

[64] Yang W, Thordarson P, Gooding JJ, Ringer SP, Braet F. Carbon nanotubes for biological and biomedical applications. Nanotechnology 2007; 18: 1-12. [http://dx.doi.org/10.1088/0957-4484/18/41/412001]

[65] Shi X, Sitharaman B, Pham QP, *et al.* Fabrication of porous ultra-short single-walled carbon nanotube nanocomposite scaffolds for bone tissue engineering. Biomaterials 2007; 28(28): 4078-90. [http://dx.doi.org/10.1016/j.biomaterials.2007.05.033] [PMID: 17576009]

[66] Zhang Y, Bai Y, Yan B. Functionalized carbon nanotubes for potential medicinal applications. Drug Discov Today 2010; 15(11-12): 428-35. [http://dx.doi.org/10.1016/j.drudis.2010.04.005] [PMID: 20451656]

[67] Liu Z, Tabakman S, Welsher K, Dai H. Carbon Nanotubes in Biology and Medicine: *In vitro* and *in vivo* Detection, Imaging and Drug Delivery. Nano Res 2009; 2(2): 85-120. [http://dx.doi.org/10.1007/s12274-009-9009-8] [PMID: 20174481]

[68] Welsher K, Liu Z, Daranciang D, Dai H. Selective probing and imaging of cells with single walled carbon nanotubes as near-infrared fluorescent molecules. Nano Lett 2008; 8(2): 586-90. [http://dx.doi.org/10.1021/nl072949q] [PMID: 18197719]

[69] Liu Z, Cai W, He L, *et al. In vivo* biodistribution and highly efficient tumour targeting of carbon nanotubes in mice. Nat Nanotechnol 2007; 2(1): 47-52. [http://dx.doi.org/10.1038/nnano.2006.170] [PMID: 18654207]

[70] Shi H, Xia T, Nel AE, Yeh JI. Part II: coordinated biosensors--development of enhanced nanobiosensors for biological and medical applications. Nanomedicine (Lond) 2007; 2(5): 599-614. [http://dx.doi.org/10.2217/17435889.2.5.599] [PMID: 17976023]

[71] Kam NW, Liu Z, Dai H. Functionalization of carbon nanotubes *via* cleavable disulfide bonds for efficient intracellular delivery of siRNA and potent gene silencing. J Am Chem Soc 2005; 127(36): 12492-3. [http://dx.doi.org/10.1021/ja053962k] [PMID: 16144388]

[72] Chen X, Tam UC, Czlapinski JL, *et al.* Interfacing carbon nanotubes with living cells. J Am Chem Soc 2006; 128(19): 6292-3. [http://dx.doi.org/10.1021/ja060276s] [PMID: 16683774]

[73] Mauter MS, Elimelech M. Environmental applications of carbon-based nanomaterials. Environ Sci Technol 2008; 42(16): 5843-59. [http://dx.doi.org/10.1021/es8006904] [PMID: 18767635]

[74] Dujardin E, Ebbesen TW, Krishnan A, Treacy MM. Wetting of Single Shell Carbon Nanotubes. Adv Mater 1998; 10: 1472-5. [http://dx.doi.org/10.1002/(SICI)1521-4095(199812)10:17<1472::AID-ADMA1472>3.0.CO;2-R]

[75] Peng XJ, Luan ZK, Ding J, Di ZH, Li YH, Tian BH. Ceria Nanoparticles Supported on Carbon Nanotubes for the Removal of Arsenate from Water. Mater Lett 2005; 59: 399-403. [http://dx.doi.org/10.1016/j.matlet.2004.05.090]

[76] Di ZC, Ding J, Peng XJ, Li YH, Luan ZK, Liang J. Chromium adsorption by aligned carbon nanotubes supported ceria nanoparticles. Chemosphere 2006; 62(5): 861-5. [http://dx.doi.org/10.1016/j.chemosphere.2004.06.044] [PMID: 16403560]

[77] Majumder M, Chopra N, Andrews R, Hinds BJ. Nanoscale hydrodynamics: enhanced flow in carbon nanotubes. Nature 2005; 438(7064): 44-4. [http://dx.doi.org/10.1038/438044a] [PMID: 16267546]

[78] Wang Z, Ci L, Chen L, Nayak S, Ajayan PM, Koratkar N. Polarity-dependent electrochemically controlled transport of water through carbon nanotube membranes. Nano Lett 2007; 7(3): 697-702. [http://dx.doi.org/10.1021/nl062853g] [PMID: 17295548]

[79] Brady-Estévez AS, Kang S, Elimelech M. A single-walled-carbon-nanotube filter for removal of viral and bacterial pathogens. Small 2008; 4(4): 481-4. [http://dx.doi.org/10.1002/smll.200700863] [PMID: 18383192]

[80] Chen Z, Zhang L, Tang Y, Jia Z. Adsorption of Nicotine and Tar from the Main Stream Smoke of Cigarettes by Oxidized Carbon Nanotubes. Appl Surf Sci 2006; 252: 2933-7. [http://dx.doi.org/10.1016/j.apsusc.2005.04.044]

[81] Kang S, Pinault M, Pfefferle LD, Elimelech M. Single-walled carbon nanotubes exhibit strong

antimicrobial activity. Langmuir 2007; 23(17): 8670-3.
[http://dx.doi.org/10.1021/la701067r] [PMID: 17658863]

[82] Li Q, Mahendra S, Lyon DY, *et al.* Antimicrobial nanomaterials for water disinfection and microbial control: potential applications and implications. Water Res 2008; 42(18): 4591-602.
[http://dx.doi.org/10.1016/j.watres.2008.08.015] [PMID: 18804836]

[83] Kotani T, Ueno M, Kawai N, Kitamoto S. Development of a New Radiation Detector Utilising Carbon Nanotube as Anode. Physica E 2007; 40(2): 422-4.
[http://dx.doi.org/10.1016/j.physe.2007.06.044]

[84] Masciangioli T, Zhang WX. Environmental technologies at the nanoscale. Environ Sci Technol 2003; 37(5): 102A-8A.
[http://dx.doi.org/10.1021/es0323998] [PMID: 12666906]

[85] Matsumoto Y, Unal U, Tanaka N, Kudo A, Kato H. Electrochemical approach to evaluate the mechanism of photocatalytic water splitting on oxide photocatalysts. J Solid State Chem 2004; 177: 4205-12.
[http://dx.doi.org/10.1016/j.jssc.2004.08.001]

[86] Sheeney-Haj-Khia L, Basnar B, Willner I. Efficient Generation of Photocurrents by Using CdS/Carbon Nanotube Assemblies on Electrodes. Angew Chem Int Ed 2005; 44: 78-83.
[http://dx.doi.org/10.1002/anie.200461666]

[87] Kongkanand A, Kuwabata S, Girishkumar G, Kamat P. Single-wall carbon nanotubes supported platinum nanoparticles with improved electrocatalytic activity for oxygen reduction reaction. Langmuir 2006; 22(5): 2392-6.
[http://dx.doi.org/10.1021/la052753a] [PMID: 16489834]

[88] Morones JR, Elechiguerra JL, Camacho A, *et al.* The bactericidal effect of silver nanoparticles. Nanotechnology 2005; 16(10): 2346-53.
[http://dx.doi.org/10.1088/0957-4484/16/10/059] [PMID: 20818017]

[89] Sozer N, Kokini JL. Nanotechnology and its applications in the food sector. Trends Biotechnol 2009; 27(2): 82-9.
[http://dx.doi.org/10.1016/j.tibtech.2008.10.010] [PMID: 19135747]

[90] Duncan TV. Applications of Nanotechnology in Food Packaging and Food Safety: Barrier Materials. Antimicrobials and Sensors
[http://dx.doi.org/10.1016/j.jcis.2011.07.017]

[91] Joseph T, Morrison M. Nanotechnology in Agriculture and Food 2006. www.nanoforum.org

[92] de Azeredo HM. Nanocomposites for Food Packaging Applications. Food Res Int 2009; 42: 1240-53.
[http://dx.doi.org/10.1016/j.foodres.2009.03.019]

[93] Evans J, Brown T, Leducq D, *et al.* Emerging Refrigeration Technologies at Laboratory Scale To Improve Food Quality And Reduce Environmental Impact And Energy Consumption http://www.icef11.org/content/papers/inm/INM1324.pdf

[94] Chinnamuthu CR, Boopathi PM. Nanotechnology and Agroecosystem. Madras Agric J 2009; 96(1-6): 17-31.

[95] Zimadahl RL. Fundamentals of weed science. 2nd ed., San Diego, CA: Academic Press 1999.

[96] http://www.smartplanet.com/blog/smart-takes/nano-rfid-tags-could-replace-barcodes-smart-groceries-bandages-coming/5481

[97] Aryasomayajula L, Wolter KJ. Carbon Nanotube Composites for Electronic Packaging Applications: A Review. J Nanotechnol 2013; 1-6.
[http://dx.doi.org/10.1155/2013/296517]

[98] Nai SM, Wei J, Gupta M. Effect of Carbon Nanotubes on the Shear Strength and Electrical Resistivity of a Lead-Free Solder. J Electron Mater 2008; 37(4): 515-22.

[http://dx.doi.org/10.1007/s11664-008-0379-6]

[99] Nanoscience and Nanotechnologies. Opportunities and Uncertainties U.K.RS/RAE.2004, http://www.nanotec.org.uk/finalReeport.htm

[100] Halliday JE. Parliament votes for tougher additives regulation http://www.foodnavigator.com/news/ng.asp?n=78139-

[101] Weiss J, Takhistov P, McClements DJ. Functional Materials in Food Nanotechnology. J Food Sci 2006; 71(9): R107-16.
[http://dx.doi.org/10.1111/j.1750-3841.2006.00195.x]

[102] Chaudhry Q, Scotter M, Blackburn J, *et al.* Applications and implications of nanotechnologies for the food sector. Food Addit Contam Part A Chem Anal Control Expo Risk Assess 2008; 25(3): 241-58.
[http://dx.doi.org/10.1080/02652030701744538] [PMID: 18311618]

[103] http://www.nanowerk.com/spotlight/spotid=23118.php

[104] Jorio A, Dresselhaus G, Dresselhaus MS. Potential Applications of Carbon Nanotubes. Top Appl Phys 2008; 111: 3-62.
[http://dx.doi.org/10.1007/978-3-540-72865-8]

[105] http://www.appliednanotech.net/tech/cnt_displays.php

[106] Rinzler AG, Hafner JH, Nikolaev P, *et al.* Unraveling nanotubes: field emission from an atomic wire. Science 1995; 269(5230): 1550-3.
[http://dx.doi.org/10.1126/science.269.5230.1550] [PMID: 17789445]

[107] Heer WA, Chatelain A, Ugarte D. A carbon nanotube field emission electron source. Science 1995; 270: 1179-80.
[http://dx.doi.org/10.1126/science.270.5239.1179]

[108] Saito Y, Uemura S. Field emission from carbon nanotubes and its applications to electron sources. Carbon 2000; 38: 169-82.
[http://dx.doi.org/10.1016/S0008-6223(99)00139-6]

[109] Fan S, Chapline MG, Franklin NR, Tombler TW, Cassell AM, Dai H. Self-oriented regular arrays of carbon nanotubes and their field emission properties. Science 1999; 283(5401): 512-4.
[http://dx.doi.org/10.1126/science.283.5401.512] [PMID: 9915692]

[110] Choi WB, Chung DS, Kang JH, *et al.* Fully sealed high brightness carbon nanotube field emission display. Appl Phys Lett 1999; 75: 3219-31.
[http://dx.doi.org/10.1063/1.125253]

[111] Bonard JM, Salvetat JP, Stockli T, Deheer WA, Forro LA, Chatelain A. Field emission from single-wall carbon nanotube film. Appl Phys Lett 1998; 73: 918-20.
[http://dx.doi.org/10.1063/1.122037]

[112] Wei Y, Weng D, Yang Y, *et al.* Efficient Fabrication of Field Electron Emitters from the Multi-Walled Carbon Nanotube Yarns. Appl Phys Lett 2006; 89(6): 063101.
[http://dx.doi.org/10.1063/1.2236465]

[113] Liu P, Wei Y, Jiang K, *et al.* Thermionic Emission and Work Function of Multi-Walled Carbon Nanotube Yarns. Phys Rev B 2006; 73(23): 235412-1-.
[http://dx.doi.org/10.1103/PhysRevB.73.235412]

[114] www.xintek.com/index.htm

[115] Cheng Y, Zhang J, Lee YZ, *et al.* Dynamic Radiography Using a Carbon Nanotube- Based Field-Emission X-ray Source. Rev Sci Instrum 2004; 75(10): 3264-7.
[http://dx.doi.org/10.1063/1.1791313]

[116] Zhang J, Yang G, Lee YZ, Chang S, Lu JP, Zhou O. Multiplexing Radiography Using a Carbon Nanotube Based X-ray Source. Appl Phys Lett 2006; 89(6)

[http://dx.doi.org/10.1063/1.2234744]

[117] Zhang JY, Cheng YZ, Lee BG, *et al.* A Nanotube-Based Field Emission X-ray Source for Microcomputed Tomography. Rev Sci Instrum 2005; 76(9): 94301-5. [http://dx.doi.org/10.1063/1.2041589]

[118] www.wtec.org/cnm/

[119] http://en.wikipedia.org/wiki/45nm

[120] Rossnagel SM, Kaun TS. Alteration of Cu Conductivity in the Size Effect Regime. J Vac Sci Technol B 2004; 22: 240-7. [http://dx.doi.org/10.1116/1.1642639]

[121] White CT, Todorov TN. Quantum electronics. Nanotubes go ballistic. Nature 2001; 411(6838): 649-51. [http://dx.doi.org/10.1038/35079720] [PMID: 11395751]

[122] Naeemi AR, Sarvari R, Meindl JD. Performance Comparison Between Carbon Nanotube and Copper Interconnects for Gigascale Integration (GSI). IEEE Electron Device Lett 2005; 26: 84-6. [http://dx.doi.org/10.1109/LED.2004.841440]

[123] http://www.nantero.com/

[124] www.eikos.com/advantages.html

[125] Dresselhaus MS, Dresselhaus G, Sugihara K, Spain IL, Goldberg HA. Graphite Fibres and Filaments. Berlin, Heidelberg: Springer 1988. [http://dx.doi.org/10.1007/978-3-642-83379-3]

[126] Hafner JH, Cheung CL, Woolley AT, Lieber CM. Structural and functional imaging with carbon nanotube AFM probes. Prog Biophys Mol Biol 2001; 77(1): 73-110. [http://dx.doi.org/10.1016/S0079-6107(01)00011-6] [PMID: 11473787]

[127] Hinds BJ, Chopra N, Rantell T, Andrews R, Gavalas V, Bachas LG. Aligned multiwalled carbon nanotube membranes. Science 2004; 303(5654): 62-5. [http://dx.doi.org/10.1126/science.1092048] [PMID: 14645855]

[128] Holt JK, Park HG, Wang Y, *et al.* Fast mass transport through sub-2-nanometer carbon nanotubes. Science 2006; 312(5776): 1034-7. [http://dx.doi.org/10.1126/science.1126298] [PMID: 16709781]

[129] Srivastava A, Srivastava ON, Talapatra S, Vajtai R, Ajayan PM. Carbon nanotube filters. Nat Mater 2004; 3(9): 610-4. [http://dx.doi.org/10.1038/nmat1192] [PMID: 15286755]

[130] Kordás K, Mustonen T, Tóth G, *et al.* Inkjet printing of electrically conductive patterns of carbon nanotubes. Small 2006; 2(8-9): 1021-5. [http://dx.doi.org/10.1002/smll.200600061] [PMID: 17193162]

[131] Ericson LM, Fan H, Peng H, *et al.* Macroscopic, neat, single-walled carbon nanotube fibers. Science 2004; 305(5689): 1447-50. [http://dx.doi.org/10.1126/science.1101398] [PMID: 15353797]

[132] Li YL, Kinloch IA, Windle AH. Direct spinning of carbon nanotube fibers from chemical vapor deposition synthesis. Science 2004; 304(5668): 276-8. [http://dx.doi.org/10.1126/science.1094982] [PMID: 15016960]

[133] Wei J, Zhu H, Wu D, Wei B. Carbon Nanotube Filaments in Household Light Bulbs. Appl Phys Lett 2004; 84: 4869-71. [http://dx.doi.org/10.1063/1.1762697]

[134] Cao A, Veedu VP, Li X, Yao Z, Ghasemi-Nejhad MN, Ajayan PM. Multifunctional brushes made from carbon nanotubes. Nat Mater 2005; 4(7): 540-5.

[http://dx.doi.org/10.1038/nmat1415] [PMID: 15951816]

[135] Yurdumakan B, Raravikar NR, Ajayan PM, Dhinojwala A. Synthetic gecko foot-hairs from multiwalled carbon nanotubes. Chem Commun (Camb) 2005; 30(30): 3799-801. [http://dx.doi.org/10.1039/b506047h] [PMID: 16041421]

[136] Subic A, Mouritz A, Troynikov O. Sustainable Design and Environmental Impact of Materials in Sports Products. Sports Technologies 2009; 2(3–4): 67-79.

[137] Allsopp M, Walters A, Santillo D. Nanotechnologies and Nanomaterials in Electrical and Electronic Goods: A Review of Uses and Health Concerns. Greenpeace Research Laboratories, Technical Note 2007.

[138] http://www.understandingnano.com/nanotubes-carbon.html

[139] http://cleantechnica.com/2008/09/15/carbon-nanotubes-might-be-used-in-future-water-filters/

[140] Nativi K, Estes B, Mullenberg S, Patel SK. Testing of solar refrigeration units using carbon nanotubes and methanol 2009.

[140] http://www.fuelcells.org

[141] Smitha B, Sridhar S, Khan A. Solid Polymer Electrolyte Membranes for Fuel Cell Applications – A Review. J Membr Sci 2005; 259: 10-26. [http://dx.doi.org/10.1016/j.memsci.2005.01.035]

[142] Stumper J, Stone C. Recent Advances in Fuel Cell Technology at Ballard. J Power Sources 2008; 176: 468-76. [http://dx.doi.org/10.1016/j.jpowsour.2007.08.071]

[143] Zhang Y, Zhang H, Zhu X, Gang L, Bi C, Liang Y. Fabrication and Characterization of a PTFE-Reinforced Integral Composite Membrane for Self-Humidifying PEMFC. J Power Sources 2007; 165: 786-92. [http://dx.doi.org/10.1016/j.jpowsour.2006.12.060]

[144] Ahluwalia R, Wang X. Fuel Cell Systems for Transportation: Status and Trends. J Power Sources 2008; 177: 167-76. [http://dx.doi.org/10.1016/j.jpowsour.2007.10.026]

[145] Sotowa C, Origi G, Takeuchi M, *et al.* The reinforcing effect of combined carbon nanotubes and acetylene blacks on the positive electrode of lithium-ion batteries. ChemSusChem 2008; 1(11): 911-5. [http://dx.doi.org/10.1002/cssc.200800170] [PMID: 18980237]

[146] http://www.researchandmarkets.com/reports/1092074/

[147] Anderman M. The challenge to fulfill electrical power requirements of advanced vehicles. J Power Sources 2004; 127(1): 2-7. [http://dx.doi.org/10.1016/j.jpowsour.2003.09.002]

[148] www.IDTechEx.com/EDLC

[149] Subash S, Chowdhury M. High Efficiency Carbon Nanotube Based Solar Cells For Electronics Devices. Proceedings of the 2009 12th International Symposium on.

[150] Bullis K. 2012. http://www.technologyreview.com/news/429170/newmethod-makes-solar-cell-production-cheaper/

[151] Dirjish M. What's The Difference Between Thin-Film And Crystalline-Silicon Solar Panels 2012. http://electronicdesign.com/power-sources/what-sdifference-between-thin-film-and-crystalline-silicon-solarpanels

[152] Klinger C, Patel Y, Postuma HW. Carbon Nanotube Solar Cells 2010. http://www.plosone.org/article/info%3Adoi%2F10.1371

[153] Bindl DJ, Wu MY, Prehn FC, Arnold MS. Efficiently harvesting excitons from electronic type-

controlled semiconducting carbon nanotube films. Nano Lett 2011; 11(2): 455-60. [http://dx.doi.org/10.1021/nl1031343] [PMID: 21166422]

[154] Jain RM, Howden R, Tvrdy K, *et al.* Polymer-free near-infrared photovoltaics with single chirality (6,5) semiconducting carbon nanotube active layers. Adv Mater 2012; 24(32): 4436-9. [http://dx.doi.org/10.1002/adma.201202088] [PMID: 22740144]

[155] Bernardi M, Lohrman J, Kumar PV, *et al.* Nanocarbon-based photovoltaics. ACS Nano 2012; 6(10): 8896-903. [http://dx.doi.org/10.1021/nn302893p] [PMID: 22953746]

[156] Xu W, Choi S, Allen M. Carbon-Fibre-Based Solar Cell. Micro Electro Mechanical Systems (MEMS). IEEE 23rd International Conference on IEEE.

[157] Chandler D. All-carbon solar cell harnesses infrared light 2010.

[158] Berger M. Novel flexible solar cell made from carbon nanotube fibres http://www.nanowerk.com/spotlight/spotid=19933.php

[159] http://www.unidym.com/press/pr_080312.html

[160] Taczak MD. A brief review of nanomaterials for aerospace applications: carbon nanotube-reinforced polymer composites. MITRE Nanosystems Group 2006.

[161] www.dtsc.ca.gov

[162] www.nasc.com

[163] www.caneus.org/devices

[164] Li J, Lu Y. Carbon Nanotube Based Chemical Sensors for Space and Terrestrial Applications. ECS Trans 2009; 19(6): 7-15.

[165] Ajayan PM, Tour JM. Materials science: nanotube composites. Nature 2007; 447(7148): 1066-8. [http://dx.doi.org/10.1038/4471066a] [PMID: 17597753]

[166] Bellucci S, Balasubramanian C, Borin P, Micciulla F, Rinaldi G. CNT Composites for Aerospace Applications. J Exp Nanosci 2007; 2(3): 193-206. [http://dx.doi.org/10.1080/17458080701376348]

[167] http://advancedtextilessource.com/2013/06/carbon-nanotube-fabrics-the-new-nonwovens/

[168] Du JH, Bai J, Cheng HM. The Present Status and Key Problems of Carbon Nanotube Based Polymer Composites. Express Polym Lett 2007; 1(5): 253-73. [http://dx.doi.org/10.3144/expresspolymlett.2007.39]

[169] http://www.advancenanocarbon.com/

[170] Transport Applications of Carbon Nanomaterials. www.cnt-ltd.co.uk/nano/information/ applications-of-carbon-nanomaterials-in-transport/

[171] Presting H, König U. Future Nanotechnology Developments for Automotive Applications. Mater Sci Eng C 2003; 23: 737-41. [http://dx.doi.org/10.1016/j.msec.2003.09.120]

[172] http://www.nanowerk.com/spotlight/spotid=18972.php

[173] Mamalis AG, Vogtländer LO, Markopoulos A. Nanotechnology and Nanostructured Materials: Trends in Carbon Nanotubes. Precis Eng 2004; 28: 16-30. [http://dx.doi.org/10.1016/j.precisioneng.2002.11.002]

[174] Sinha N, Ma J, Yeow JT. Carbon nanotube-based sensors. J Nanosci Nanotechnol 2006; 6(3): 573-90. [http://dx.doi.org/10.1166/jnn.2006.121] [PMID: 16573108]

[175] Schlapbach L, Züttel A. Hydrogen-storage materials for mobile applications. Nature 2001; 414(6861): 353-8.

[http://dx.doi.org/10.1038/35104634] [PMID: 11713542]

[176] Bakshi SR, Agarwal A. An Analysis of the Factors Affecting Strengthening in Carbon Nanotube Reinforced Aluminium Composites. Carbon 2011; 49: 533-44. [http://dx.doi.org/10.1016/j.carbon.2010.09.054]

[177] http://www.nanocyl.com/Products-Solutions/Sectors/Automotive

[178] www.advancenanocarbon.com/default.asp?tag=5

[179] Ruoff RS, Lorents DC. Mechanical and Thermal Properties of Carbon Nanotubes. Carbon 1995; 33(7): 925-30. [http://dx.doi.org/10.1016/0008-6223(95)00021-5]

[180] Kovaleva AA. Carbon Nanotubes in the Construction Industry conf.sfu-kras.ru/sites/mn2012/thesis/s023/s023-009.pdf

[181] http://www.nanocyl.com/

[182] www.riba-insight.com/

[183] www.riba-insight.com

[184] Gian H, Greenhalgh ES, Shaffer MS, Bismarck A. Carbon Nanotube-Based Hierarchical Composites: A Review. J Mater Chem 2010; 20: 4751-62. [http://dx.doi.org/10.1039/c000041h]

[185] http://www.innovationintextiles.com/

[186] Behabtu N, Green MJ, Pasquali M. Carbon Nanotube-Based Neat Fibres. Nano Today 2008; 3(5-6): 24-34. [http://dx.doi.org/10.1016/S1748-0132(08)70062-8]

[187] Zhang M, Atkinson KR, Baughman RH. Multifunctional carbon nanotube yarns by downsizing an ancient technology. Science 2004; 306(5700): 1358-61. [http://dx.doi.org/10.1126/science.1104276] [PMID: 15550667]

[188] Coyle S, Wu Y, Lau KT, De Rossi D, Wallace G, Diamond D. Smart Nanotextiles: A Review of Materials and Applications. MRS Bull 2007; 32(05): 434-42. [http://dx.doi.org/10.1557/mrs2007.67]

[189] www.safenano.org

[190] Watch W. Textile recycling information sheet http://www. wasteonline.org.uk2006.

[191] Köhler AR, Som C, Helland A, Gottschalk F. Studying the Potential Release of Carbon Nanotubes Throughout the Application Life Cycle. J Clean Prod 2008; 16(8-9): 927-37. [http://dx.doi.org/10.1016/j.jclepro.2007.04.007]

[192] Laboratories DO. www.nanosra.nrl.navy.mil/laboratories.php2007.

[193] www.nrl.navy.mil/nanoscience/publications.html2007.

[194] www.globalsecurity.org/military/systems/ground/ofw.htm2007.

[195] Machines NM. www.eurekalert.org/pub_releases/2006-04/nsf-ms041306.php2007.

[196] Carafano JJ, Gudgel A. Nanotechnology and National Security: Small Changes, Big Impact. Heritage Foundation Backgrounder 2007; p. 2071.

CHAPTER 2

# Methods for the Synthesis of Carbon Nanotubes

**Jagjiwan Mittal***

*Amity Institute of Nanotechnology, Amity University, NOIDA, Uttar Pradesh, India*

**Abstract:** Carbon nanotubes (CNTs) have received the attention of researchers for the past 25 years. It is one of the most investigative material in the history of Nanoscience and mainly responsible for current Nanotechnology boom. This chapter describes the various techniques of synthesis of CNTs such as arc discharge, laser ablation and chemical vapor deposition. Special techniques like use of liquid solution in arc discharge, lithographic techniques, solar, electrolysis and hydrothermal for CNTs synthesis are also included. The chapter also discusses the mechanism proposed for the development of carbon nanotubes in these methods and summarizes the removal of impurities from these synthesized CNTs by chemical, physical, and multi-steps purifications.

**Keywords:** Arc discharge, Carbon nanotubes (CNTs), Chemical Purifications, Dry oxidation, Electrolysis, Hydrothermal, Laser ablation: CVD, Mechanisms, MWCNTs, Physical purifications, Solar, Synthesis of CNTs, SWCNTs, Wet oxidation.

## 2.1. INTRODUCTION

After first description in 1952 [1], carbon nanotubes (CNTs) were observed in 1976 [2]. However, Iijima got the fame of discovering CNTs preparation in 1991 [3]. Since then, a number of techniques were reported for the synthesis of CNTs. Among these techniques, three most preferable methods are arc discharge, laser ablation and chemical vapor deposition. First two are "bottom down" processes, whereas third is the "bottom up" process. Although techniques using very high temperatures such as arc discharge or laser ablation were first reported but the method mostly used to synthesize CNTs now a days is the low temperature chemical vapor deposition (CVD) ($<800°C$), because the orientation, alignment, length, diameter, purity and density of CNTs can be precisely controlled in this technique [4]. This method can produce large quantities of nanotubes on industrial-scale to make them economically viable. Along with these three, other

* **Corresponding author Jagjiwan Mittal:** Amity Institute of Nanotechnology, Amity University, NOIDA, Uttar Pradesh, India; Tel: +911204392130; E-mail: jmittal@amity.edu

techniques such solar, hydrothermal, electrochemical methods are also described herein.

Purification of CNTs has been a puzzling work, since their discovery. It is a must requirement for the applications of CNTs in various fields [5 - 9]. Soot formed during CNTs synthesis in different processes usually contains a number of impurities such as metal particles, amorphous carbon, and carbon shells. In Arc discharge, the impurity problem is much more severe than in CVD. Separation between nano graphite shells, carbon onions, amorphous carbons, metal particles, carbon nanotubes is still challenging. It is difficult due to insolubility of CNT and other carbon materials except fullerenes in different solvents and the limitation of liquid chromatography. The chapter discusses the various methods of synthesizing CNTs particularly using arc discharge, laser ablation and CVD techniques in detail as well as the different techniques reported for their purifications from the soot.

## 2.2. SYNTHESIS OF CNTS

### 2.2.1. Electric Arc Discharge Method

Arc discharge is an electrical phenomenon, in which gas electrically breaksdown and creates an electrical discharge [10]. High amount of current flow through a normally nonconductive medium *e.g.* air produces a plasma, which may produce visible light. Although the arc discharge method is same in synthesizing the multi-walled or single walled carbon nanotubes but there is a slight difference in the type of raw material being used.

#### *2.2.1.1. Synthesis of Multiwalled Carbon Nanotubes (MWCNTs)*

A schematic diagram for the arc discharge technique for the synthesis of carbon nanotubes is shown in Fig. (**1**). Chamber shown is a vacuum chamber and synthesis is carried out using two electrodes of diameter between 6 mm to 12 mm. One of them, anode is the high purity graphite rod and cathode may be either pure graphite or metal rod. Both electrodes are enclosed in a chamber having water cooled surface. At the start of the synthesis, electrodes are brought very close to each other (1-2 mm) until the flow of current reaches between 50-150A. This produces the plasma between the electrodes. The synthesis is carried out at low pressure of inert and/or reactant gas. When temperature is reached at 5000°C, it sublimes the graphite anode (positive electrode) and consumed in the process. To continue the plasma for longer periods, gap between the electrodes must be maintained. Since anode is consumed in the process, either of the electrode should be moved manually or electronically for keeping the constant gap between the electrodes. Depending on the conditions, the reaction time varies from 30 s to 10 min. During consumption of graphite in the arc, various products are formed and collected in the different parts of the reactor. Large quantities of rubbery soot comprises mostly fullerenes are formed on the reactor walls. Grey color material containing MWCNTs and graphite nanoparticles is deposited at the cathode.

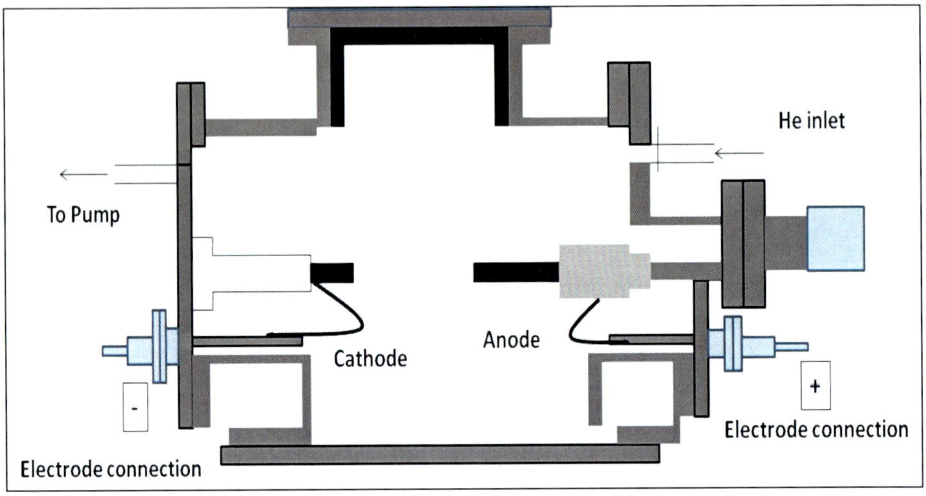


**Fig. (1).** Sketch showing electric arc discharge method for the synthesis of carbon nanotubes (CNTs).

Type, quantity and quality of CNTs depend on the various parameters employed in the arc reactor. Some studies on the effect of different parameters in arc discharge on the nanotubes development are provided below:

### *2.2.1.2. Effect of Gaseous Atmosphere*

Generally, He or Ar atmosphere is used in arc discharge but carbon nanotubes are successfully synthesized in the other atmospheres also. Effect of different atmospheres on the final morphology of carbon nanotubes was studied by Wang *et al.* [11]. In the high pressure of $CH_4$, thick CNTs embellished with carbon nanoparticles were obtained using the high current DC arc discharge of graphite electrodes. However, thin and long MWNTs were obtained, when $CH_4$ gas pressure of 50 Torr and an arc current of 20 A was used. Zhao *et al.* [12] observed the high variations in CNT morphology, while using $CH_4$ gas than He gas. These authors [13] also used hydrogen gas atmosphere for preparation of fine and long MWNTs.

Shimotani *et al.* [14] synthesized MWCNTs under four different atmospheres namely He, ethanol, acetone and hexane atmosphere at various pressures from 150 to 500 Torr and concluded that ethanol, acetone and hexane atmospheres produce more than double MWCNTs than in the He atmosphere. This increase may be due to the ionization and decomposition of organic gases into hydrogen and carbon atoms, which produces more carbon atom. Influence of $NH_3$ atmosphere for producing high amount of the carbon nanotubes was shown by Jiang *et al.* [15].

Ebbesen and Ajayan [16] found the dependence of the gas pressure in the reaction vessel on the purity and yield of the carbon nanotubes. In one study, it was observed that the purity of the obtained material was influenced by the current density and $H_2$gas pressure variation [17].

### 2.2.1.3. Liquid Atmosphere

Several studies reported the synthesis of carbon nanotubes using arc discharge in liquid solutions *e.g.* liquid nitrogen was used in arc discharge by Jung *et al.* [18] and Sornsuwit and Maaithong [19] for synthesizing MWCNTs in high yield and purity. Montoro *et al.* [20] prepared MWCNTs in $H_3VO_4$ aqueous solution. In another work, MWCNTs were produced using water between pure graphite electrodes [21]. The production of carbon nanomaterials under water or liquid nitrogen was also achieved by Xing *et al.* [22].

### 2.2.1.4. Pulsed Techniques

Mostly DC arc discharge is applied to synthesize carbon nanotubes, but few researchers have used pulsed techniques. Parkansky *et al.* [23] used single-pulse arc for depositing vertically oriented MWNTs using a graphite counter-electrode in ambient air. Tsai *et al.* [24] prepared MWCNTs using single-pulse discharge arc in air with peak current of 2.5A and a discharging time of 1000ms. Imasaka *et al.* [25] used intermittent arc discharge process in water producing carbon nano-onions and nanotubes. Lee *et al.* [26] used plasma rotating arc discharge process and found the dependence of nanotubes yield on the rotation speed of the anode.

### 2.2.1.5. Mechanism of MWCNTs Growth by Arc-Discharge

Four models are proposed for the synthesis of MWCNTs during arc discharge method. Main aspects of these models are described below.

#### 2.2.1.5.1. Vapor Phase Growth

Mechanisms proposed for the formation of MWCNTs in arc discharge method during the early periods were based on the assumptions that the nucleation and growth of MWCNTs happened by the direct condensation from the vapors formed in the arc. After analyzing the density and velocity distribution of vapors near the cathode surface region, Gamaly and Ebbesen [27] proposed the formation of two groups of carbon particles with different velocity distributions. One group of carbon particles has the thermal velocity, because of 3700°C temperature in the arc, while other group is composed of carbon ions. The velocity of carbon ions is much higher than thermal particles. Nanotubes and nanoparticles are formed in three stages. Open structures are formed in the first stage, which lead to rapid tube growth in the second stage. In last stage, nanotube growth terminates abruptly and cap is formed. Another vapor phase growth model by Louchev *et al.* [28 - 30]. proposed the adsorption of atoms onto a nanotube surface, followed by surface diffusion to the growth edge. Liu *et al.* [31]suggested the formation of heptagons formations at the growing edge.

#### *2.2.1.5.2. Liquid Phase Growth*

This is another model proposed by De Heeret *et al.* in 2005 [32]. This model is based on the observation of the beads of amorphous carbon on the surface of CNTs, which suggests the formation of solidified liquid droplets during the arc. At the start of arc-discharge, the carbon anode is locally bombarded by the electrons from cathode. This results in the formation and ejection of liquid carbon globules on the surface of the anode. Due to the high vapor pressure of liquid carbon, the surface of a carbon globule cools very rapidly but the interior cools slowly, which causes the liquid carbon to supercool. CNTs and nanoparticles nucleate and grow within this super-cooled liquid carbon.

#### *2.2.1.5.3. Solid Phase Growth*

Solid phase model of MWCNTs growth by arc discharge was proposed by Harris *et al.* in 1994 [33]. Initially carbon in the vapor phase consisting largely of C2 species is formed in the arc and condensed on the cathode as a fullerene soot-like material. Because of high temperature, condensed carbon results in the formation of multiwalled nanotubes. Growth of nanotubes terminates when carbon supply is exhausted or arcing process is intercepted. Setlur *et al.* [34]. demonstrated that the heating a non-graphitizing boron doped carbon to 2200–2400°C produced multiwalled nanotubes, similar to the arc-discharge. Since the temperature is less than the melting point of carbon 3527°C in arc discharge, these authors proposed a mechanism on the basis of solid state growth. According to the mechanism, pentagonal and heptagonal rings are formed in the early stages, which later on converted into hexagons and result in nanotube growth. Solid phase model was also proposed by Chadderton and Chen [35]on the basis of carbon nanotubes synthesis by thermal annealing of mechanically milled graphite powder at ~1400°C.

#### *2.2.1.5.4. Crystallization Model*

As an alternate model to vapor, liquid and solid phase models, Zhou and Chow in

2003, proposed a crystallization model [36]. According to this model, formation of the tubes is a crystallization process, which starts at the surface and goes on toward the center. Synthesis is a two-stage process occurs in solid state. In the first stage, amorphous carbon called assemblies having variety of shapes, depending on their surface energy and the local discharge conditions are formed on the surface of the cathode. In the second stage, graphitization of the assemblies during cooling occurs from the surface toward the interior region. Huang *et al.* [37] supported the crystallization model by forming the amorphous carbon nanowires *in situ* by electron beam deposition inside a HRTEM and synthesizing the graphitized structure similar to MWCNTs after heating at >2000°C.

### *2.2.1.6. Single Wall Nanotubes*

Single walled carbon nanotubes (SWCNTs) were first synthesized in two separate studies from Iijima *et al.* and Bethune *et al.* in 1993 [4, 38]. Main aspect of preparation of the SWCNTs in the arc discharge is the use of metals catalysts. MWCNTs are formed in arc discharge without any catalyst but metal catalyst is an essential component for SWCNTs.

In the process, metal catalysts either individually or in combinations with other metals, are mixed with the graphite powder and filled in the hole made in the anode center. After this, same procedure of arc discharge is applied as in the case of MWCNTs. When no catalyst is used, only the soot and the cathode deposit are formed. With catalyst, the spongy collaret around the cathodic deposit and web like structure between the cathode and chamber walls is also formed. The collaret mainly contains the SWCNTs. SWCNTs exist in bundles and free from metal catalyst. Tips of most SWCNTs are closed, free of catalyst exist in bundles. The diameters of SWCNTs formed in arc discharge are in the range of 1.1–1.4 nm and several microns long.

Various metal catalysts were used for the synthesis of CNTs. Bethune *et al.* [38] used Co as catalyst and synthesized SWCNTs of 1.2 nm. Ajayan *et al.* [39] also employed Co catalyst for synthesis of SWNTs of 1–2 nm diameter using arc discharge in He atmosphere. Another catalysts mostly used for SWCNTs synthesis is nickel. Seraphin *et al.* and Saito *et al.* [40, 41] synthesized SWCNTs using Ni. Zhou *et al.* [42] reported prepared SWCNTs using yttrium carbide. In a separate study, Saito *et al.* [43] showed Rh, Pd, and Pt formed SWCNTs, but the Ru, Os, Ir not. These authors also synthesized SWCNTs using Fe, Co, Ni, F/Ni, La, and Ce as catalysts [44]. Chen *et al.* reported synthesis of zing SWCNTs using 1% Fe catalyst in $H_2$–Ar mixture gas [45]. Fan *et al.* successfully produced cheap SWCNTs in argon DC arc discharge from charcoal as carbon source and FeS (20 wt %) as catalyst [46]. Shi *et al.* [47] witnessed the formation of high density of

SWCNT bundles using He pressure of 700 Torr. Not only SWCNTs, unwanted products such as MWCNTs, fullerenes, graphite nanoparticles, and amorphous carbons are also formed in this arc discharge process. Therefore, the purification of soot obtained in arc discharge method to get pure SWCNTs is still a challenge.

#### *2.2.1.7. Double Walled Carbon Nanotubes (DWCNTs) Synthesis*

Usually SWCNTs or MWCNTs are formed in the arc discharge process but studies have been undertaken to synthesize DWCNTs. Hutchison *et al.* [48] first reported the synthesis of DWCNTs in an arc discharge technique using graphite rod filled with a mixture of Ni, Co, Fe and S in the mixed atmosphere of argon and hydrogen. Sugai *et al.* [49] prepared high-quality DWNTs by high-temperature pulsed arc discharge method using Y/Ni alloy Catalyst. Qiu *et al.* [50] synthesized DWCNTs using potassium chloride in trace amount as a promoter in an iron sulfide catalyst. Qiu *et al.* [51] showed that the dependence of DWCNTs formation on the composition of the catalyst and $H_2$ pressure. Liu *et al.* reported the preparation of DWCNTs using nickel formate dehydrate [52].

### 2.2.2. Laser Ablation Method

Guo *et al.* [53] at Rice University for the first time used pulsed laser-ablation process for the synthesis of SWCNTs. Hafner *et al.* successfully developed a laser ablation method for the mass production of SWCNTs [54]. In the method, metal loaded sample is placed in a quartz tube, heated to >1000°C and inert gas is flowed through the tube. Sample is exposed to the laser and produces the soot, which is flown away by the inert gas to water cooled side of the tube. The procedure was improved later using double beam laser by Thess *et al.* [55] and Rao *et al.* [56] and prepared SWCNTs with < 90% purity with better graphitized structure. In the method, a laser beam (532 nm), is focused onto a metal-graphite composite target, which is placed in a high-temperature furnace ~1200°C. These targets are made by baking a paste at higher temperature after mixing high-purity catalyst metal or metal-oxide with graphite powder and carbon cement at room temperature. Targeted laser beam vaporizes the surface of composite and inert gas takes away the soot formed from the high-temperature zone to a water-cooled zone and deposited there. The major drawback of the method is the small carbon deposit. The major impurities in the product obtained in the soot are graphite particles, catalysts, amorphous carbon and fullerenes. In addition, it is an expensive process. The quantity and the quality of produced SWCNTs material depends on the type of metal catalysts and their ratio [57 - 59], gas and its pressure [60 - 62], the temperature of the reaction furnace [63 - 65], and the laser parameters [66 - 68]. Catalysis metals are either used as individual or in combinations in the target, some of them are Co, Ni, Pt, Cu, Co/Ni, Ni/Pt, Co/Pt,

Co/Cu, Nb,. It was observed that target having 1% (at %) each of Ni and Co, gave best purity and yield of SWCNTs. Usually Ar with or without $H_2$ is used as carrier gas in laser ablation process. However, other atmospheres like $N_2$, $O_2$, *etc*. were also studied. It was found that increasing the gas pressure reduces the formation of SWCNT and lighter gases are more efficient for rapid cooling than heavier ones.

#### *2.2.2.1. Mechanism of SWCNTs Growth in Arc Discharge and Laser Ablation*

Both arc discharge and laser ablation techniques are top down methods for the synthesis of CNTs. Starting materials are graphite–metal mixtures. Vaporization of this mixture followed by condensation in an inert atmosphere are the same processes in both the methods. Carbon soot produced by both methods contains bundles of SWCNTs together with disordered carbon and metal particles. Because of these similarities, it is assumed that the mechanisms of SWCNTs formation in the arc-discharge and laser-ablation processes are similar. It was observed during HRTEM studies that the metal particles do not exist at the tips of SWCNTs after producing in arc-discharge and laser-ablation techniques [69]. Further, diameter of metal particles is much higher than individual SWCNT in the soot. Therefore, formation mechanism of SWCNTs in these methods is the "base growth" rather than "tip growth". It means that metal catalysts exist at the base and carbon is supplied from the base for the growth of CNTs. Saito in 1995 [70] proposed a vapor–liquid–solid (VLS) model. According to this model, first stage is the co-condensation of carbon and metal atoms from the vapor phase to form liquid metal carbide. When this liquid is supersaturated with carbon, solid phase nanotubes begin to grow. However, other studies [71] showed that SWCNTs growth may be a solid state phenomenon. Kataura *et al*. [72]. proposed that in the first phase in the gas phase at 2000–3000°C, small carbon clusters rich in pentagonal rings are formed. During cooling, metal atoms condense, forming droplets and supersaturated with carbon. These composite particles then act as precursors for SWCNTs growth, with carbon being supplied by precipitation or from the carbon in the particles.

### 2.2.3. Chemical Vapor Deposition Method

Although arc discharge and laser ablation methods were discovered first to synthesize the CNTs but they are least used now a days. Arc process is simple and easy to perform but separation of CNTs from the soot is a major problem. Difficulty in purifying the carbon nanotubes from the rest of the soot particularly graphite nanoparticles is an area of concern. Same difficulty is observed in the laser ablation method. Besides this, laser ablation is a costly method. Another problem in both the methods is the difficulties in setting the parameters to synthesize the carbon nanotubes having the desired characteristics. Efforts were made but did not show the encouraging results. Therefore, research was undertaken to develop a suitable method to produce carbon nanotubes to get rid of the problems in these two methods. The chemical vapor deposition (CVD) method solved these problems up to some extent. The important aspect of this method is that it can produce carbon nanotubes with predefined properties [73]. It is already known for more than half century that carbon fibers can be prepared by decomposition of carbon containing gases over metals surface using CVD method. However, Yacaman *et al.* [74] reported the synthesis of carbon nanotubes using this method for the first time. CVD uses the catalytic decomposition of gases on the catalysts at pre-adjusted temperature in a furnace. The hydrocarbons are decomposed using supported transition metal catalysts for synthesis of CNTs. Fig. (**2**) depicts a typical diagrams showing the main elements of the process. The experiment is usually carried out in a high temperature furnace at atmospheric pressure. Mostly horizontal furnace is used for the continuous mass production, but, vertical furnace can also be used. Catalyst is placed on a ceramic or quartz support and reaction is carried out in a quartz tube. Mixture of hydrocarbon and an inert gas is passed over the catalyst at temperatures ranging from 500 °C to 1100 °C. After the reaction, system is then cooled down to room temperature. The catalyst and carbon source is injected from the one side of the furnace and the resultant carbon nanotubes grow during decomposition of hydrocarbon at the catalysis. Nanosize metal catalysts are either kept in the reactor before heating the furnace or prepared *in situ* using their precursors. Application of fluidized bed reactor is another option in CVD. Catalysts are usually kept in the furnace center and an upward flow of carbon feedstock gases is used.

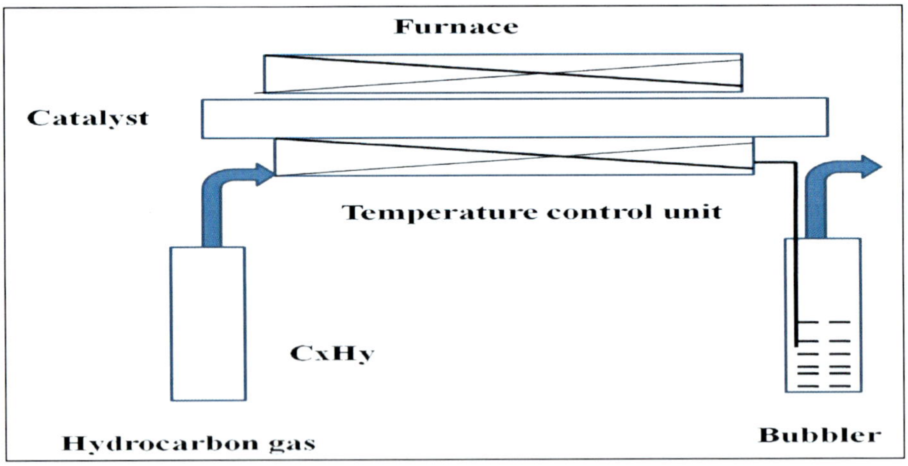


**Fig. (2).** Sketch showing the chemical vapor deposition (CVD) method for the synthesis of carbon nanotubes.

This technique can control the length and diameter of the nanotubes by controlling the reaction time, catalyst and other parameters. Using this method, it is possible to synthesize CNTs in the forms of powder [75, 76], aligned forest [77, 78], in various shapes like straight, curved, planar-spiral, and helix. Planar arrays of aligned carbon nanotubes over large areas with high density and ordering have been grown using CVD on single-crystal substrates of sapphire or quartz. The CNTs often have an amorphous carbon coating, and sometimes metal particles at their tips.

Fig. (**3**) shows an example of MWCNTs synthesized using CVD method. The characteristics of the CNTs synthesized by CVD method depend on the temperature and the pressure in the process, volume, concentration and kind of hydrocarbon, the nature, size of metal catalyst and its pretreatment, the nature of the support and the reaction time.

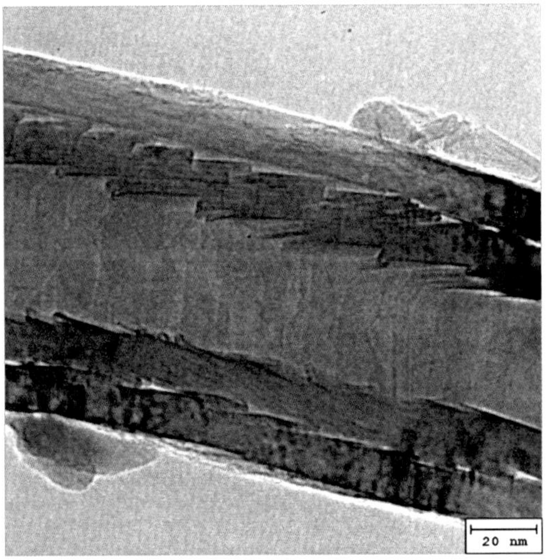


**Fig. (3).** Synthesis of multiwalled carbon nanotubes by chemical vapor deposition method(CVD).

### *2.2.3.1. Carbon Source and Inert Gas*

Various hydrocarbons as source of carbon are used to produce carbon nanotubes using CVD. It may be a gas such as $C_2H_4$, $CH_4$, $C_2H_2$ or aromatic liquid *i.e.* xylene or benzene [73, 79 - 82]. One of the inert gas such as nitrogen, helium and argon is used to purge along with hydrocarbon source. This serves two purposes: (1) It removes the oxygen from the furnace and avoid the oxidation of carbon and (2) it takes the liquid carbon source, wherever it is used. As an alternative to gas and liquid carbon sources, polymers were also tried for the synthesis of CNTs.

### 2.2.3.2. Catalyst

Various catalysts [83 - 85] and combinations were used to improve the yield and the quality of CNTs production. Fe/Mo supported on MgO was found to be effective catalyst combination in the synthesis of SWCNTs and DWCNTs [86 - 89]. Co and Co/Fe metal catalyst supported on zeolite and $CaCO_3$ doped with different kinds of K-containing chemicals [90] were tried in the CVD synthesis of MWCNTs. Studies showed that bimetallic K doped catalyst have the higher activity than mono metallic. Natural minerals such as forsterite, diopside, quartz, magnesite ($MgCO_3$) and brucite ($Mg(OH)_2$) crystals [91] were used as catalyst for the production of CNTS, because of the presence of Fe in the materials as impurity. Magnesite crystal was used to synthesize SWCNTs, which demonstrates the existence of naturally occurring SWCNTs. Aligned CNTs were formed within the pores of an alumina membrane using ethylene without any metal catalyst [92]. CNTs showed a narrow diameter distribution.

### 2.2.3.3. Carbon Source and Catalysts Combinations

Different combinations of catalysis and carbon sources are used for the development of carbon nanotubes. Wei *et al.* [93] synthesized aligned CNTs using Co on $SiO_2$support and a mixture of $NH_3/C_2H_2$ gases. High purity MWNTs in large quantities were produced by methane on Co-Mo/MgO catalysts [94] or Ni/Mo, Y/Mo and Ni-Y/Mo systems [95]. It was observed that inclusion of Cu in the catalyst promote the forming of multi-branched carbon nanotubes. Borowiak and Rümmeli [96] published the formation of Y junction CNTs using Cu on MgO support using alcohol.

Carbon nanofibers (CNFs) were prepared using $CO/Ar/O_2$ system at low-temperature without any metal catalyst by microwave plasma-enhanced CVD. Addition of a small amount of $O_2$ suppresses the isotropic deposition of amorphous carbon and assists the anisotropic linear growth of crystallized carbon deposit [97]. Cheng *et al.* [98] developed partially filled MWCNTs using titanate-modified palygorskite as catalyst and acetylene as carbon source by CVD at high temperature.

### 2.2.3.4. Substrate

Usually the substrates used in CVD process are silicon, glass and alumina. Porous silicon can be used for growing self-oriented nanotubes on large surface. Hexagonal close packed nano-channel alumina template with a Co-Ni catalyst is useful for synthesizing well-graphitized free-standing nanotube arrays. Diameter-controlled and well defined nanotubes [99] are prepared using this method. Metal filled and bamboo-like CNTs and carbon onions were synthesized on the new copper-supported Ni/Y composite catalyst by decomposition. Jayatissa and Guo [100] observed the dependence of the diameter of the CNTs on the thickness of the deposited metal film on the substrate.

### 2.2.3.5. Gas Phase Metal Catalyst

Another method to use the catalyst is in gas phase. In this method, both the catalyst and the hydrocarbon precursor gas are fed into a furnace in gas phase. This method is suitable for large-scale synthesis. Spray-pyrolysis or Injection CVD uses spraying or pumping the catalyst–precursor mixture into the furnace. Horvath *et al.* [101] found that maximum yield was obtained when ferrocene –nickelocene catalyst mixture was used with xylene as carbon source. Singh *et al.* [102] synthesized aligned MWNT films on the quartz substrate by an injecting ferrocene/toluene solution into the furnace.

### 2.2.3.6. Different Types of CVD

Usually thermal-CVD is used for synthesizing nanotubes. However, other types of CVD techniques such as hot-filament (HF-CVD), plasma-enhanced (PE-CVD), microwave-plasma (MP-CVD) or radiofrequency-CVD technique [103, 104] are also used. Thermal-CVD and PE-CVD are commonly used to grow aligned MWCNTs and SWCNTs on various substrates including Ni, Si, SiO2, Cu/Ti/Si, stainless steel and glass. CNTs with different orientation can be produced by microwave plasma-CVD using the same substrates as for PE-CVD. Chaisitsak *et al.* [105]. studied the influence of the Si support, gas ratio and the growth pressure during hot-filament enhanced CVD method in the synthesis of CNTs. High yield growth of vertically aligned SWCNTs and DWCNTs films was reported using a water-assisted [106], an oxygen-assisted plasma-enhanced [107], and microwave plasma-enhanced CVD techniques (MWP-CVD) [108], Alcohol-CVD is another CVD process, which is used for high purity SWCNTs using ferrocene and ethanol in different ratios.

### 2.2.3.7. Uniform Vertically Aligned Synthesis of CNTs

The growth of vertically aligned carbon nanotubes forests is studied extensively, because of the possible application in industry. The aim is to grow vertically aligned CNTs with homogeneous distribution on the surfaces having high uniformity. Nanolithography is used to create pattern of the catalyst on the surface, on which CNTs are formed using CVD method. Kim *et al.* [109] used Si wafer as a substrate and patterned Ni dots catalyst and grew aligned CNTs using CVD. Yamada *et al.* [110] synthesized vertically aligned DWCNTs forests using water-assisted CVD synthesis and Fe catalyst. Chang *et al.* [111] grew vertically aligned MWCNTs with a bamboo-like structure and high density on a large area of on a TiN coated quartz plate with Fe catalysts patterned by UV nano-imprint lithography (NIL) using PECVD. Ammonia and acetylene were used as the etchant gases and the carbon source, respectively. In non- lithographic techniques, Kim *et al.* [112] prepared aligned CNTs on AAO/Si substrate using DC-PECVD at 500°C using Co layer as catalyst, acetylene gas as a carbon source and ammonia gas as carrier gas.

#### 2.2.3.8. Liquid State Pyrolysis Chemical Vapor Deposition

Aerosol pyrolysis CVD is the pyrolysis of mixed liquid aerosols of both liquid hydrocarbon and catalyst in CVD. Byeon *et al.* [113] synthesized vertically aligned CNTs arrays of 4.38 mm height using in-situ formation of metal catalysis *via* pyrolysis of ferrocene–ethanol aerosol without supporting material. Jeong *et al.* produced aligned and clean CNTs using ultrasonic evaporator atomizing the mixed liquid solution by the thermal pyrolysis process [114]. Khatri *et al.* reported SWCNTs synthesis using ethanol, cobalt and molybdenum acetates by an ultrasonic spray pyrolysis on silicon substrates at 850°C [115]. Camarena *et al.* reported the synthesis of MWCNTs by spray pyrolysis using toluene as the carbon source and ferrocene as the catalyst [116]. In a similar study, Sadeghian synthesized MWCNTs using hexane as a carbon source and ferrocene as a catalyst [117]. During study of CNTs growth *via* catalytic pyrolysis of acetylene, Nebol'sin and Vorob'ev observed that the surface free energy determines the catalytic activity of the liquid droplet on the tip of the CNT and is responsible for the constant nanotube diameter [118]. Natural products like pinenefrom turpentine, neem oil extracted from the seeds of the neem-Azadirachtaindica and coconut oil were used for MWCNTs production by spray pyrolysis [119 - 121]. Quan *et al.* reported the CNTs synthesis through waste pyrolysis oil [122].

#### 2.2.3.9. Solid State Pyrolysis Chemical Vapor Deposition

Kucukayan *et al.* synthesized MWCNTs through pyrolysis of the sulfuric acid-carbonized byproduct of sucrose [123]. Kuang *et al.* prepared straight CNTs in large scale through thermal CVD by pyrolysis of two mixed metal phthalo-cyanines with a certain amount of S at 800–950°C. Two kinds of mixed transition metal phthalo-cyanines (M (II) Pc)) were used [124] as the carbon source as well as catalysts favoring the growth of the straight CNTs. Here the metal atoms were Fe and Co. Du *et al.* [125] prepared MWCNTs through the solid-phase transformation of metal-containing glass-like carbon nanoparticles by heating at temperatures between 800–1000°C

A new method known as V-type pyrolysis flame for mass CNTs production was developed by Liu *et al.* [126]. They prepared the CNTs using penta carbonyl iron as the catalyst and carbon monoxide as the carbon source.

## *2.2.3.10. CNT Growth Mechanism in CVD*

Extensive studies have been undertaken during past two decades to understand the CNT growth mechanism in CVD but still debatable. Basic mechanism involves the dissociation of hydrocarbon molecules catalyzed by the transition metal, and the saturation of carbon atoms in the metal nanoparticle. The precipitation of carbon from the metal particle leads to the formation of carbon nanotubes. Two types of detailed mechanisms are proposed based on observations outlined below:

When hydrocarbon (a carbon source) vapors contact with the metal catalyst nanoparticles at high temperature, it decomposes into carbon and hydrogen species in which hydrogen is removed with inert gas and carbon gets dissolved into the metal. When carbon solubility in metal saturates, as-dissolved carbon precipitates out and crystallizes as cylindrical network. This network is energetically stable because of no dangling bonds. Since decomposition of hydrocarbon is an exothermic process, it provides the heat to the metal's exposed zone, which is used in the endothermic processes of precipitation and crystallization of carbon. This precise thermal gradient inside the metal particle keeps the process on.

There are two models proposed for growing carbon nanotubes on metal catalyst. First model is "tip growth" and other is "base growth". Tip growth occurs when the catalyst-substrate interaction is weak and metal forms an acute contact angle with the substrate. Hydrocarbon decomposes on the top surface of the metal and carbon diffuses down through the metal. Therefore, CNT precipitates out across the metal bottom, pushing the whole metal particle off the substrate. As long as the metal's top is open and available for fresh hydrocarbon decomposition, the concentration gradient in the metal allows carbon diffusion. This results in the growth of CNT. When metal is fully covered with excess carbon, its catalytic activity ceases and the CNT growth is stopped. The evidence of this model came from observation of metal presence at the tip of nanotubes.

Base growth model happens when the catalyst-substrate interaction is strong and metal forms an obtuse contact angle with the substrate. Early hydrocarbon decomposition and carbon diffusion occurs like in the tip-growth, but the CNT precipitation from metal carbon solution fails to push the metal particle up because of strong interaction between substrate and metal. Therefore, the precipitation is obliged to emerge out from the solution on the top of the metal. Carbon diffuses out as a hemisphere and extends vertically up in the form of cylinder. Successive decomposition of hydrocarbon at the surface of the metal results in the diffusion of carbon upward. Thus CNT grows up having the metal catalyst particle at its base.

### 2.2.4. Other Methods of CNTs Synthesis

Few other methods are also capable to synthesize CNTs other than above mentioned three methods.

#### *2.2.4.1. Hydrothermal Synthesis*

Hydrothermal technique is a technique, which can be used for the synthesis of CNTs. This process has many advantages over other methods such as the starting materials are more common to obtain and are stable in ambient temperature conditions. It is mostly a low temperature process and does not require the flow of hydrocarbon or carrier gases. Gogosti *et al.* prepared both closed and open ends of MWCNTs having >100 walls by heating a mixture of polyethylene and water with a Ni catalyst at 700-800 °C under 60–100 MPa pressure [127]. Same authors produced graphitic CNTs having long and wide internal channels by heating ethylene glycol ($C_2H_4O_2$) with Ni catalyst at 730–800 °C under 60–100 MPa pressure [128]. Formation of large quantity of CNTs by heating with NaOH, $CH_2Cl_2$ and Li *et al.* 150–160 °C for 24 h was shown by Manafi *et al.* [129]. Nanotubes obtained were ~2–5 µm long with diameter of 60 nm. Moreno *et al.* formed the MWCNTs by hydrothermal treatment to amorphous carbon, at temperatures between 600°C- 800 °C and 100 MPa pressure, in the absence of metal catalysts [130]. The spontaneous formation of short nanotubes from nanoporous carbon in the presence of elemental Cs at 50 °C was shown by Stevens *et al.* [131].

#### *2.2.4.2. Electrolysis*

Electrolysis is a less common method for CNT production. It was developed by Hsu *et al.* in 1995 [132]. Electrolysis method uses electrowinning of chloride salts of alkali or alkaline-earth metals on a graphite cathode. This method has several advantages over others such as possibility to control the synthesis process by the electrolysis modes, cheap raw materials and possibility to control the product structures and morphologies. Bai *et al.* synthesized SWCNTs [133] by electrolytic conversion of graphite to CNTs in fused NaCl at 810 °C using Ar as inert gas.

Novoselova *et al.* [134] produced CNTs by cathodic reduction of $CO_2$andternary mixture of alkali metal chlorides (NaCl: KCl: CsCl) as base electrolyte. CNTs were also synthesized from $C_2H_2$ by electrolysis at -40 °C in liquid $NH_3$ without a metal catalyst [135].

#### *2.2.4.3. Solar Technique*

The process of carbon nanotubes production using concentrated solar light was described by Laplaze *et al.* [136]. It involved the use of sun light to induce the vaporization of the target. A high-flow vacuum pump maintained the pressure using Ar or He as buffer gas. The vaporization temperature remained in the range 2627–2727 °C. It is possible to control nanotubes parameters in this process by different synthesis parameters. Luxembourg *et al.* [137] produced SWCNTs in bulk using a 50 kW solar reactor.

## 2.3. PURIFICATION OF SOOT CONTAINING CNTS

Purification of CNTs is the getting the desired nanotubes including the other types of undesired nanotubes in the raw soot. Since there are different types of impurities, no single step can remove all of them. Therefore, various techniques are combined with each other to purify and to remove different impurities at the same time. Following steps are used for purification of CNTs.

### 2.3.1. Chemical Purification

Most preferred method of purifying the black soot obtained after the synthesis is the use of chemicals. Among these mostly used are the oxidizing agents. There are two types of oxidations: wet oxidation and dry oxidation. Wet oxidation involves the use of the chemicals in liquid state particularly concentrated acids or oxidizing agents. On the other hand, dry oxidation refers to oxidation by air, $O_2$ and other gases at suitable temperature. Oxidation removes the amorphous carbon, carbon nanoparticles and metal particles from soot. These materials have high oxidation activity than nanotube. Higher activities of amorphous carbon are due to presence of high energy hanging bonds and carbon nanoparticles having big curvature and pentagonal carbon rings [138]. Metals are also more than reactive than carbon nanotubes.

There is some damage to CNTs structure also during oxidation such as tips opening of the nanotubes because pentagonal rings at the tip of CNTs are less stable than hexagonal rings in other part of CNTs. Efficiency and yield of the separating carbon nanotubes from the soot depends on metal content, oxidation time, environment, oxidizing agent and temperature. Among various oxidizing agents use of acids are more common but other reagents are also studied.

#### *2.3.1.1. Wet Oxidation*

Refluxing the soot in mineral acids is effective in oxidizing metal particles and amorphous carbon. Acids commonly used are hydrochloric acid (HCl), nitric acid ($HNO_3$) and sulfuric acid ($H_2SO_4$). Metal particles are dissolved and removed from the soot during treatment for short durations with nitric acid [139 - 141]. Oxidation by $HNO_3$ is also found to open the tip of carbon nanotubes [140].

However, some distractive effect due to oxidation of graphenes of CNTs by $HNO_3$ was also observed [141]. Silica matrix used to support the catalyst in CVD process can be easily dissolved in hydrofluoric acid (HF) [142] and removed smaller amorphous carbon. Zeolite support was removed using HF in other studies [143, 144]. However, over oxidation of carbon nanotubes using acids destroy the morphology of nanotubes and form the carbon nanoparticles [145].

Chemicals other than mineral acids such as potassium permanganate ($KMnO_4$), hydrogen peroxide ($H_2O_2$) and sodium hydroxide (NaOH) [146] are also used. Since the byproduct of $KMnO_4$ oxidation is difficult to remove, it appears as a less suitable oxidation agent. $H_2O_2$ is more suitable reagent. Suzuki *et al.* [147] shown in earlier study that the capabilities of oxidation by $H_2O_2$with less impurities in the final product with water as byproduct. Fe metal particles and amorphous carbon were removed by refluxing the soot with $H_2O_2$ solution Scientists used NaOH for purification of CNTs [148, 149].

Mixtures of acids or oxidizing agents were used for obtaining better yield of purified carbon nanotubes. It was observed in one study [150] that purification of CNTs increased many folds by using mixtures of $HNO_3$ and HCl than $HNO_3$. The dissolution of the metal particles led to the metal oxides and HCl removed from these oxides. Ziegler *et al.* [151] used $H_2SO_4/H_2O_2$ mixture for purification. Zhang *et al.* [152] carried out the treatment using the nitration mixture ($H_2SO_4/HNO_3$). This solution is able to cut CNTs into fragments with opened ends. It was observed that concentrated acid can absorb microwave rapidly by quickly dissolving the metal particles and thus preventing damages to the tubes. Amorphous carbon metal catalyst can be removed with higher efficiency microwave [153].

#### *2.3.1.2. Dry Oxidation*

Dry oxidations are carried out using gases. The commonly used gases for this purpose include dry/wet air [154 - 156], $CO_2$ [157], $H_2$ [158], and $NH_3$ [159]. Oxidation of carbon soot at moderate temperature in air is able to remove the amorphous carbon and other carbonaceous impurities. During heating in air, the reactivity of amorphous carbon is higher than the CNTs. It is converted into carbon dioxide ($CO_2$) during oxidation. Because of this, the amorphous carbon can be easily eliminated from the CNTs by selecting the suitable temperature (300-400°C) for heating the soot. However, heating at very high temperatures (>400C) is detrimental to tube structures. Oxidation by $CO_2$ is shown to be selectively eradicating the disordered amorphous carbon. Smith *et al.* [157] removed the reactive carbon species *e.g.* amorphous carbon using oxidation with $CO_2$ at 600°C. Vivek chand *et al.* [158] eliminated amorphous carbon and carbon coating on the metal catalyst particles by heating in hydrogen gas at high temperature. Wang *et al.* [159] used $NH_3$ to eradicate residual carbon impurities and repaired the damaged CNTs sidewalls. In one study, Hou *et al.* [160] purified CNTs by removing the amorphous carbon and catalyst particles by intercalation of Br followed by oxidation in air at 500°C

### 2.3.2. Physical Purification

It was observed that oxidation by acids or chemicals caused some damage to the tube structure. Therefore, to reduce this damage, physical methods such as filtration, ultrasonication and chromatography are used to purify CNTs. Abatemarco *et al.* [161] filtered the soot using membranes with different pore diameters in sequence from the largest to smallest in the cross- flow system. It showed that all large aggregates were retained by the larger pore size membranes, whereas CNTs were retained on the smaller pore size membranes. Therefore, CNTs, polyhedral nanoparticles, and large aggregates were separated from each other during the filtration. In addition, this made possible for fractionation of the CNTs according to the length. This was further supported by the study of Tagmatachis *et al.* [162], who separated water-insoluble CNTs according to the length or size using field-flow fractionation.

Sonication was found to be useful eliminate amorphous impurities using solvents such as dichloromethane and *O*-dichlorobenzene [163, 164]. Solvent molecules interacted with CNTs and solubilized them. Ultrasonic treatment normally separates MWCNTs from carbon nanoparticles [165].

High performance liquid chromatography (HPLC) [166]and size exclusion chromatography (SEC) [167, 168] are the two techniques, which can separate nanotubes having different lengths. CNTs tend to self-associate into aggregates, therefore, use of surfactants can separate them into individual tubes [169]. SDS [170] and dodecyl-benzene sodium sulfonate (NaDDBS) [171, 172] were used to prevent the aggregation of carbon nanotubes in water. Heating to high temperature melts the metal and can be removed.

### 2.3.3. Multi-Step Purification

Multi-step purification is the combination of chemical and physical treatments. It is used when a single treatment is not sufficient to remove all the impurities simultaneously which are present in the CNTs. In the recent work of Montoro *et al.* [169], a multi-step purification consisting four treatments were used. Fullerene and soluble impurities were first removed by sohxlet extraction with toluene. In the second step liquid phase oxidation with $H_2O_2$was conducted to get rid of the amorphous carbon. The metallic particles were eliminated by acid treatment.

Finally, physical separation in sodium dodecyl sulfate (SDS) solution was carried out to remove the graphite and metallic particles.

## 2.4. CONCLUSION

Although lot of research has been carried on the synthesis, purification *etc.* of CNTs but many aspects are still to be investigated. Even after 25 years of production cost of CNTs specially SWCNTs are very high. New ideas of synthesis of carbon nanotubes are required to make affordable. Recently the focus of the research in materials laboratories has been changed from carbon nanotubes to graphene. Therefore, more research is required for the application of carbon nanotubes which can attract scientists again to focus on synthesis of carbon nanotubes.

## CONFLICT OF INTEREST

The authors declare no conflict of interest, financial or otherwise.

## ACKNOWLEDGEMENTS

Author is grateful to Dr. Marc Monthioux, CNRS, France and Prof. K.L. Lin, NCKU, Taiwan for providing opportunities for his research on carbon nanotubes.

## REFERENCES

[1] Radushkevich LV, Lukyanovich VM. On the structure of carbon, formed by the thermal decomposition of carbon monoxide on iron surface. Zh Fiz Khim 1952; 26: 88-95.

[2] Oberlin A, Endo M, Koyama T. Filamentous growth of carbon through benzene decomposition. J Cryst Growth 1976; 32: 335-49. [http://dx.doi.org/10.1016/0022-0248(76)90115-9]

[3] Iijima S. Helica l microtubules of graphitic carbon. Nature 1991; 354: 56-8. [http://dx.doi.org/10.1038/354056a0]

[4] Iijima S, Ichihashi T. Single-shell carbon nanotubes of 1-nm diameter. Nature 1993; 363: 603-5. [http://dx.doi.org/10.1038/363603a0]

[5] Mittal J, Monthioux M, Allouche H, Stephan O. Room temperature filling of single-wall carbon nanotubes with chromium oxide in open air. Chem Phys Lett 2001; 339: 311-8. [http://dx.doi.org/10.1016/S0009-2614(01)00354-2]

[6] Mittal J, Lin KL. The formation of electric circuits with carbon nanotubes and copper using tin solder. Carbon 2011; 49: 4385-91. [http://dx.doi.org/10.1016/j.carbon.2011.06.029]

[7] Mittal N, Jain S, Mittal J. Application of Electron Energy Loss Spectroscopy for Single Wall Carbon Nanotubes. J Appl Spectrosc 2015; 82: 1-12. [Review]. [http://dx.doi.org/10.1007/s10812-015-0056-5]

[8] Lota G, Frackowiak E, Mittal J, Monthioux M. High performance supercapacitor from chromium oxide-nanotubes based electrodes. Chem Phys Lett 2007; 434: 73-7. [http://dx.doi.org/10.1016/j.cplett.2006.11.089]

[9] Mittal J, Monthioux M, Allouche H. Synthesis of SWNT-based hybrid nanomaterials from photolysis-enhanced chemical processes Twenty-Fifth Biennial Conference on Carbon, Lexington, KY Novel/142.

[10] Zajickova L, Elias M, Jasek O, Kudrle V, Frgala Z. Matejkova, Bursik J, Kadlecikova M, Atmospheric pressure microwave torch for synthesis of carbon nanotubes. Plasma Phys Contr Fusion 2005; 47: B655-66.
[http://dx.doi.org/10.1088/0741-3335/47/12B/S48]

[11] Wang M, Zhao XL, Ohkohchi M, Ando Y. Carbon nanotubes grown on the surface of cathode deposit by arc discharge Fullerene Sci. Technol 1996; 4: 1027-39.

[12] Zhao XL, Wang M, Ohkohchi M, Ando Y. Morphology of carbon nanotubes prepared by carbon arc Jpn. J. Appl. Phys, Part 1 1996; 35: 4451-6.

[13] Zhao XL, Ohkohchi M, Wang M, Iijima S, Ichihashi T, Ando Y. Preparation of high-grade carbon nanotubes by hydrogen arc discharge. Carbon 1997; 35: 775-81.
[http://dx.doi.org/10.1016/S0008-6223(97)00033-X]

[14] Shimotani K, Anazawa K, Watanabe H, Shimizu M. New synthesis of multi-walled carbon nanotubes using an arc discharge technique under organic molecular atmospheres. Appl Phys, A Mater Sci Process 2001; 73: 451-4.
[http://dx.doi.org/10.1007/s003390100821]

[15] Jiang Y, Wang H, Shang XF, Li ZH, Wang M. Influence of $NH_3$ atmosphere on the growth and structures of carbon nanotubes synthesized by the arc-discharge methodInorg. Mater 2009; 45: 1237-9.

[16] Ebbesen TW, Ajayan PM. Large-scale synthesis of carbon nanotubes. Nature 1992; 358: 220-2.
[http://dx.doi.org/10.1038/358220a0]

[17] Cadek M, Murphy R, McCarthy B, *et al.* Optimization of the arc-discharge production of multi-wall carbon nanotubes. Carbon 2002; 40: 923-8.
[http://dx.doi.org/10.1016/S0008-6223(01)00221-4]

[18] Jung SH, Kim MR, Jeong SH, *et al.* High-yield synthesis of multi-walled carbon nanotubes by arc discharge in liquid nitrogen. Appl Phys, A Mater Sci Process 2003; 76: 285-6.
[http://dx.doi.org/10.1007/s00339-002-1718-8]

[19] Sornsuwit N, Maaithong W. Study of Multi-Walled Carbon Nanotube Synthesis Using Liquid Nitrogen and Post-Process Filtration. Int J Precis Eng Manuf 2008; 9: 18-21.

[20] Montoro LA, Lofrano RC, Rosolen JM. Synthesis of single-walled and multi walled carbon nanotubes by arc-water method. Carbon 2005; 43: 200-3.
[http://dx.doi.org/10.1016/j.carbon.2004.08.014]

[21] Guo J, Wang X, Yao Y, Yang X, Liu X, Xu B. Structure of nanocarbons prepared by arc discharge in water. Mater Chem Phys 2007; 105: 175-8.
[http://dx.doi.org/10.1016/j.matchemphys.2007.04.022]

[22] Xing G, Jia SL, Shi ZQ. The production of carbon nano-materials by arc discharge under water or liquid nitrogen. N Carbon Mater 2007; 22: 337-41.
[http://dx.doi.org/10.1016/S1872-5805(08)60005-0]

[23] Parkansky N. carbon nanotubes by plasma rotating arc discharge technique. Diamond Rel Mat 2002; 11: 914-7.
[http://dx.doi.org/10.1016/S0925-9635(01)00639-2]

[24] Tsai YY, Su JS, Su CY, He WH. J Mater Process Technol 2009; 209: 4413-6.
[http://dx.doi.org/10.1016/j.jmatprotec.2008.10.049]

[25] Imasaka K, Kanatake Y, Ohshiro Y, Suehiro J, Hara M. Production of carbon nano onions and nanotubes using an intermittent arc discharge in water. Thin Solid Films 2006; 506–507: 250-4.
[http://dx.doi.org/10.1016/j.tsf.2005.08.024]

[26] Lee SJ, Baik HK, Yoo J, Han JH. Large scale synthesis of carbon nanotubes by plasma rotating arc discharge technique. Diamond Rel Mat 2002; 11: 914-7. [http://dx.doi.org/10.1016/S0925-9635(01)00639-2]

[27] Gamaly EG, Ebbesen TW. Mechanism of carbon nanotube formation in the arc discharge. Phys Rev B Condens Matter 1995; 52(3): 2083-9. [http://dx.doi.org/10.1103/PhysRevB.52.2083] [PMID: 9981282]

[28] Louchev OA. Transport-kinetical phenomena in nanotube growth. J Cryst Growth 2002; 237(1): 65-9. [http://dx.doi.org/10.1016/S0022-0248(01)01852-8]

[29] Louchev OA, Sato Y, Kanda H. Morphological stabilization, destabilization, and open-end closure during carbon nanotube growth mediated by surface diffusion. Phys Rev E Stat Nonlin Soft Matter Phys 2002; 66(1 Pt 1): 011601. [http://dx.doi.org/10.1103/PhysRevE.66.011601] [PMID: 12241366]

[30] Louchev OA, Sato Y, Kanda H. Mechanism of thermos kinetical selection between carbon nanotube and fullerene-like nanoparticle formation. J Appl Phys 2002; 91(12): 10074-80. [http://dx.doi.org/10.1063/1.1479469]

[31] Liu YW, Wang L, Zhang H. A possible mechanism of uncatalyzed growth of carbon nanotubes. Chem Phys Lett 2006; 427(1–3): 142-6. [http://dx.doi.org/10.1016/j.cplett.2006.05.122]

[32] de Heer WA, Poncharal P, Berger C, *et al.* Liquid carbon, carbon-glass beads, and the crystallization of carbon nanotubes. Science 2005; 307(5711): 907-10. [http://dx.doi.org/10.1126/science.1107035] [PMID: 15705847]

[33] Harris PJ, Tsang SC, Claridge JB, Green ML. High-resolution electron microscopy studies of a microporous carbon produced by arc-evaporation. J Chem Soc, Faraday Trans 1994; 90(18): 2799-802. [http://dx.doi.org/10.1039/ft9949002799]

[34] Setlur AA, Doherty SP, Dai JY, Chang RP. A promising pathway to make multiwalled carbon nanotubes. Appl Phys Lett 2000; 76(21): 3008-10. [http://dx.doi.org/10.1063/1.126561]

[35] Chadderton LT, Chen Y. Nanotube growth by surface diffusion. Phys Lett A 1999; 263(4 6): 401-5. [http://dx.doi.org/10.1016/S0375-9601(99)00731-8]

[36] Zhou D, Chow L. Complex structure of carbon nanotubes and their implications for formation mechanism. J Appl Phys 2003; 93(12): 9972-6. [http://dx.doi.org/10.1063/1.1573733]

[37] Huang JY, Chen S, Ren ZF, Chen G, Dresselhaus MS. Real-time observation of tubule formation from amorphous carbon nanowires under high-bias Joule heating. Nano Lett 2006; 6(8): 1699-705. [http://dx.doi.org/10.1021/nl0609910] [PMID: 16895359]

[38] Bethune DS, Klang CH, De Vries MS, *et al.* Cobalt-catalysed growth of carbon nanotubes with single-atomic-layer walls. Nature 1993; 363: 605-7. [http://dx.doi.org/10.1038/363605a0]

[39] Ajayan PM, Lambert JM, Bernier P, Barbedette L, Colliex C, Planeix JM. Growth morphologies during cobalt-catalyzed single-shell carbon nanotube synthesis. Chem Phys Lett 1993; 215: 509-17. [http://dx.doi.org/10.1016/0009-2614(93)85711-V]

[40] Seraphin S, Zhou SD, Jiao J, *et al.* Catalytic role of nickel, palladium, and platinum in the formation of carbon nanoclusters. Chem Phys Lett 1994; 217: 191-8. [http://dx.doi.org/10.1016/0009-2614(93)E1376-R]

[41] Saito Y, Okuda M, Fujimoto N, Yoshikawa T, Tomita M, Hayashi T. Single-wall carbon nanotubes growing radially from Ni fine particles formed by arc evaporation. Jpn. J. Appl. Phys, Part 2 1994; 33:

L526-9.

[42] Zhou D, Seraphin S, Wang S. Single□walled carbon nanotubes growing radially from $YC_2$ particles. Appl Phys Lett 1994; 65: 1593-5.
[http://dx.doi.org/10.1063/1.112924]

[43] Saito Y, Nishikubo K, Kawabata K, Matsumoto T. Carbon nanocapsules and single□layered nanotubes produced with platinum□group metals (Ru, Rh, Pd, Os, Ir, Pt) by arc discharge. J Appl Phys 1996; 80: 3062-7.
[http://dx.doi.org/10.1063/1.363166]

[44] Saito Y, Okuda M, Koyama T. Carbon nanocapsules and single-wall nanotubes formed by arc evaporation. Surf Rev Lett 1996; 3: 863-7.
[http://dx.doi.org/10.1142/S0218625X96001558]

[45] Chen B, Zhao X, Inoue S, Ando Y. Fabrication and dispersion evaluation of single-wall carbon nanotubes produced by FH-arc discharge method. J Nanosci Nanotechnol 2010; 10(6): 3973-7.
[http://dx.doi.org/10.1166/jnn.2010.2015] [PMID: 20355400]

[46] Fan WW, Zhao J, Lv YK, Bao WR, Liu XG. J. Synthesis of SWNTs from charcoal by arc discharging, Wuhan Univ. Technol. Mater Sci Ed 2010; 25: 194-6.

[47] Shi Z, Lian Y, Liao FH, *et al.* Large scale synthesis of single-wall carbon nanotubes by arc-discharge method. J Phys Chem Solids 2000; 61: 1031-6.
[http://dx.doi.org/10.1016/S0022-3697(99)00358-3]

[48] Hutchison JL, Kiselev NA, Krinichnaya EP, *et al.* Double-walled carbon nanotubes fabricated by a hydrogen arc discharge methodCarbon 2001; 39: 761-70.

[49] Sugai T, Yoshida H, Shimada T, Okazaki T, Shinohara H. New synthesis of high-quality double-walled carbon nanotubes by high-temperature pulsed arc discharge. Nano Lett 2003; 3: 769-73.
[http://dx.doi.org/10.1021/nl034183+]

[50] Qiu JS, Wang ZY, Zhao ZB, Wang TH. Synthesis of double-walled carbon nanotubes from coal in hydrogen-free atmosphere. Fuel 2007; 86: 282-6.
[http://dx.doi.org/10.1016/j.fuel.2006.05.024]

[51] Qiu HX, Shi ZJ, Guan LH, *et al.* High-efficient synthesis of double-walled carbon nanotubes by arc discharge method using chloride as a promoter. Carbon 2006; 44: 516-21.
[http://dx.doi.org/10.1016/j.carbon.2005.08.021]

[52] Liu QF, Ren WC, Li F, Cong HT, Cheng HM. Synthesis and high thermal stability of double-walled carbon nanotubes using nickel formate dihydrate as catalyst precursor. J Phys Chem C 2007; 111: 5006-13.
[http://dx.doi.org/10.1021/jp068672k]

[53] Guo T, Nikolaev P, Rinzler AG, Tomanek D, Colbert DT, Smalley RE. Self-assembly of tubular fullerenes. J Phys Chem 1995; 99: 10694-7.
[http://dx.doi.org/10.1021/j100027a002]

[54] Hafner JH, Bronikowski MJ, Azamian BR, *et al.* Catalytic growth of single-wall carbon nanotubes from metal particles. Chem Phys Lett 1998; 296: 195-202.
[http://dx.doi.org/10.1016/S0009-2614(98)01024-0]

[55] Thess A, Lee R, Nikolaev P, *et al.* Fischer, J.E.; Smalley, R.E. Crystalline ropes of metallic carbon nanotubes. Science 1996; 273(5274): 483-7.
[http://dx.doi.org/10.1126/science.273.5274.483] [PMID: 8662534]

[56] Rao AM, Richter E, Bandow S, *et al.* Diameter-selective Raman scattering from vibrational modes in carbon nanotubes. Science 1997; 275(5297): 187-91.
[http://dx.doi.org/10.1126/science.275.5297.187] [PMID: 8985007]

[57] Yudasaka M, Yamada R, Sensui N, Wilkins T, Ichihashi T, Iijima S. Mechanism of the effect of NiCo,

Ni and Co catalysts on the yield of single-wall carbon nanotubes formed by pulsed Nd:YAG laser ablation. J Phys Chem B 1999; 103: 6224-9.
[http://dx.doi.org/10.1021/jp9908451]

[58] Zhang M, Yudasaka M, Iijima S. Single-wall carbon nanotubes: A high yield of tubes through laser ablation of a crude-tube target. Chem Phys Lett 2001; 336: 196-200.
[http://dx.doi.org/10.1016/S0009-2614(01)00101-4]

[59] Kataura H, Kumazawa Y, Maniwa Y, *et al.* Diameter control of single-walled carbon nanotubes. Carbon 2000; 38: 1691-7.
[http://dx.doi.org/10.1016/S0008-6223(00)00090-7]

[60] Nishide D, Kataura H, Suzuki S, Tsukagoshi K, Aoyagi Y, Achiba Y. High-yield production of single-wall carbon nanotubes in nitrogen gas. Chem Phys Lett 2003; 372: 45-50.
[http://dx.doi.org/10.1016/S0009-2614(03)00352-X]

[61] Yudasaka M, Komatsu T, Ichihashi T, Achiba Y, Iijima S. Pressure dependence of the structures of carbonaceous deposits formed by laser ablation on targets composed of carbon, nickel, and cobalt. J Phys Chem B 1998; 102: 4892-6.
[http://dx.doi.org/10.1021/jp973446j]

[62] Gorbunov AA, Friedlein R, Jost O, Golden MS, Fink J, Pompe W. Gas-dynamic consideration of the laser evaporation synthesis of single-wall carbon nanotubes. Appl Phys A (Suppl) 1999; 69: S593-6.
[http://dx.doi.org/10.1007/s003390051483]

[63] Braidy N, El Khakani MA, Botton GA. Effect of laser intensity on yield and physical characteristics of single wall carbon nanotubes produced by the Nd:YAG laser vaporization method. Carbon 2002; 40: 2835-42.
[http://dx.doi.org/10.1016/S0008-6223(02)00260-9]

[64] Kokai F, Takahashi K, Yudasaka M, Yamada R, Ichihashi T, Iijima S. Growth dynamics of single-wall carbon nanotubes synthesized by CO2 laser vaporization. J Phys Chem B 1999; 103: 4346-51.
[http://dx.doi.org/10.1021/jp990065s]

[65] Bandow S, Asaka S, Rao YA, Grigorian L, Richter E, Eklund PC. Effect of the growth temperature on the diameter distribution and chirality of single-wall carbon nanotubes. Phys Rev Lett 1998; 80: 3779-82.
[http://dx.doi.org/10.1103/PhysRevLett.80.3779]

[66] Dillon AC, Parilla PA, Alleman JL, Perkins JD, Heben MJ. Controlling single-wall nanotube diameters with variation in laser pulse power. Chem Phys Lett 2000; 316: 13-8.
[http://dx.doi.org/10.1016/S0009-2614(99)01259-2]

[67] Yudasaka M, Ichihashi T, Komatsu T, Iijima S. Single-wall carbon nanotubes formed by a single laser-beam pulse. Chem Phys Lett 1999; 299: 91-6.
[http://dx.doi.org/10.1016/S0009-2614(98)01219-6]

[68] Zhang H, Ding Y, Wu C, *et al.* The effect of laser power on the formation of carbon nanotubes prepared in $CO_2$ continuous wave laser ablation at room temperature. Physica B 2003; 325: 224-9.
[http://dx.doi.org/10.1016/S0921-4526(02)01528-4]

[69] Saito Y, Okuda M, Fujimoto N, Yoshikawa T, Tomita M, Hayashi T. Single-wall carbon nanotubes growing radially from Ni fine particles formed by arc evaporation. Jpn J Appl Phys Part 2-Lett 1994; 33(4A): L526-9.
[http://dx.doi.org/10.1143/JJAP.33.L526]

[70] Saito Y. Nanoparticles and filled nanocapsules. Carbon 1995; 33(7): 979-88.
[http://dx.doi.org/10.1016/0008-6223(95)00026-A]

[71] Kataura H, Kumazawa Y, Maniwa Y, Ohtsuka Y, Sen R, Suzuki S, *et al.* Diameter control of single-walled carbon nanotubes. Carbon 2000; 38(11–12): 1691-7.
[http://dx.doi.org/10.1016/S0008-6223(00)00090-7]

[72] Sen R, Suzuki S, Kataura H, Achiba Y. Growth of single-walled carbon nanotubes from the condensed phase. Chem Phys Lett 2001; 349(5-6): 383-8. [http://dx.doi.org/10.1016/S0009-2614(01)01208-8]

[73] Sinnott SB, Andrews R. Carbon Nanotubes: Synthesis, properties and applications. Crit Rev Solid State Mater Sci 2001; 26: 145-249. [http://dx.doi.org/10.1080/20014091104189]

[74] Yacamàn MJ, Yoshida MM, Rendon L, Santiesteban JG. Catalytic growth of carbon microtubules with fullerene structure. Appl Phys Lett 1993; 62: 202-4. [http://dx.doi.org/10.1063/1.109315]

[75] Flahaut E, Govindaraj A, Peigney A, Laurent C, Rousset A, Rao CN. Synthesis of single-walled carbon nanotubes using binary (Fe, Co, Ni) alloy nanoparticles prepared *in situ* by the reduction of oxide solid solutions. Chem Phys Lett 1999; 300: 236-42. [http://dx.doi.org/10.1016/S0009-2614(98)01304-9]

[76] Couteau E, Hernadi K, Seo JW, *et al.* CVD synthesis of high-purity multiwalled carbon nanotubes using CaCO3 catalyst support for large-scale production. Chem Phys Lett 2003; 378: 9-17. [http://dx.doi.org/10.1016/S0009-2614(03)01218-1]

[77] Ren ZF, Huang ZP, Xu JW, *et al.* Synthesis of large arrays of well-aligned carbon nanotubes on glass. Science 1998; 282(5391): 1105-7. [http://dx.doi.org/10.1126/science.282.5391.1105] [PMID: 9804545]

[78] Fan S, Chapline MG, Franklin NR, Tombler TW, Cassell AM, Dai H. Self-oriented regular arrays of carbon nanotubes and their field emission properties. Science 1999; 283(5401): 512-4. [http://dx.doi.org/10.1126/science.283.5401.512] [PMID: 9915692]

[79] Dai HJ, Rinzler AG, Nikolaev P, Thess A, Colbert DT, Smalley RE. Single-wall nanotubes produced by metal-catalyzed disproportionation of carbon monoxide. Chem Phys Lett 1996; 260: 471-5. [http://dx.doi.org/10.1016/0009-2614(96)00862-7]

[80] Hernadi K, Fonseca A, Nagy JB, Siska A, Kiricsi I. Production of nantoubes by the catalytic decomposition of different carbon-containing compounds. Appl Catal A Gen 2000; 199: 245-55. [http://dx.doi.org/10.1016/S0926-860X(99)00561-X]

[81] Nikolaev P, Bronikowski MJ, Bradley RK, *et al.* Gas-phase catalytic growth of single-walled carbon nanotubes from carbon monoxide. Chem Phys Lett 1999; 313: 91-7. [http://dx.doi.org/10.1016/S0009-2614(99)01029-5]

[82] Maruyama S, Kojima R, Miyauchi R, Chiashi S, Kohno M. Low-temperature synthesis of high-purity single-walled carbon nanotubes from alcohol. Chem Phys Lett 2002; 360: 229-34. [http://dx.doi.org/10.1016/S0009-2614(02)00838-2]

[83] Su M, Zheng B, Liu J. A scalable CVD method for the synthesis of single-walled carbon nanotubes with high catalyst productivity. Chem Phys Lett 2000; 322: 321-6. [http://dx.doi.org/10.1016/S0009-2614(00)00422-X]

[84] Li Y. Liu J, Wang QY, Wang ZL. Preparation of monodispersed Fe–Mo nanoparticles as the catalyst for CVD synthesis of carbon nanotubes. Chem Mater 2001; 13: 1008-14. [http://dx.doi.org/10.1021/cm000787s]

[85] Ciambelli P, Sannino D, Sarno M, Fonseca A, Nagy JB. Selective formation of carbon nanotubes over Co-modified beta zeolite by CCVD. Carbon 2005; 43: 631-40. [http://dx.doi.org/10.1016/j.carbon.2004.10.027]

[86] Ago H, Uehara N, Yoshihara N, *et al.* Gas analysis of the CVD process for high yield growth of carbon nanotubes over metal-supported catalysts. Carbon 2006; 44: 2912-8. [http://dx.doi.org/10.1016/j.carbon.2006.05.049]

[87] Qingwen L, Hao Y, Yan C, Jin Z, Zhongfan L. A scalable CVD synthesis of high-purity single-walled

carbon nanotubes with porous MgO as support material. J Mater Chem 2002; 12: 1179-83. [http://dx.doi.org/10.1039/b109763f]

[88] Liu BC, Lyu SC, Jung SI, *et al.* Single-walled carbon nanotubes produced by catalytic chemical vapor deposition of acetylene over Fe–Mo/MgO catalyst. Chem Phys Lett 2004; 383: 104-8. [http://dx.doi.org/10.1016/j.cplett.2003.10.134]

[89] Ago H, Imamura S, Okazaki T, Saito T, Yumura M, Tsuji M. CVD growth of single-walled carbon nanotubes with narrow diameter distribution over Fe/MgO catalyst and their fluorescence spectroscopy. J Phys Chem B 2005; 109(20): 10035-41. [http://dx.doi.org/10.1021/jp050307q] [PMID: 16852214]

[90] Balogh Z, Halasi G, Korbely B, Hernadi K. CVD synthesis of MWNTs over K-doped supported catalysts. Appl Catal A Gen 2008; 344: 191-7. [http://dx.doi.org/10.1016/j.apcata.2008.04.019]

[91] Kawasaki S, Shinoda M, Shimada T, Okino F, Touhara H. Single-walled carbon nanotubes grown on natural minerals. Carbon 2006; 44: 2139-41. [http://dx.doi.org/10.1016/j.carbon.2006.03.015]

[92] Ciambelli P, Sannino D, Sarno M, Fonseca A, Nagy JB. Hydrocarbon decomposition in alumina membrane: an effective way to produce carbon nanotubes bundles. J Nanosci Nanotechnol 2004; 4(7): 779-87. [http://dx.doi.org/10.1166/jnn.2004.108] [PMID: 15570960]

[93] Wei S, Kang WP, Davidson JL, Huang JH. Aligned carbon nanotubes fabricated by thermal CVD at atmospheric pressure using Co as catalyst with $NH_3$ as reactive gas. Diamond Rel Mat 2006; 15: 1828-33. [http://dx.doi.org/10.1016/j.diamond.2006.09.010]

[94] Valles C, Mendoza MP, Maser WK, *et al.* Effects of partial and total methane flows on the yield and structural characteristics of MWCNTs produced by CVD. Carbon 2009; 47: 998-1004. [http://dx.doi.org/10.1016/j.carbon.2008.11.047]

[95] Mendoza MP, Valles C, Maser WK, *et al.* Ni-Y/Mo catalyst for the large-scale CVD production of multi-wall carbon nanotubes. Carbon 2005; 43: 3034-7. [http://dx.doi.org/10.1016/j.carbon.2005.05.048]

[96] Palen EB, Rümmeli MH. Activated Cu catalysts for alcohol CVD synthesized non-magnetic bamboo-like carbon nanotubes and branched bamboo-like carbon nanotubes. Superlattices Microstruct 2009; 46: 374-8. [http://dx.doi.org/10.1016/j.spmi.2008.10.029]

[97] Mori S, Suzuki M. Catalyst-free low-temperature growth of carbon nanofibers by microwave plasma-enhanced CVD. Thin Solid Films 2009; 517: 4264-7. [http://dx.doi.org/10.1016/j.tsf.2009.02.009]

[98] Cheng J, Zhang X, Liu F, *et al.* Synthesis of carbon nanotubes filled with $Fe_3C$ nanowires by CVD with titanate modified palygorskite as catalyst. Carbon 2003; 41: 1965-70. [http://dx.doi.org/10.1016/S0008-6223(03)00185-4]

[99] Mamalis AG, Vogtländer LO, Markopoulos A. Nanotechnology and nanostructured materials: Trends in carbon nanotubes. Precis Eng 2004; 28: 16-30. [http://dx.doi.org/10.1016/j.precisioneng.2002.11.002]

[100] Jayatissa AH, Guo K. Synthesis of carbon nanotubes at low temperature by filament assisted atmospheric CVD and their field emission characteristics. Vacuum 2009; 83: 853-6. [http://dx.doi.org/10.1016/j.vacuum.2008.08.009]

[101] Horvath ZE, Kertesz K, Petho L, *et al.* Darabont Al, Incze, PN, Sarkozi Z, Biro LP. Inexpensive, upscalable nanotube growth methods. Curr Appl Phys 2006; 6: 135-40. [http://dx.doi.org/10.1016/j.cap.2005.07.026]

[102] Singh C, Shaffer M, Kinloch I, Windle A. Production of aligned carbon nanotubes by the CVD injection method. Physica B 2002; 323: 339-40. [http://dx.doi.org/10.1016/S0921-4526(02)01045-1]

[103] Qi JL, Wang X, Tian HW, Peng YS, Liu C, Zheng WT. Syntheses of carbon nanomaterials by radio frequency plasma enhanced chemical vapor deposition. J Alloys Compd 2009; 486: 265-72. [http://dx.doi.org/10.1016/j.jallcom.2009.06.122]

[104] Zhu M, Wang J, Outlaw RA, Hou K, Manos DM, Holloway BC. Synthesis of carbon nanosheets and carbon nanotubes by radio frequency plasma enhanced chemical vapor deposition. Diamond Rel Mat 2007; 16: 196-201. [http://dx.doi.org/10.1016/j.diamond.2006.05.007]

[105] Chaisitsak S, Yamada A, Konagi M. Hot filament enhanced CVD synthesis of carbon nanotubes by using a carbon filament. Diamond Rel Mat 2004; 13: 438-44. [http://dx.doi.org/10.1016/S0925-9635(03)00572-7]

[106] Hata K, Futaba DN, Mizuno K, Namai T, Yumura M, Iijima S. Water-assisted highly efficient synthesis of impurity-free single-walled carbon nanotubes. Science 2004; 306(5700): 1362-4. [http://dx.doi.org/10.1126/science.1104962] [PMID: 15550668]

[107] Zhang G, Mann D, Zhang L, *et al.* Ultra-high-yield growth of vertical single-walled carbon nanotubes: Hidden roles of hydrogen and oxygen. Proc Natl Acad Sci USA 2005; 102(45): 16141-5. [http://dx.doi.org/10.1073/pnas.0507064102] [PMID: 16263931]

[108] Hiramatsu M, Nagao H, Taniguchi M, Amano H, Ando Y, Hori M. High-rate growth of films of dense, aligned double-walled carbon nanotubes using microwave plasma-enhanced chemical vapor deposition. Jpn J Appl Phys 2005; 44: 693-5. [http://dx.doi.org/10.1143/JJAP.44.L693]

[109] Kim D, Lim SH, Guilley AJ, *et al.* Growth of vertically aligned arrays of carbon nanotubes for high field emission. Thin Solid Films 2008; 516: 706-9. [http://dx.doi.org/10.1016/j.tsf.2007.06.089]

[110] Yamada T, Namai T, Hata K, *et al.* Size-selective growth of double-walled carbon nanotube forests from engineered iron catalysts. Nat Nanotechnol 2006; 1(2): 131-6. [http://dx.doi.org/10.1038/nnano.2006.95] [PMID: 18654165]

[111] Kim MJ, Choi JH, Park JB, Kim SK, Yoo JB, Park CY. Growth characteristics of carbon nanotubes *via* aluminum nanopore template on Si substrate using PECVD. Thin Solid Films 2003; 435: 312-7. [http://dx.doi.org/10.1016/S0040-6090(03)00339-0]

[112] Chang WS, Kim JW, Choi DC, Han CS. Fabrication of free-standing carbon nanotube electrode arrays on a quartz wafer. Thin Solid Films 2010; 518: 6624-9. [http://dx.doi.org/10.1016/j.tsf.2010.03.139]

[113] Byeon H, Kim SY, Koh KH, Lee S. Growth of ultra long multiwall carbon nanotube arrays by aerosol-assisted chemical vapor deposition. J Nanosci Nanotechnol 2010; 10(9): 6116-9. [http://dx.doi.org/10.1166/jnn.2010.2574] [PMID: 21133158]

[114] Jeong N, Seo Y, Lee J. Vertically aligned carbon nanotubes synthesized by the thermal pyrolysis with an ultrasonic evaporatorDiamond Relat. Mater 2007; 16: 600-8.

[115] Camarena JP, Gomez HE, Somanathan R, *et al.* Vertically aligned carbon nanotubes synthesized by the thermal pyrolysis with an ultrasonic evaporator. J Nanosci Nanotechnol 2011; 11: 5539-45. [http://dx.doi.org/10.1166/jnn.2011.3430] [PMID: 21770216]

[116] Sadeghian Z. Large-scale production of multi-walled carbon nanotubes by low-cost spray pyrolysis of hexane. N Carbon Mater 2009; 24: 33-8. [http://dx.doi.org/10.1016/S1872-5805(08)60034-7]

[117] Pinault M, L'Hermite MM, Reynaud C, Beyssac O, Rouzaud JN, Clinard C. Carbon nanotubes

produced by aerosol pyrolysis: growth mechanisms and post-annealing effectsDiamond Relat. Mater 2004; 13: 1266-9.

[118] Nebol'sin VA, Vorob'ev AV. Role of surface energy in the growth of carbon nanotubes *via* catalytic pyrolysis of hydrocarbons. Inorg Mater 2011; 47: 128-32. [http://dx.doi.org/10.1134/S0020168511020105]

[119] Romero JL, Yanez JC, Tinoco JL, Nunez GA, Sandoval SA, Delgado FP. Fuller Nanotub Carbon Nanostruct 2011; 19: 483-96. [http://dx.doi.org/10.1080/1536383X.2010.494785]

[120] Kumar R, Tiwari RS, Srivastava ON. Scalable synthesis of aligned carbon nanotubes bundles using green natural precursor: neem oil. Nanoscale Res Lett 2011; 6(1): 92. [http://dx.doi.org/10.1186/1556-276X-6-92] [PMID: 21711585]

[121] Paul S. Samdarshi, A green precursor for carbon nanotube synthesis. N Carbon Mater 2011; 26: 85-8. [http://dx.doi.org/10.1016/S1872-5805(11)60068-1]

[122] Quan C, Li A, Gao N. Synthesis of carbon nanotubes and porous carbons from printed circuit board waste pyrolysis oil. J Hazard Mater 2010; 179(1-3): 911-7. [http://dx.doi.org/10.1016/j.jhazmat.2010.03.092] [PMID: 20400225]

[123] Kucukayan G, Ovali R, Ilday S, *et al.* An experimental and theoretical examination of the effect of sulfur on the pyrolytically grown carbon nanotubes from sucrose-based solid state precursors. Carbon 2011; 49: 508-17. [http://dx.doi.org/10.1016/j.carbon.2010.09.050]

[124] Kuang YH, Li KZ, Li HJ, Xu ZW, Wang YJ. A Solid-state Hybrid Method to Synthesize Straight Carbon Nanotubes by Pyrolysis of Two Mixed Metal Phthalocyanines Chin. J Inorg Chem 2009; 25: 951-5.

[125] Du G, Song C, Zhao J, Feng S, Zhu Z. Solid-phase transformation of glass-like carbon nanoparticles into nanotubes and the related mechanism. Carbon 2008; 46: 92-8. [http://dx.doi.org/10.1016/j.carbon.2007.10.029]

[126] Liu YC, Zheng NN, Huang JD, Sun BM. Advanced Polymer Science and Engineering. Stafa-Zurich: Trans Tech Publications Ltd 2011; pp. 99-103.

[127] Gogotsi Y, Libera JA, Yoshimura M. Hydrothermal synthesis of multiwall carbon nanotubes. J Mater Res 2000; 15: 2591-4. [http://dx.doi.org/10.1557/JMR.2000.0370]

[128] Gogotsi Y, Naguib N, Libera J. *In situ* chemical experiments in carbon nanotubes. Chem Phys Lett 2002; 365: 354-60. [http://dx.doi.org/10.1016/S0009-2614(02)01496-3]

[129] Manafi S, Nadali H, Irani HR. Low temperature synthesis of multi-walled carbon nanotubes *via* a sonochemical/hydrothermal method. Mater Lett 2008; 62: 4175-6. [http://dx.doi.org/10.1016/j.matlet.2008.05.072]

[130] Moreno JM, Swamy SS, Fujino T, Yoshimura M. Carbon nanocells and nanotubes grown in hydrothermal fluids. Chem Phys Lett 2000; 329: 317-22. [http://dx.doi.org/10.1016/S0009-2614(00)01017-4]

[131] Stevens MG, Subramoney S, Foley HC. Spontaneous formation of carbon nanotubes and polyhedra from cesium and amorphous carbon. Chem Phys Lett 1998; 292: 352-6. [http://dx.doi.org/10.1016/S0009-2614(98)00671-X]

[132] Hsu WK, Hare JP, Terrones M, Kroto HW, Walton DR, Harris PJ. Condensed-phase nanotubes. Nature 1995; 377: 687. [http://dx.doi.org/10.1038/377687a0]

[133] Bai JB, Hamon AL, Marraud A, Jouffrey B, Zymla V. Synthesis of SWNTs and MWNTs by a molten

salt (NaCl) method. Chem Phys Lett 2002; 365: 184-8.
[http://dx.doi.org/10.1016/S0009-2614(02)01447-1]

[134] Novoselova IA, Oliinyk NF, Volkov SV, *et al.* Electrolytic synthesis of carbon nanotubes from carbon dioxide in molten salts and their characterization. Physica E 2008; 40: 2231-7.
[http://dx.doi.org/10.1016/j.physe.2007.10.069]

[135] Matveev AT, Golberg D, Novikov VP, Klimkovich LL, Bando Y. Synthesis of carbon nanotubes below room temperature. Carbon 2001; 39: 155-8.
[http://dx.doi.org/10.1016/S0008-6223(00)00209-8]

[136] Laplaze D, Bernier P, Maser WK, Flamant G, Guillard T, Loiseau A. Carbon nanotubes: The solar approach. Carbon 1998; 36: 685-8.
[http://dx.doi.org/10.1016/S0008-6223(98)00025-6]

[137] Luxemburg D, Flamant G, Laplaze D. Solar synthesis of single-walled carbon nanotubes at medium scale. Carbon 2005; 43: 2302-10.
[http://dx.doi.org/10.1016/j.carbon.2005.04.010]

[138] Moon JM, An KH, Lee YH, Park YS, Bae DJ, Park GS. High-yield purification process of single walled carbon nanotubes. J Phys Chem B 2001; 105: 5677-81.
[http://dx.doi.org/10.1021/jp0102365]

[139] Mohanapriya S, Lakshminarayanan V. Simultaneous purification and spectrophotometric determination of nickel present in as-prepared single-walled carbon nanotubes (SWCNT). Talanta 2007; 71(1): 493-7.
[http://dx.doi.org/10.1016/j.talanta.2006.04.001] [PMID: 19071332]

[140] Delpeux S, Szostak K, Frackowiak E, Beguin F. An efficient two-step process for producing opened multi-walled carbon nanotubes of high purity. Chem Phys Lett 2005; 404: 374.
[http://dx.doi.org/10.1016/j.cplett.2005.01.108]

[141] Zhao N, He C, Li J, Jiang Z, Li Y. Study on purification and tip-opening of CNTs fabricated by CVD. Mater Res Bull 2006; 41: 2204.
[http://dx.doi.org/10.1016/j.materresbull.2006.04.029]

[142] Engel-Herbert R, Pforte H, Hesjedal T. CVD synthesis and purification of single-walled carbon nanotubes using silica-supported metal catalyst. Mater Lett 2007; 61(11): 2589-93.
[http://dx.doi.org/10.1016/j.matlet.2006.10.004]

[143] Igarashi H, Murakami H, Murakami Y, Maruyama S, Nakashima N. Purification and characterization of zeolite-supported single-walled carbon nanotubes catalytically synthesized from ethanol. Chem Phys Lett 2004; 392: 529-32.
[http://dx.doi.org/10.1016/j.cplett.2004.05.104]

[144] Colomer JF, Piedigrosso P, Fonseca A, Nagy JB. Different purification methods of carbon nanotubes produced by catalytic synthesis. Synth Met 1999; 103: 2482-3.
[http://dx.doi.org/10.1016/S0379-6779(98)01066-2]

[145] Kushwaha N, Mittal J. Over-oxidation of multi-walled carbon nanotubes and formation of fluorescent carbon nanoparticles. Mater Lett 2015; 145: 37-40.
[http://dx.doi.org/10.1016/j.matlet.2015.01.059]

[146] Hernadi K, Siska A, Thien-Nga L, Forro L, Kiricsi I. Reactivity of different kinds of carbon during oxidative purification of catalytically prepared carbon nanotubes. Solid State Ion 2001; 141: 203-9.
[http://dx.doi.org/10.1016/S0167-2738(01)00789-5]

[147] Suzuki T, Suhama K, Zhao X, Inoue S, Nishikawa N, Ando Y. Purification of single-wall carbon nanotubes produced by arc plasma jet method. Diamond and related mater 2007; 16: 1116-20.
[http://dx.doi.org/10.1016/j.diamond.2006.11.006]

[148] Pinero ER, Cacciaguerra T, Simon P, Beguin F. A single step process for the simultaneous purification and opening of multiwalled carbon nanotubes. Chem Phys Lett 2005; 412: 184-9.

[http://dx.doi.org/10.1016/j.cplett.2005.06.099]

[149] Ramesh P, Okazaki T, Sugai T, *et al.* Purification and characterization of double-wall carbon nanotubes synthesized by catalytic chemical vapor deposition on mesoporous silica. Chem Phys Lett 2006; 418: 408-12.
[http://dx.doi.org/10.1016/j.cplett.2005.11.018]

[150] Salernitano E, Giorgi L, Makris TD, *et al.* Purification of MWCNTs grown on a nanosized unsupported Fe-based powder catalyst. Diamond Relat mater 2007; 16(8): 1565-70.
[http://dx.doi.org/10.1016/j.diamond.2006.12.038]

[151] Ziegler KJ, Gu Z, Peng H, Flor EL, Hauge RH, Smalley RE. Controlled oxidative cutting of single-walled carbon nanotubes. J Am Chem Soc 2005; 127(5): 1541-7.
[http://dx.doi.org/10.1021/ja044537e] [PMID: 15686387]

[152] Zhang Y, Shi Z, Gu Z, Iijima S. Structure modification of single-wall carbon nanotubes. Carbon 2000; 38(15): 2055-9.
[http://dx.doi.org/10.1016/S0008-6223(00)00047-6]

[153] Chen CM, Chen M, Peng YW, Yu HW, Chen CF. High efficiency microwave digestion purification of multi-walled carbon nanotubes synthesized by thermal chemical vapor deposition. Thin Solid Films 2006; 498(1): 202-5.
[http://dx.doi.org/10.1016/j.tsf.2005.07.088]

[154] Chen CM, Chen M, Peng YW, Yu HW, Chen CF. High efficiency microwave digestion purification of multi-walled carbon nanotubes synthesized by thermal chemical vapor deposition. Thin Solid Films 2006; 498: 202-5.
[http://dx.doi.org/10.1016/j.tsf.2005.07.088]

[155] Valentini L, Cantalini C, Lozzi L, Armentano I, Kenny JM, Santucci S. Reversible oxidation effects on carbon nanotubes thin films for gas sensing applications. Mater Sci Eng C 2003; 23: 523-9.
[http://dx.doi.org/10.1016/S0928-4931(02)00356-9]

[156] Wiltshire JG, Khlobystov AN, Li LJ, Lyapin SG, Briggs GA, Nicholas RJ. Comparative studies on acid and thermal based selective purification of HiPCO produced single-walled carbon nanotubes. Chem Phys Lett 2004; 386: 239-43.
[http://dx.doi.org/10.1016/j.cplett.2004.01.067]

[157] Smith MR, Hedges SW, LaCount R, *et al.* Selective oxidation of single-walled carbon nanotubes using carbon dioxide. Carbon 2003; 41: 1221-30.
[http://dx.doi.org/10.1016/S0008-6223(03)00054-X]

[158] Vivekchand SR, Govindaraj A, Seikh MM, Rao CN. New method of purification of carbon nanotubes based on hydrogen treatment. J Phys Chem B 2004; 108: 6935-7.
[http://dx.doi.org/10.1021/jp048737o]

[159] Wang Y, Gao L, Sun J, *et al.* An integrated route for purification, cutting and dispersion of single-walled carbon nanotubes. Chem Phys Lett 2006; 432: 205-8.
[http://dx.doi.org/10.1016/j.cplett.2006.10.054]

[160] Hou PX, Bai S, Yang QH, Liu C, Cheng HM. Multi-step purification of carbon nanotubes. Carbon 2002; 40: 81-5.
[http://dx.doi.org/10.1016/S0008-6223(01)00075-6]

[161] Abatemarco T, Stickel J, Belfort J, Frank BP, Ajayan PM, Belfort G. Fractionation of multiwalled carbon nanotubes by cascade membrane microfiltration. J Phys Chem B 1999; 103: 3534-8.
[http://dx.doi.org/10.1021/jp984020n]

[162] Tagmatarchis N, Zattoni A, Reschiglian P, Prato M. Separation and purification of functionalised water-soluble multi-walled carbon nanotubes by flow field-flow fractionation. Carbon 2005; 43: 1984-9.
[http://dx.doi.org/10.1016/j.carbon.2005.03.011]

[163] Lu KL, Lago RM, Chen YK, Green ML, Harris PJ, Tsang SC. Mechanical damage of carbon nanotubes by ultrasound. Carbon 1996; 34: 814-6.
[http://dx.doi.org/10.1016/0008-6223(96)89470-X]

[164] Heller DA, Barone PW, Strano MS. Sonication-induced changes in chiral distribution: A complication in the use of single-walled carbon nanotube fluorescence for determining species distribution. Carbon 2005; 43: 651-3.
[http://dx.doi.org/10.1016/j.carbon.2004.10.010]

[165] Yang C, Hu X, Wang D, *et al.* Ultrasonically treated multi-walled carbon nanotubes (MWCNTs) as PtRu catalyst supports for methanol electrooxidation. J Power Sources 160: 187-93.
[http://dx.doi.org/10.1016/j.jpowsour.2006.05.015]

[166] Farkas E, Anderson ME, Chen Z, Rinzler AG. Length sorting cut single wall carbon nanotubes by high performance liquid chromatography. Chem Phys Lett 2002; 363: 111-6.
[http://dx.doi.org/10.1016/S0009-2614(02)01203-4]

[167] Yang Y, Xie L, Chen Z, Liu M, Zhu T, Liu Z. Purification and length separation of single-walled carbon nanotubes using chromatographic method. Synth Met 2005; 155: 455-60.
[http://dx.doi.org/10.1016/j.synthmet.2005.05.012]

[168] Duesberg GS, Blau W, Byrne HJ, Muster J, Burghard M, Roth S. Chromatography of carbon nanotubes. Synth Met 1999; 103: 2484-5.
[http://dx.doi.org/10.1016/S0379-6779(98)01068-6]

[169] Montoro LA, Rosolen JM. A multi-step treatment to effective purification of single-walled carbon nanotubes. Carbon 2006; 44: 3293-301.
[http://dx.doi.org/10.1016/j.carbon.2006.06.018]

[170] Jiang L, Gao L, Sun J. Production of aqueous colloidal dispersions of carbon nanotubes. J Colloid Interface Sci 2003; 260(1): 89-94.
[http://dx.doi.org/10.1016/S0021-9797(02)00176-5] [PMID: 12742038]

[171] Ago H, Kugler T, Cacialli F, *et al.* Work function of purified and oxidised carbon nanotubes. Synth Met 1999; 103: 2494-5.
[http://dx.doi.org/10.1016/S0379-6779(98)01062-5]

[172] Matarredona O, Rhoads H, Li Z, Harwell JH, Balzano L, Resasco DE. Dispersion of single-walled carbon nanotubes in aqueous solutions of the anionic surfactant NaDDBS. J Phys Chem B 2003; 107: 13357-67.
[http://dx.doi.org/10.1021/jp0365099]

CHAPTER 3

# Carbon Nanotubes: Synthesis Methods

**Ranu Nayak***

*Amity Institute of Nanotechnology, Sec 125, Amity University, Noida, UP, INDIA*

**Abstract:** This chapter is focused on the unique electrical, optical, mechanical, thermal and vibrational properties of carbon nanotubes. Reader's get an insight of how these unique properties of carbon nanotubes can be functional to various applications. Each of these properties are being emphasized with some basic definitions, equations, and references from published literatures. This chapter also discusses the property changes observed in CNTs due to their structural differences as in single walled and multiwalled CNTs.

**Keywords:** CNT, Electrical properties, Mechanical properties, MWNT, Optical properties, Properties, SWNT, Thermal properties.

## 3.1. INTRODUCTION

Widely researched carbon nanotubes (CNT), is a versatile material for its unique mechanical, thermal, optical, electrical and vibrational properties. A blend of numerous unique properties makes it an attractive choice for multiple applications. The unlimited fascination for this material and its unique properties are primarily from its unusual structure and aspect ratio.

This chapter focuses on properties of both single walled (SWCNTs) and multi-walled carbon nanotubes (MWCNT). Structurally, SWCNTs consist of a single layer of graphite rolled into a cylindrical shape and were discovered in 1993 [1]. SWCNTs are prototypes of quantum wires composed of a single element (carbon),with walls only one atom thick and tons of atoms in circumference. Every carbon atom lies on the surface of the nanotubes and forms part of a strong covalently bonded structure. SWNTs are typically about 1-3 nm in diameter but have lengths from hundreds of nanometers to several centimeters, with long-range crystalline periodicity retained along the tube axis. The unique structure of SWNTs have generated excellent mechanical properties, including an elastic modulus near 1000 GPa and tensile strength of several tens of GPa, that are

* **Corresponding author Ranu Nayak:** Amity Institute of Nanotechnology, Sec 125, Amity University, Noida, Uttar Pradesh, India; Tel: +919818936883; E-mails: rnayak@amity.edu; ranunayakbose@gmail.com

**Jagriti Narang & Chandra Shekhar Pundir (Eds.)**

comparable to those predicted for singles sheets of graphite. It is the $sp^2$ bonding between carbon atoms in CNT that makes it mechanically stronger than steel, and even stronger than the $sp^3$ bonded diamond. SWCNTs also possess remarkable electronic properties. Extensive research on SWCNTs clearly demonstrates its potential role in the field of molecular electronics, in which the active part of a device is composed of either a single or a few molecules. The electronic structures of SWCNTs strongly depend on their chiral/non-chiral form nanotubes and also lead to unusual and potentially useful optical properties.

MWCNTs, discovered in 1991 by Ijima [2] are multiple layers of graphite superimposed and rolled in on themselves to form a tubular shape. There are two structural models for MWNTs, namely, Russian doll model and Parchment model. In the Russian Doll model, sheets of graphite are arranged in concentric cylinders, while in the Parchment model, a single sheet of graphite is rolled in around itself, like a scroll of parchment or a rolled newspaper. MWCNTs and SWCNTs have similar properties. In MWCNTs, since there are multiple layers, the outer-walls can not only shield the inner carbon nanotubes from chemical interactions with outside substances but also present higher tensile strength properties than SWCNTs [3]. The inner diameter of MWCNTs varies from 0.4 nm to a few nm and outer diameter varies from 2 to 30 nm. The length can vary from 1 μm to few cm. Amazingly CNTs also possess excellent elasticity. It can be easily bent to a large angle, twisted, and buckled without any physical deformation. However, defects in the structure of the nanotube can weaken its strength. Overall, unique combination of high electrical conductivity (as good as Copper), extraordinary mechanical strength (20 times stronger and simultaneously 5 times lighter than steel) and efficient thermal conductivity (as good as diamond) makes CNTs open with wide range of industrial applications.

## 3.2. ELECTRICAL PROPERTIES

Carbon nanotubes (SWNTs) show metallic or semiconducting properties depending on their atomic geometry, *i.e.* diameter and chirality. The wrapping of graphene sheet into a cylinder configures into three different forms of SWCNTs, namely, Armchair, Chiral, and Zigzag. The three different structural forms of SWCNT are distinguished by a pair of indices (n, m) that describe the chiral vector $\vec{C}_h$ as mentioned in equation 1. $a_1$ and $a_2$ are base vectors of graphite and also determine the diameter (d) of the nanotubes as represented in equation 2.

$$\vec{C}_h = n\vec{a}_1 + m\vec{a}_2 \equiv (n, m) \qquad \text{where} 0 \leq m \leq n \qquad \textbf{(1)}$$

$$d = \frac{a\sqrt{m^2+mn+n^2}}{\pi} \quad (2)$$

where a = 1.42 × √3 Å corresponds to the lattice constant.

The number of unit vectors in the honeycomb crystal lattice of graphene along two directions is determined by the integers n and m. As a norm, when m= 0, the CNTs are called zigzag nanotubes; when n =m, the nanotubes are called armchair nanotubes, and for all other sets of n and m, they are called chiral type nanotubes. The Arm chair form having n=m, is metallic in nature, *i.e.* highly conductive. More clearly, the condition for a SWCNT to be metallic is given by the equation 3.

$$2n+ m =3q, \text{ where q in an integer} \quad (3)$$

All other combination of n, m leads to semiconductor type property of CNTs. The band gap for semiconductor nanotubes is inversely proportional to the nanotube diameter. Introduction of structural defects change the chirality of nanotubes. Overall, it is quite apparent that electrical properties of CNTs are strongly depend on their atomic geometry and chirality.

CNTs are quasi 1D quantum wire structures. When electron flows through theses nanotubes, it behaves like an electron wave (De-Broglie wave). These electron waves can interfere with another electron wave and also with themselves. Each collision of electron wave acts as a scattering point that adds to increase the resistivity of the nanotube. Number of collisions is estimated by the mean free path of electrons. The mean free path is the average length that an electron travels before it is scattered by a scattering center. Nanotubes having long mean free path, much larger than the length and diameter of the nanotube experience a ballistic flow of electrons, where no scattering takes place. These are ideal conductors, also known as ballistic conductor. In a ballistic conductor, the wave-function of an electron is determined by solution of Schrodinger's equation; whereas, classical conductor satisfies the Ohm's law.

In order to observe conductance in nanotubes having metallic as well as semiconducting property, two electrodes need to be fabricated at the ends of a nanotube, where the contacts should be perfect ohmic. For ballistic transport between two electrodes, 1 and 2; if Vis the voltage between the electrodes, the contact resistance of a ballistic conductor is given by equation 4.

$$R_C = \frac{h}{2e^2} \cdot \frac{1}{M} \qquad \textbf{(4)}$$

where M is the number of conduction channel (an integer); $h/2e^2$ is the quantized resistance with a value of 12.9 kΩ. The inverse of equation 4 gives the contact conductance G, is given by

$$G_c = G_0 M \qquad \textbf{(5)}$$

where $G_0 = \frac{2e^2}{h}$ is the quantum conductance.

Thus, in a wire with no scattering the conductance is proportional to M. In a metal of finite length L it is expected to find two conduction channels (M=2). Thus in the ballistic limit, the conductance of metallic nanotubes is $2G_0$.

When conductance occur in a non-ballistic regime, *i.e.* with scattering centers, conductance is given by equation 6 also known as Landauer formalism

$$G_0 = \frac{2e^2}{h} M\mathcal{T} \qquad \textbf{(6)}$$

where T is the transmission probability of a channel to go from electrode 1 to electrode 2 . T = 1, implies a ballistic conduction. Electrical contacts are very crucial for controlling the transmission of electron through the nanotubes. These are major scattering centers if not properly fabricated. Besides the electrode contacts semiconductor nanotubes have other defects and doping which also contributes towards lowering of conductance. The quantized resistance and conductance can be observed in pure semiconductors at very low temperature on samples, which have a small M [4].

The mean free path of a metallic SWCNT is of the order of 1 μm, whereas semiconducting nanotubes have mean free path several orders less. Hence, room temperature resistivity of a metallic nanotube is quite low, about $10^{-4}$-$10^{-3}$Ω-cm, while resistivity at room temperature in a semiconducting nanotube is as high as $10^{1}$Ω-cm.

Transparent electrical contacts made to metallic single-walled SWCNTs have revealed to be ballistic conductors exhibiting two units of quantum conductance *i.e.* $4e^2/h$ [5 - 7]. Semiconducting SWCNTs have been successfully used for fabrication of miniaturized field-effect transistor, which can act as a switch. By applying a voltage to a gate electrode, the nanotube can be switched from a conducting (ON state) to an insulating state (OFF state). Palladium metal contacts across a semiconductor SWCNTs has been shown to substantially reduce the barriers for electron transport. Palladium (Pd) is a noble metal with high work function and good wetting interactions with nanotubes. Using Pd contacts, room temperature conductance could be near the ballistic transport limit of $4e^2/h$ [8] along with high current-carrying capability of 25 μA at low temperatures. Authors here reported that for Pd-contacted, long, semiconducting SWCNTs devices (length L = 3 mm, diameter d < 3 nm) the room-temperature conductance under ON state ($G_{ON}$) was observed to be $0.1 \times 2G_0$ ($R_{ON}$ = 60 kΩ). Short semiconducting SWCNTs device with L = 300 nm, showed $G_{ON} = 0.5 \times 2G_0$ ($R_{ON}$ = 13 kΩ). The long SWCNTs showed metallic behavior between room temperature to 200K. With further lowering in temperature $G_{ON}$ is found to decrease. The short nanotubes do not show such behavior and the average $G_{ON}$ is found to monotonically increase with decrease in temperature upto 50K (Fig. **1**).

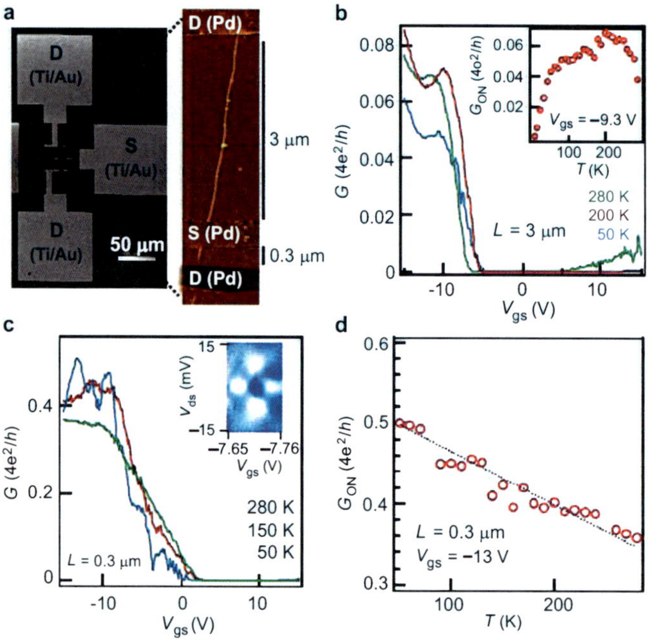


**Fig. (1).** Pd-contacted long (L = 3 mm) and short (L = 300 nm) back-gated SWNT devices formed on the same nanotubes on $SiO_2$/Si. (**a**) A SEM image (left) and AFM image (right) of a representative device. (**b**) G (at low S–D bias $V_{ds}$)*vs.* gate voltage $V_{gs}$ for a 3-mm-long SWNT (d < 3.3 nm) device recorded at various T.Inset, GON *versus* T for the device. (**c**) G *vs.* $V_{gs}$ for a 300-nm-long tube section on the same tube as for (b) at various T. (**d**) $G_{ON}$*vs.* T for the L = 300 nm semiconducting tube down to 50 K.(Source: Ref [8]).

Another interesting fact is that SWCNTs having impurities can dramatically change resistance on application of an external transverse applied field. Thus, electrical properties (resistance) of SWCNTs can be tuned by applied gate voltages or electric fields [9]. Whereas, in metallic SWCNTs electrical resistance remain unaltered with varying gate voltage or transverse electric fields [10 - 12]. However, resistance of metallic SWCNTs with impurity or structural defect is strongly dependent on the applied electric field, especially on its strength and direction [13]. When transverse electric fields are applied, the field induced change of the screened electrostatic potential near the impurity sites changes the quasibound-state energies and thereby the resonant backscattering and the resistance of the CNTs. Here is an example: When two carbon atoms on opposite sides of a (10, 10) CNT are replaced by a nitrogen and a boron atom, the resistance of the CNT is 6.7 kΩ, "ON-state" when $E_{ext} = 0$. When an electric field $E_{ext} = 0.4V/Å$ is applied pointing from the nitrogen-doped side to boron-doped side, the resistance of the CNT increases to 221 kΩ, switching to "OFF-state". With further increase in the electric field strength, the resistance reduces back to 20 kΩ. When the direction of the electric field is reversed, the doped CNT remains in the low-resistive "ON-state" (less than 10 kΩ) at all values of field strength. Thus, electrical switching is possible in both semiconducting as well as metallic SWCNTs (with impurities). In a larger-diameter SWCNTs, the electrical switching is possible with a weaker electric field. As a negative gate voltage is applied, the electron motion through the gates is constrained. At a threshold gate voltage, the current in the channel goes to zero. This is known as Coulomb blockade when electrons tunnel step-by-step. If the gate voltage is further increased (positively) above threshold voltage, a series of periodic coulombic oscillations are observed. This phenomenon become dominant a very low temperatures, as shown in Fig. (**2**). At this low temperature, electrons are delocalized over the length of the tube. Each coulombic oscillation indicate the flow of a single electron. This phenomenon is used to realize Single electron transistors.

Fig. (**3**) indicate the conductance (G) variation for a semiconductor SWCNTs [14]. At room temperature, G increases with decrease in $V_g$ (towards negative value) thereafter, at large negative gate voltage a saturation in the value of conductance is observed, which may be due to the contact resistance developed during tunneling into the nanotubes [15, 16]. As temperature is lowered, *G* is reduced and breaks up into a series of periodic coulombic oscillations as a function of $V_g$.

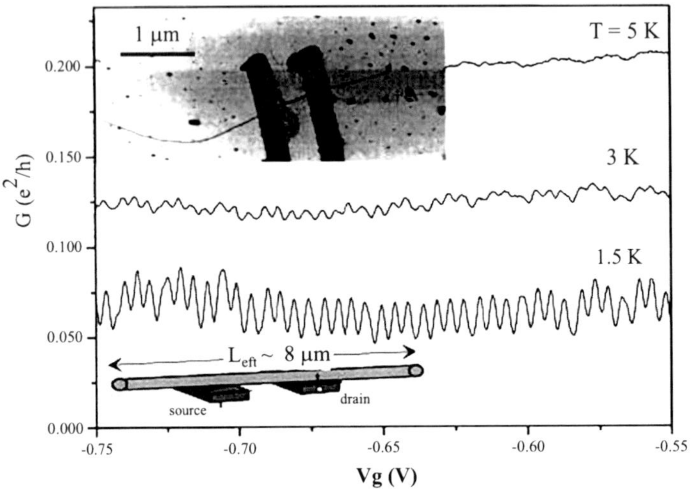


**Fig. (2).** Conductance *versus* gate voltage at different temperatures for the metallic nanotube device shown in the upper inset. The 3 nm diameter and 8 mm long nanotube rope is draped over two contacts that make tunnel contact to a metallic tube in the rope. A voltage applied to the doped substrate is used to adjust the carrier density. The appearance of the CB oscillations only at very low temperatures 1.5 K indicates that the electrons are delocalized over the entire length of the tube, as indicated in the lower inset. (Source: Ref [14]).

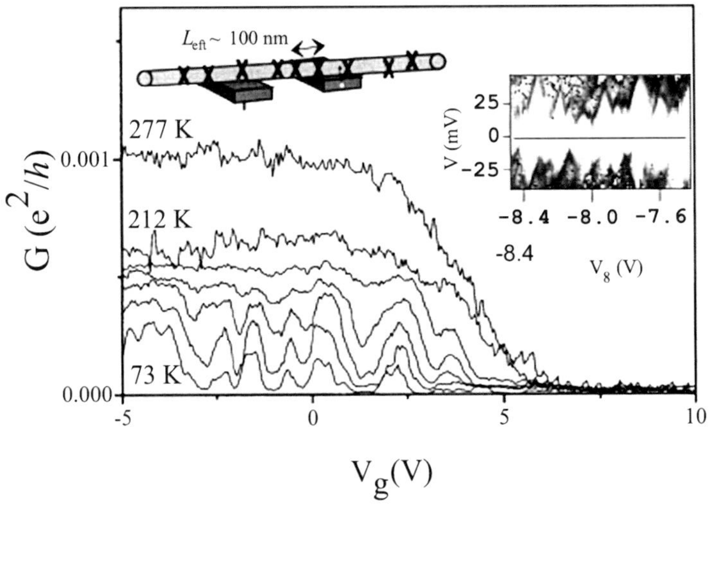


**Fig. (3).** *G* vs$V_g$for a semiconducting nanotube device with contacts separated by 0.5 mm. Irregular Coulomb oscillations are observed below $T$= 150 K. (Source Ref [14]).

Thus, in semiconducting nanotubes, disorders effectively breaks the tube into a series of quantum dots separated by potential barriers, the conductance is thus determined by tunneling through a series of quantum dots. However, the fact as observed by comparing Fig. (**2** and **3**) clearly indicate that for semiconductor nanotubes, G $<<e^2/h$, hence, these are more intensely influenced by disorderness than metallic nanotubes.

## 3.3. OPTICAL PROPERTIES

Optical properties of CNTs can be referred by UV-vis-IR absorption, photoluminescence and Raman scattering.

### 3.3.1. UV-vis and FTIR Spectroscopy

Optical absorption spectroscopy is the measurement of absorption of *em* radiation by a sample of interest. As the material absorbs light at a given wavelength it interacts with the sample and intensity of light deceases exponentially as it propagates through an absorbing medium. This phenomenon is explained by the Beer-Lambert law as shown in equation 7.

$$I(L)=I_0 e^{-\alpha Lc} \quad \textbf{(7)}$$

where $I_0$ is the intensity of the incident light, α is the absorption coefficient, $L$ is the optical path length, and $c$ is the concentration of an absorbing species in the material. By measuring the transmitted intensity relative to the incident intensity, the absorbance (at a given wavelength λ) is given by equation 8.

$$A=-\ln\left(\frac{I}{I_0}\right) \quad \textbf{(8)}$$

The optical response of SWCNTs is dominated by transitions between valence and conduction bands, where momentum conservation only allows transition pairs of singularities that are symmetrically placed with respect to the Fermi energy level. These vertical transitions are labeled as $E_{ii}$ ($i$= 1,2,3,...). Light is absorbed, when resonance occurs with the $E_{ii}$ values for the (n, m) nanotubes in a sample. SWCNTs have structure in the visible and near infrared spectrum and large reflectance in the infrared at wavelengths greater than 4μm [15]. The absorbance of MWCNTs is characterized by a π plasmon with a peak near 248 nm. However, this is not a characteristic peak for MWCNTs, as it is also observed in SWCNTs and graphite as well. The reflectance in MWCNTs is quite uniform in the visible through the infrared wavelength range. It is to be noted that the optical spectrum in the visible and near infrared range for MWCNTs does not depict the chiral angle of the outer shell or radius or length of the CNTs. Since many forms of CNTs have anisotropic structure, the optical properties strongly depend on the polarization orientation with respect to applied field or nanotube position [16]. Optical absorption of SWCNTs is often used to evaluate the mean diameter of nanotubes. A typical optical absorption of SWCNTs is shown in Fig. (**4**).

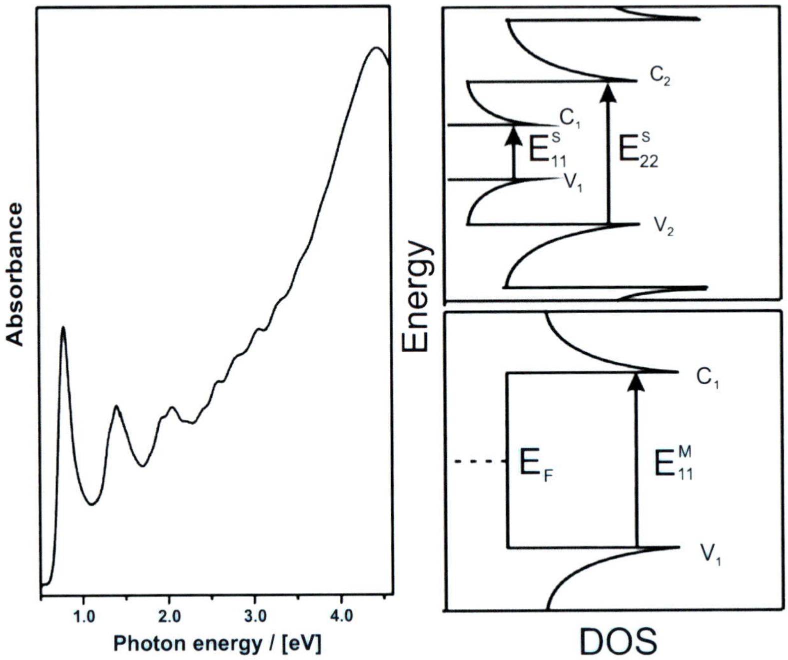


**Fig. (4).** (a) A typical optical absorption spectrum of SWNT sample showing three broad peaks that are associated with the first ($E^S_{11}$) and second ($E^S_{22}$) optical transitions for single SWNT and the first transition($E^M_{11}$) for bundled SWNT. $E^F$ is the Fermi level, while c and v are the conduction and valence bands, respectively.

FTIR can commonly detect functionalizations on CNTs *i.e.* any defects in the nanotube structure. The defects/functionalizations can alter the surface wettability of the nanotubes and may thus also alter its toxicity. Peaks related to O-H bond were observed at1445 cm$^{-1}$,1600 and 3450 cm$^{-1}$indicating oxidation in MWNTs [17] as shown in Fig. (**5**). IR-active peaks for 60 nm diameter MWNT were detected at 1725 cm$^{-1}$ for COOH groups,1584 cm$^{-1}$ for G-band, 1200 cm$^{-1}$ for D-band and several peaksin the range of 3000 cm$^{-1}$ range corresponding to CHx groups [18].

Optical properties of graphite describe it as a birefringent material. Hence, optical properties of CNTs are also be constrained by tube orientation with respect to the direction of propagating beam propagation. Another interesting physical parameter is the dark black color of the CNTs that essentially indicates a very high optical absorbance. Mizuno *et al.* have demonstrated this fact (black color) from visible wavelengths to 200 μm [19].

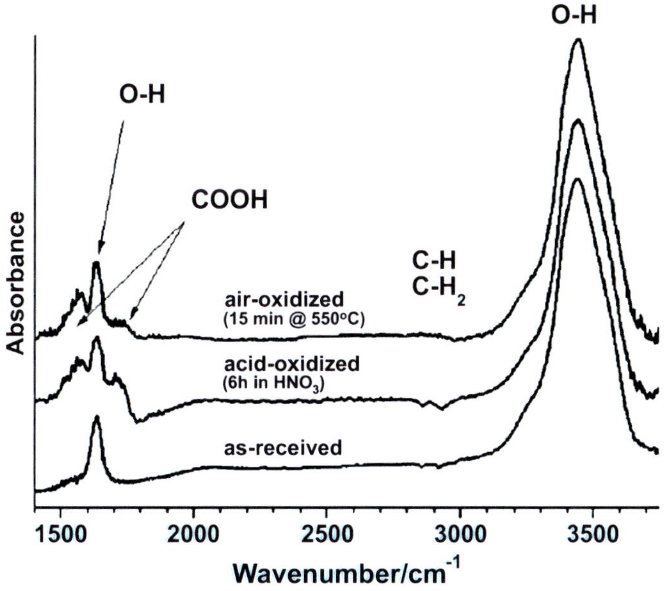


**Fig. (5).** FTIR spectra of three types of MWCNTs with functional groups assigned. (Source: Ref [17]).

### 3.3.2. Raman Spectroscopy

Infrared and Raman spectroscopies are complementary to the extent that one relates to dipole moment, while the other relates to polarizability; however, the infrared spectroscopy generates less information than the Raman spectroscopy features.

Raman scattering refers to the inelastic scattering of a photon from of molecule. Most scattering events are elastic (Rayleigh scattering is an elastic scattering) processes. However, a small fraction of incident photons can be in-elastically scattered, by exchanging energy with the molecule as a result of interacting with optical phonons. There are two types of Raman scattering, known as Stokes scattering and anti-Stokes scattering (Fig. **6**). In both cases, the scattering process is described by three steps consisting of (i) absorption of incident photon ($\hbar\nu$) into an electronic excitation of the molecule, (ii) generation or absorption of a phonon ($\hbar\omega$), and (iii) emission of aphoton ($\hbar\nu \pm \hbar\omega$).

The probability for the Raman scattering process is very low (~one in $10^7$ incident photons), as it involves an excitation to a virtual level that is only allowed for a very short time. However, the probability is strongly increased, by a factor of $10^2$-$10^4$, if the transition energy for either the absorbed or the emitted photon coincides with an electronic transition energy of the molecule. Observation of this process is called resonant Raman spectroscopy. In the case of SWCNTs, the cross section for a resonant Raman process is so large that it is even possible to analyze the vibrational modes of an individual SWCNTs. Furthermore, the Raman signals detected for different excitation energies can be used to identify SWCNTs with different chiralities at resonance.

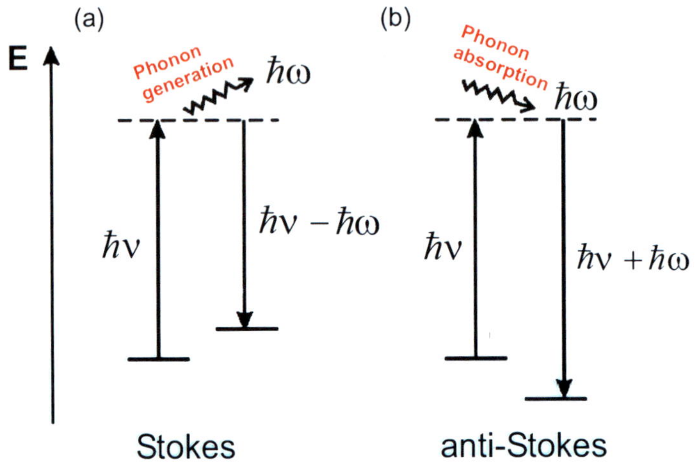


**Fig. (6).** Schematic diagram of energy levels showing (a) Stokes and (b) anti-Stokes Raman scattering process.

Thus, Raman spectroscopy is the most important optical characterization for CNTs that yields information about its purity, defects and tube alignment, and gives signature peak distinction for the presence of SWCNTs and MWCNTs relative to other carbon allotropes [20]. Important to note that the interpretation of Raman spectra for MWCNT is quite complicated, because of its multiple wall structural complexity and hence has not yielded the same level of yield as Raman spectroscopy for SWCNTs [20]. Since Raman spectra analysis for SWCNTs are well established, hence the interpretation of the MWNTs are based on the SWCNT Raman analysis but with some limitations *i.e.* the effects that are missing in SWCNTs due the presence of single graphite layer, thus avoiding the interaction of multiple layers of graphite with each other.

A typical Raman spectrum (Fig. 7) of a SWNT has the following key features:

1. A low-frequency peak (<200 $cm^{-1}$), assigned to the $A_{1g}$symmetry radial breathing mode (RBM). This peak is absent in graphite and is a signature peak for SWCNT. The frequency of this mode depends only on the tube diameter. Important to note that RBM frequency is inversely proportional to the tube diameter, hence the RBM is limited to tube diameter > 2 nm.
2. A complete description of SWCNTs chiral indices can be inferred from

resonant Raman measurements as demonstrated in the Kataura plot for allowed excitations [21].

3. A group of peaks around 1340 $cm^{-1}$, also known as the "D-band", is assigned to the presence of disorderness in graphitic materials. It has been observed that the D-band is present for nanotubes satisfying a certain chirality only [22]. Further, D-band can be split into two bands whose separation depends upon the incident laser energy.
4. A group of peaks in the 1550–1600 $cm^{-1}$ range are called the G-band (graphite-like band). The G-band is further split into two features (G+ and G-) due to the curvature of SWNTs, which induces an energy difference between the axial and transverse in-plane vibrational modes. The G+ band corresponds to atomic displacements along the nanotube axis, and the G- band corresponds to circumferential atomic displacements, dependent on the diameter of the tube. The G- band has a different line-shape for semiconducting and metallic nanotubes. The signature of a single-shell vibration is expressed by the presence of a low-frequency RBM and the splitting of the G-band.
5. A peak 2600 $cm^{-1}$ corresponds to second-order harmonic of the D mode (disorder band). This band is indicative of long range ordering in a nanotube, and it originates from a two-phonon, 2nd order scattering process that can occur when excited phonons are scattered by defects or inconsistencies in the crystal lattice. The ratio of the *G*/*D* modes in Raman spectrais often used to quantify the structural quality of carbon nanotubes.

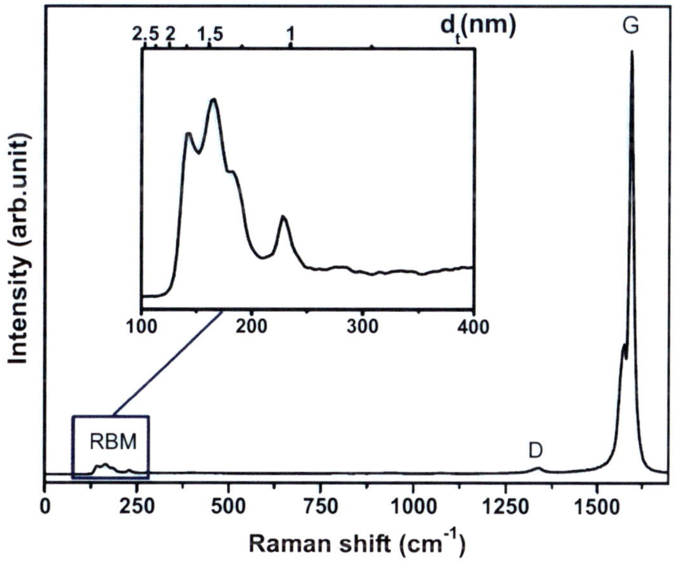


**Fig. (7).** A typical Raman spectrum of a SWCNT thin film excited by a 533 nm laser. The inset shows the enlarged RBM peaks.

Raman spectroscopy on MWCNTs has been focused on investing the presence of the above mentioned peaks/bands together important electronic and structural information. Some of the important features of MWCNT's Raman spectrum are shown in Fig. (**8**).

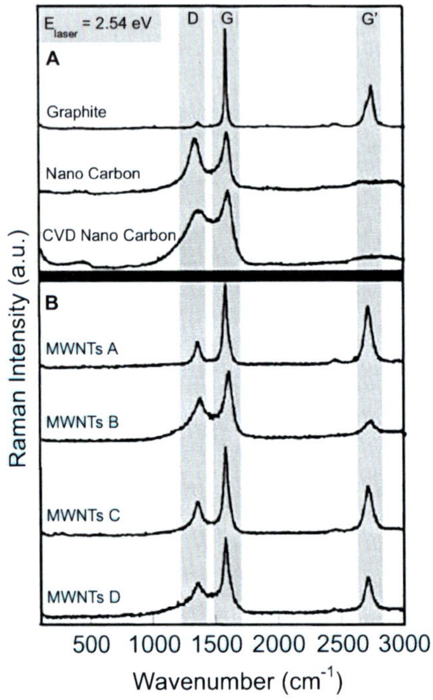


**Fig. (8).** Revealing features of MWCNTs with respect to purity compared to other carbon structures. (Source: Ref [24]).

As a variation from Raman spectra of SWCNTs, many samples of MWCNTs have observed the G-band's frequency and line-shape comparable to those of the $E_{2g}$ phonon of graphite. However, the line-width cannot be considered a consistent standard to distinguish MWCNTs from graphite, since the line-width may be affected by a number of factors such as tube diameter. In MWCNTs the low frequency RBM and the splitting of G band is often not observed. The main reason being larger diameter of the innermost shell in MWCNTs in excess of 2 nm. Absence of G band splitting in MWCNTs can also be used to distinguish between MWCNTs and SWCNTs. However MWCNTs with thin inner shell have shown RBM at low frequency region between 100-600 cm$^{-1}$ [23].

## 3.4. MECHANICAL PROPERTIES

CNTs possess unique mechanical properties as in excellent resilience, ability to withstand cross-sectional and twisting distortions, large stretchability, and ability to withstand compression without fracture. In fact, CNTs are known to be the strongest material. In CNTs, three kinds of forces between carbon atoms produce their characteristic elastic properties. The fundamental atomic forces consist of strong σ-bonding and π-bonding forces between intra layer C=C bonds and weak interlayer interactions. The π-electron contribution to the elastic energy is relatively small compared to that from the σ electrons, but the π-electron contribution is still important for obtaining the lattice distortion more commonly known as the Peierls distortion. The Peierls instability is known to cause a metal insulator transition, by opening an energy gap at the Fermi energy to benefit from the electronic energy gain, which is introduced by the distortion of the lattice into a doubled unit cell.

Material parameters of interest to reveal the unusual mechanical properties of CNTs are Young's Modulus, Resilience, non-linear elastic parameters and strength.

### 3.4.1. Measurement of Young's Modulus

The first study on Young's Modulus was performed by recording the amplitude of thermal vibrations at the free ends of an anchored CNT as a function of temperature [25]. In this experiment CNTs were connected like a cantilever and heated at one end. The thermal vibrations generated due to heat were principally elastic relaxed phonons in equilibrium with the environment; therefore the amplitude of vibration changed with time. The amplitude of the oscillations was recorded by using TEM for several nanotubes. An average value of 1.8 TPa for the Young's modulus was obtained, but with large errors. Nevertheless these values were indicative of the excellent axial stiffness of CNTs. AFM has also been utilized to analyze the mechanical properties of CNTs where an AFM tip has been used to bend an anchored CNT and record the force exerted by the tube as a function of the displacement from its equilibrium position. An average value of Young's Modulus measured by this technique was 1.28±0.5 TPa. This technique showed no dependence on the tube diameters [26]. Another technique based on resonant electrostatic deflection of MWCNTs in presence of an external ac-field was demonstrated for measurement of elastic bending modulus [27]. An ac voltage was applied to the CNTs and the frequency of the signal was adjusted to resonate the vibration of the bending modes of the nanotube. The frequency of resonance was correlated to the Young modulus of the sample. For small diameter CNTs this modulus was found to be ~ 1TPa. However, the modulus was found to decrease by one order of magnitude when the nanotube diameter increased from 8 nm to 40 nm. This decrease in modulus value was linked to the origin of different bending modes for the nanotube.

### 3.4.2. Measurement of Non-Linear Mechanics and Resilience

CNTs have very unusual properties to get large amplitude distortion violating Hooke's law, which reveal its non-linear mechanical properties. Several experimental evidence and theoretical model confirms that nanotubes can sustain significant nonlinear elastic deformations under external forces without irreversible atomic arrangement. CNTs can get kinks or ripples on compression and bending; or flatten into deflated ribbons under torsion, and still can reversibly restore their original shape.These distorted forms of MWNTs as shown in Fig. (**9**) are caused by the van der waals inter-tube interaction (Fig. **9d**), and their stability can be explained by the competition between the van der Waals and elastic restoring energies.

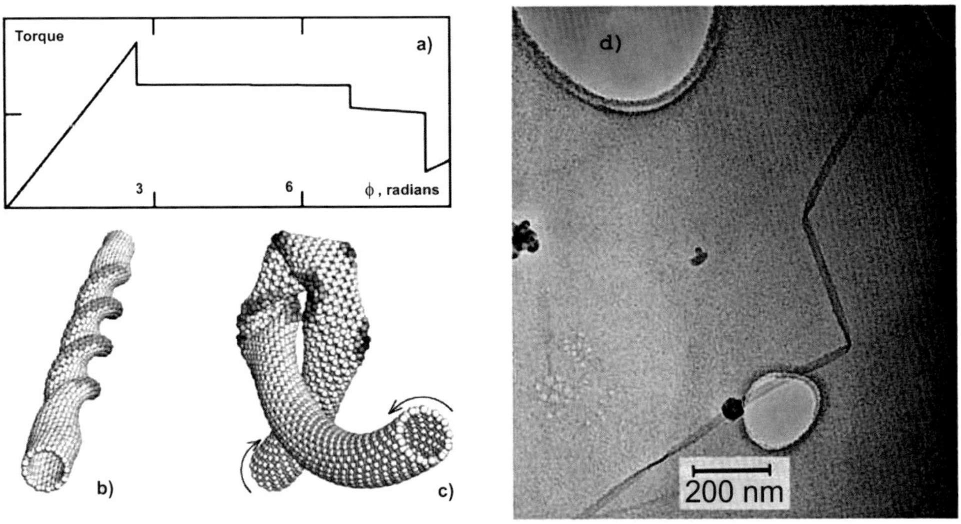


**Fig. (9).** Simulation of torsion and collapse (Source: Ref [28]). (**a**) The strain energy of a 25nm long (13, 0) tube as a function of torsion angle. (**b**) At the first bifurcation the cylinder. flattens into a straight spiral (**c**) and then the entire helix buckles sideways, and coils in a forced tertiary structure. (**d**) Collapsed tube as observed in experiment (Source: Ref [29]).

Strong evidence of large angle fully reversible bending has been demonstrated [30, 31] as shown in Fig. (**10**). Despite the appearance of kinks, this superb flexibility is primarily due to the $sp^2$ network in CNTs and its ability to re-hybridize when deformed out of plane. The degree of $sp^2$–$sp^3$ re-hybridization is proportional to the local curvature during bending [32]. Thus, the strength of the carbon–carbon bond does not guarantee resistance to radial, deformations *i.e.* graphitic sheets of the nanotubes [33], though difficult to stretch are quite easy to bend and to deform.

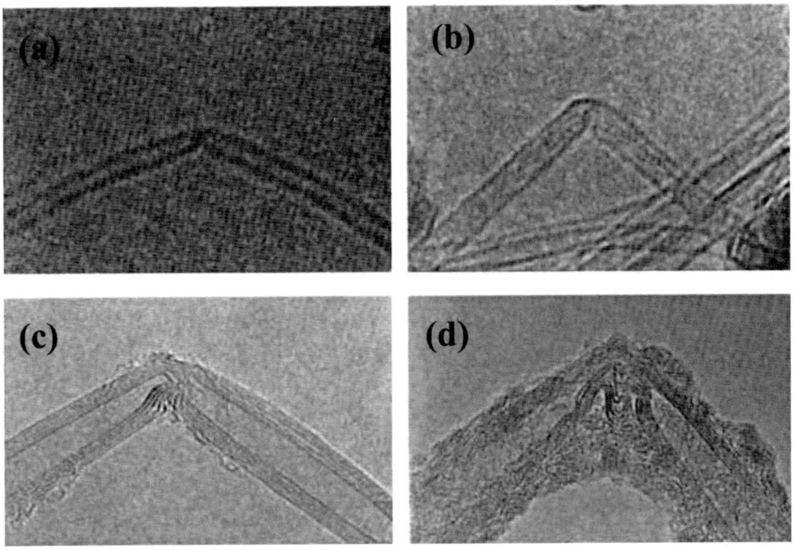


**Fig. (10).** HREM images of bent nanotubes under mechanical duress. (**a**) and (**b**) single kinks in the middle of SWNT with diameters of 0.8 and 1.2 nm, respectively. (**c**)and (**d**) MWNT of about 8nm diameter showing a single and a two-kink complex, respectively (Source: Ref [31]).

### 3.4.3. Strength Measurements

It needs extremely sophisticated devices to measure the strength of CNTs, as a nanotube is too small to be pulled apart with standard tension devices, and too strong for nanophotonic "optical tweezer" devices.

Good estimation of CNT's strength can be made by adhering the nanotube in a matrix material and applying a tensile load. These experiments, are successfully performed with stress-induced fragmentation of carbon nanotube in a polymer matrix The estimated strength on the CNT is 45 GPa, based on a simple iso-strain model of the carbon nanotube-matrix [34]. Yu *et al.* demonstrated direct tensile strength measurement of MWCNT by pulling the two ends of a tube using AFM tips [35]. In this work tensile strengths of MWCNTs were reported in the range of 11 to 63 GPa with very less dependence on the outer shell diameter (Fig. **11**).

In general, calculations and experimental evidences of the elastic properties of CNTs confirm that they are extremely rigid in the axial direction (high tensile) and can be readily distorted in the perpendicular direction (radial deformations), due to their high aspect ratio. Also chemical functionalization of CNTs has been predicted to degrade the mechanical properties by more than 15% [36]. Also, composite materials reinforced by either SWCNT/MWCNT have been fabricated and significant enhancement in mechanical properties has been reported [37]. In fact, the high modulus and the low weight of CNT material makes it ideal reinforcing agent in a variety of composite materials that can be used in the aircraft and sports industries. Further details on composite CNT materials are presented in the later chapter of this book.

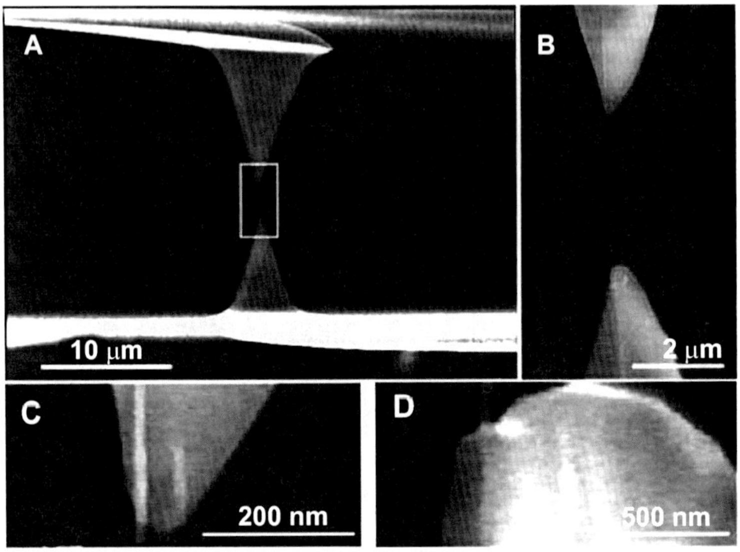


**Fig. (11).** **A**: SEM image of two oppositely aligned AFM tips holding a MWCNT, which is attached at both ends on the AFM silicon tip surface by electron beam deposition of carbonaceous material. The lower AFM tip in the image is on a soft cantilever whose deflection is used to determine the applied force on the MWCNT. **B–D**: Large magnification SEM image of the indicated region in (A) and the weld of the MWCNT on the top AFM tip (Source: Ref [35]).

## 3.5. THERMAL AND VIBRATIONAL PROPERTIES

Thermal properties of CNTs mainly refer to the phonon vibration and specific heat.In general, the specific heat $C$ comprises of phonon $C_{ph}$ and electron $C_e$ contributions, but for CNTs, the dominant contribution to the specific heat is from the phonons. The phonon contribution is obtained by integrating over the phonon density of states with a convolution factor that reflects the energy and occupation of each phonon state (equation 9):

$$C_{ph} = \int_0^{\omega_{max}} k_B \left(\frac{\hbar\omega}{k_B T}\right)^2 \frac{e^{\left(\frac{\hbar\omega}{k_B T}\right)} \rho(\omega)\, d\omega}{\left(e^{\left(\frac{\hbar\omega}{k_B T}\right)} - 1\right)^2} \quad \textbf{(9)}$$

where $\rho(\omega)$ is the phonon density of states and $\omega_{max}$ is the highest phonon energy. For T≠0, the convolution factor is 1at $\omega = 0$, and decreases smoothly to a value of

~0.1 at $\hbar\omega = k_B T/6$, so that the specific heat rises with $T$ as more phonon states are occupied.

At low temperatures, $C_{ph}$ probes only the lowest energy phonons and the above equation modifies to equation 10.

$$C_{ph} \propto T^{d/\alpha}$$

where d is the dimensionality of the system. For a linearly dispersing mode ($\alpha$ =1), the specific heat is linear proportional to temperature for a 1-D system (*e.g.* CNTs)

### 3.5.1. Phonon Density of States

The phonon band structure of isolated nanotubes has been calculated [38 - 40] as shown in Fig. (**12**) wherefrom the phonon density of states can be calculated. It can be found from the Fig. (**12**) that six phonon bands of the graphene sheet has been transformed into 66 sub-bands in ID SWCNT. The appearance of large number of sub-bands with sharp peaks is due to the 1D van Hove singularities in CNTs. It is interesting to note from Fig. (**12**), that at low energies, the geometry of the CNTs causes the phonon structure to substantially differ from that of the parent graphene sheet. Whereas, at high energies the density of states for SWNT and grapheme sheet is similar, thus, indicating that at high temperature specific heat of CNT should be approximately equal to that of graphene.

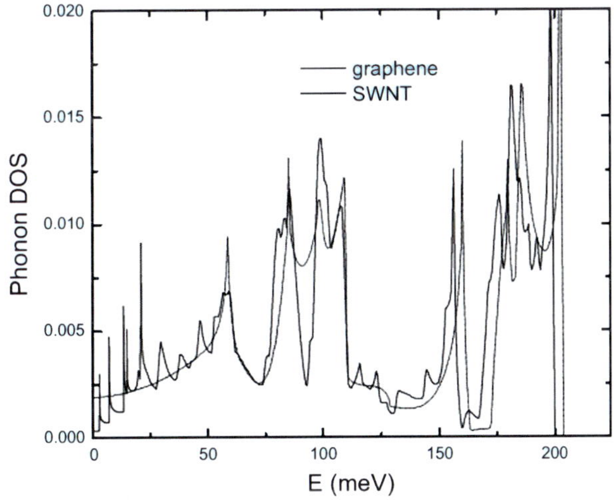


**Fig. (12).** Phonon density of states of an isolated SWCNT (solidline), compared to a 2-D graphene sheet. (Source: Ref [38]).

Fig. (**13**) indicate the variation of overall specific heat with temperature for 1-D CNT structure, 2-D graphene sheet and 3-D graphite structure. At higher temperatures above 100K, the specific heat of all the materials is similar. However, at lower temperatures, the specific heats diverge substantially. The isolated graphene sheet displays the largest specific heat, which shows a linear behavior with temperature because of the 2-D quadratic sheet-rolling mode. The isolated nanotube specific heat is smaller, as a result of the absence of the rolling mode.

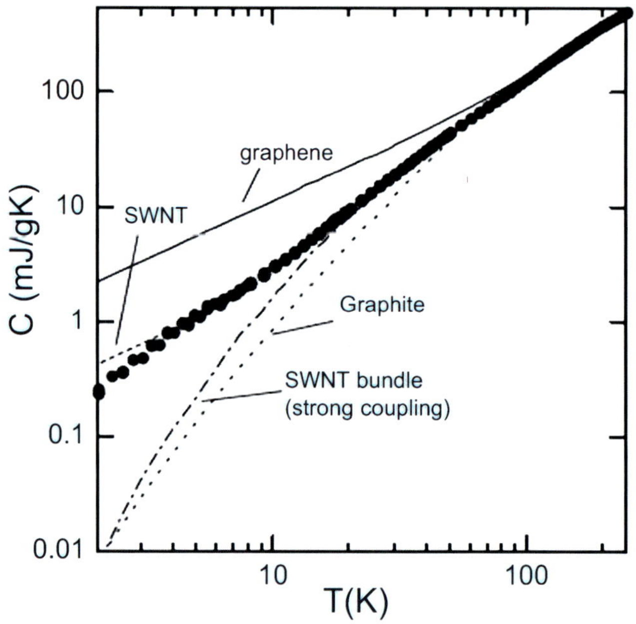


**Fig. (13).** Calculated specific heat (lines) of graphene, isolated SWNTs, graphite, and strongly coupled SWNT bundles. The solid points represent the measured specific heat of a bulk sample of SWNTs. (Source: Ref [41]).

### 3.5.2. Specific Heat at Low Temperature

At low temperatures specific heat (C) behaves linearly with temperature. The contribution of each acoustic phonon mode to specific heat is can be expressed as equation 10:

$$C_{ph} = \frac{\pi k_B^2 T}{\hbar v_{ph} \rho_m} \tag{10}$$

where $\rho_m$ is the linear mass density.

A metallic SWNT is a 1-D structure with constant density of states near the Fermi energy level. At low temperatures, it will have an electronic heat capacity that is linearly proportional to temperature, and can be expressed as equation 11:

$$C_{el} = \frac{4\pi k_B^2 T}{3\hbar \upsilon_F \rho_m} \quad \textbf{(11)}$$

where $\upsilon_F$ is the Fermi velocity.

It can be obtained from equation (10) and equation (11)

$$\frac{C_{ph}}{C_{el}} = \frac{v_{ph}}{v_F} \cong 100 \quad \textbf{(12)}$$

Equation 12 indicates that the phonon contribution is considerably more the electron contribution at low temperatures.

### 3.5.3. Thermal Conductivity

It is the phonon vibration that generates the thermal conductivity behaviour in CNTs. Thermal conductivity tensor (κ) can be expressed by equation 13:

$$\kappa_{zz} = \Sigma\, C\, v_z^2\, \tau \quad \textbf{(13)}$$

where $C$, $\upsilon$, and $\tau$ are the specific heat, group velocity, and relaxation time of a given phonon state, and the sum is overall the phonon states. The major contribution is usually from the acoustic phonons.

Fig. (**14**) represents the thermal conductivity variation of SWCNTs with varying temperature [42]. The thermal conductivity decreases smoothly from 350K to 8K, and displays a change of slope near 30K. It is to be noted, that the thermal conductivity observed in this report is quite different from that of graphite [43]; although both the materials are composed of graphite sheets. The dissimilarity in thermal conductivity of CNTS from graphite has been explained in terms of equation 13 [44]. The reasons given are- 1$^{st}$ first, additional phonon modes in graphite, due to inter planar vibrations, which is missing in CNT; 2$^{nd}$ – Difference in scattering processes ingraphite and tubes;3$^{rd}$ - The "rolling up" of a grapheme sheet to make a nanotube has significant effects on the phonon spectrum.

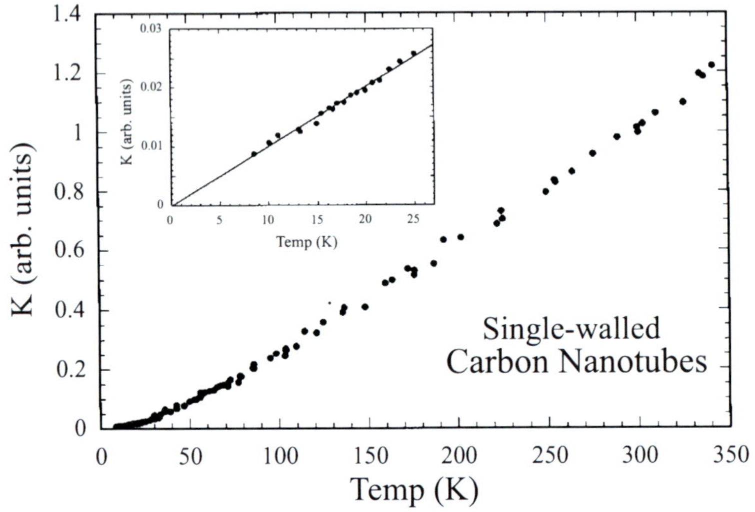


**Fig. (14).** Temperature-dependent thermal conductivity of SWNTs. The inset shows the low-temperature behavior in greater detail. The solid line is a linear fit to the data below 25 K, and extrapolates to zero at 0K. (Source: Ref [42]).

It is already discussed in the previous Section 3.1. on electrical properties that for 1-D structures like CNTs, the electronic conductance is quantized. Similarly, the thermal conductance of CNTs is independent of material parameters, and there exists a quantum of thermal conductance, that behaves linearly with temperature and can be expressed as equation 14:

$$G_{th} = \frac{\pi^2 T\, k_B^2}{3h} \quad (14)$$

Quantization of thermal conductance was first observed by Rego and Kirczenow [45] and later was confirmed by Schwab *et al.* [46].

Overall, theoretical calculations of thermal conductivity of CNTs indicate phonon thermal conductivity of isolated nanotubes as high as 37,000 W/m K at 100K whish is quite comparable to the highest measured thermal conductivity of a pure diamond (41,000 W/mK at 104 K) [47]. Even at room temperature, the thermal conductivity is quite high (6600 W/m K), higher than reported room temperature thermal conductivity of pure diamond (by a factor of 2). Experiments show a room temperature thermal conductivity over 200 W/mK for bulk samples of SWCNTs, and over 3000W/mK for individual MWCNTs [42, 48, 49]. Moreover, CNT nanocomposites can significantly increase the thermal conductivity. Further details on CNT nanocomposites are presented in the next chapter.

## 3.6. CONCLUSION

This book chapter discusses the unique properties of carbon nanotubes using basic ideas of the lattice, electronic and phonon structures. It is clear that the physical properties of different geometries of carbon nanotubes provide a new dimension for research expansion in various fields of science. This book chapter is meant for readers from any field of science, who wants to learn the basic properties of carbon nanotubes.

## CONFLICT OF INTEREST

The authors declare no conflict of interest, financial or otherwise.

## ACKNOWLEDGEMENTS

The author would like to acknowledge the scientists whose work has been referred in this chapter. Author would also like to thank Amity University for generous scientific and moral support. The author especially thanks Dr Jagriti Narang and Dr C S Pundir for their constant support and encouragement for writing this book chapter.

## REFERENCES

[1] Iijima S, Ichihashi T. Single-shell carbon nanotubes of 1-nm diameter. Nature 1993; 363: 603-5. [http://dx.doi.org/10.1038/363603a0]

[2] Iijima S. Helical microtubules of graphitic carbon. Nature 1991; 354: 56-8. [http://dx.doi.org/10.1038/354056a0]

[3] Vander Wal RL, Berger GM, Ticich TM. Carbon nanotube synthesis in a flame using laser ablation for *in situ* catalyst generation. Appl Phys, A Mater Sci Process 2003; 77: 885-9. [http://dx.doi.org/10.1007/s00339-003-2196-3]

[4] Beenakker CW, Williamson JG, Kouwenhoven LP, Foxon CT, Kouwenhoven LP, van der Marel D. van Wees BJ; van Houten H; van der Marel D. Quantized conductance of point contacts in a two-dimensional electron gas. Phys Rev Lett 1988; 60(9): 848-50. [http://dx.doi.org/10.1103/PhysRevLett.60.848] [PMID: 10038668]

[5] White CT, Todorov TN. Carbon nanotubes as long ballistic conductors. Nature 1998; 393: 240-2. [http://dx.doi.org/10.1038/30420]

[6] Liang W, Bockrath M, Bozovic D, Hafner JH, Tinkham M, Park H. Fabry - Perot interference in a nanotube electron waveguide. Nature 2001; 411(6838): 665-9. [http://dx.doi.org/10.1038/35079517] [PMID: 11395762]

[7] Kong J, Yenilmez E, Tombler TW, *et al.* Quantum interference and ballistic transmission in nanotube electron waveguides. Phys Rev Lett 2001; 87(10): 106801. [http://dx.doi.org/10.1103/PhysRevLett.87.106801] [PMID: 11531494]

[8] Javey A, Guo J, Wang Q, Lundstrom M, Dai H. Ballistic carbon nanotube field-effect transistors. Nature 2003; 424(6949): 654-7. [http://dx.doi.org/10.1038/nature01797] [PMID: 12904787]

[9] McEuen PL, Fuhrer MS. Single-walled carbon nanotube electronics. IEEE Trans NanoTechnol 2002; 1: 78-85.

[http://dx.doi.org/10.1109/TNANO.2002.1005429]

[10] McEuen PL, Bockrath M, Cobden DH, Yoon Y-G, Louie SG. Disorder, pseudospins, and backscattering in carbon nanotubes. Phys Rev Lett 1999; 83: 5098-101. [http://dx.doi.org/10.1103/PhysRevLett.83.5098]

[11] Nygård J, Cobden DH, Bockrath M, Mceuen PL, Lindelof PE. Electrical transport measurements on single-walled carbon nanotubes. Nano Lett 1999; 3: 183-7. [http://dx.doi.org/10.1007/s003390051004]

[12] Li Y, Rotkin SV, Ravaioli U. Electronic response and bandstructure modulation of carbon nanotubes in a transverse electrical field. Nano Lett 2003; 3: 183-7. [http://dx.doi.org/10.1021/nl0259030]

[13] Son Y-W, Ihm J, Cohen ML, Louie SG, Choi HJ. Electrical switching in metallic carbon nanotubes. Phys Rev Lett 2005; 95(21): 216602. [http://dx.doi.org/10.1103/PhysRevLett.95.216602] [PMID: 16384166]

[14] McEuen P, Bockrath M, Cobden D, Yoon Y-G, Louie S. Disorder, Pseudospins, and Backscattering in Carbon Nanotubes. Phys Rev Lett 1999; 83: 5098-101. [http://dx.doi.org/10.1103/PhysRevLett.83.5098]

[15] Kamarás, Katalin and Pekker, Áron and Bruckner M and B. Wide-range optical spectra of carbon nanotubes: A comparative study. Phys Status Solidi Basic Res 2008; 245: 2229-32. [http://dx.doi.org/10.1002/pssb.200879647]

[16] Lin MF, Shyu FL, Chen RB. Optical properties of well-aligned multiwalled carbon nanotube bundles. Phys Rev B 2000; 61: 14114-8. [http://dx.doi.org/10.1103/PhysRevB.61.14114]

[17] Osswald S, Havel M, Gogotsi Y. Monitoring oxidation of multiwalled carbon nanotubes by Raman spectroscopy. J Raman Spectrosc 2007; 38: 728-36. [http://dx.doi.org/10.1002/jrs.1686]

[18] Kouklin N, Tzolov M, Straus D, Yin A, Xu JM. Infrared absorption properties of carbon nanotubes synthesized by chemical vapor deposition. Appl Phys Lett 2004; 85: 4463. [http://dx.doi.org/10.1063/1.1812837]

[19] Mizuno K, Ishii J, Kishida H, *et al.* A black body absorber from vertically aligned single-walled carbon nanotubes. Proc Natl Acad Sci USA 2009; 106(15): 6044-7. [http://dx.doi.org/10.1073/pnas.0900155106] [PMID: 19339498]

[20] Jorio A, Pimenta MA, Filho AG, Saito R, Dresselhaus G, Dresselhaus MS, *et al.* Characterizing carbon nanotube samples with resonance Raman scattering. New J Phys 2003; 5: 139-9. [http://dx.doi.org/10.1088/1367-2630/5/1/139]

[21] Rao AM, Richter E, Bandow S, *et al.* Diameter-selective raman scattering from vibrational modes in carbon nanotubes. Science 1997; 275(5297): 187-91. [http://dx.doi.org/10.1126/science.275.5297.187] [PMID: 8985007]

[22] Maultzsch J, Reich S, Thomsen C. Chirality-selective Raman scattering of the D mode in carbon nanotubes. Phys Rev B 2001; 64: 121407. [http://dx.doi.org/10.1103/PhysRevB.64.121407]

[23] Zhao X, Ando Y, Qin L-C, Kataura H, Maniwa Y, Saito R. Radial breathing modes of multiwalled carbon nanotubes. 2002; Vol. 361.

[24] DiLeo RA, Landi BJ, Raffaelle RP. Purity assessment of multiwalled carbon nanotubes by Raman spectroscopy. J Appl Phys 2007; 101: 064307. [http://dx.doi.org/10.1063/1.2712152]

[25] Treacy M, Ebbesen TW, Gibson JM. Exceptionally high Young's modulus observed for individual carbon nanotubes. Nature 1996; 381: 678-80.

[http://dx.doi.org/10.1038/381678a0]

[26] Wong EW, Sheehan PE, Lieber CM. Nanobeam Mechanics. Elasticity, Strength, and Toughness of Nanorods and Nanotubes

[27] Poncharal P, Wang ZL. D. Ugarte WA de H. Electrostatic Deflections and Electromechanical Resonances of Carbon Nanotubes

[28] Yakobson BI, Brabec CJ, Bernholc J. Structural mechanics of carbon nanotubes: From continuum elasticity to atomistic fracture. J Comput Mater Des 1996; 3: 173-82. [http://dx.doi.org/10.1007/BF01185652]

[29] Chopra NG, Zettl A. Measurement of the elastic modulus of a multi-wall boron nitride nanotube. Solid State Commun 1998; 105: 297-300. [http://dx.doi.org/10.1016/S0038-1098(97)10125-9]

[30] Desperes JF, Daguerre E, Lafdi K. Flexibility of graphene layers in carbon nanotubes. Carbon N Y 1995; 33: 87-92. [http://dx.doi.org/10.1016/0008-6223(95)91118-Q]

[31] Iijima S, Brabec C, Maiti A, Bernholc J. Structural flexibility of carbon nanotubes. J Chem Phys 1996; 104: 2089. [http://dx.doi.org/10.1063/1.470966]

[32] Haddon RC. Chemistry of the fullerenes: The manifestation of strain in a class of continuous aromatic molecules. Science 1993; 261: 1545-50.

[33] Hiura H, Ebbesen TW, Fujita J, Tanigaki K, Takada T. Role of sp3 defect structures in graphite and carbon nanotubes. Nature 1994; 367: 148-51. [http://dx.doi.org/10.1038/367148a0]

[34] Wagner HD, Lourie O, Feldman Y, Tenne R. Stress-induced fragmentation of multiwall carbon nanotubes in a polymer matrix. Appl Phys Lett 1998; 72: 188. [http://dx.doi.org/10.1063/1.120680]

[35] Yu MF, Lourie O, Dyer MJ, Moloni K, Kelly TF, Ruoff RS. Strength and breaking mechanism of multiwalled carbon nanotubes under tensile load. Science 2000; 287(5453): 637-40. [http://dx.doi.org/10.1126/science.287.5453.637] [PMID: 10649994]

[36] Garg A, Sinnott SB. Effect of chemical functionalization on the mechanical properties of carbon nanotubes. Chem Phys Lett 1998; 295: 273-8. [http://dx.doi.org/10.1016/S0009-2614(98)00969-5]

[37] Mamedov AA, Kotov NA, Prato M, Guldi DM, Wicksted JP, Hirsch A. Molecular design of strong single-wall carbon nanotube/polyelectrolyte multilayer composites. Nat Mater 2002; 1(3): 190-4. [http://dx.doi.org/10.1038/nmat747] [PMID: 12618809]

[38] Saito R, Takeya T, Kimura T, Dresselhaus G, Dresselhaus MS. Raman intensity of single-wall carbon nanotubes. Phys Rev B 1998; 57: 4145-53. [http://dx.doi.org/10.1103/PhysRevB.57.4145]

[39] Saito R, Dresselhaus G, Dresselahus M. Physical Properties of Carbon Nanotubes 1998.

[40] Sánchez-Portal D, Artacho E, Soler J, Rubio A, Ordejón P. Ab initio structural, elastic, and vibrational properties of carbon nanotubes. Phys Rev B 1999; 59: 12678-88. [http://dx.doi.org/10.1103/PhysRevB.59.12678]

[41] Hone J, Batlogg B, Benes Z, Johnson AT, Fischer JE. Quantized phonon spectrum of single-wall carbon nanotubes. Science 2000; 289: 1730. [http://dx.doi.org/10.1126/science.289.5485.1730]

[42] Hone J, Whitney M, Piskoti C, Zettl A. Thermal conductivity of single-walled carbon nanotubes. Phys Rev B 1999; 59: R2514-6. [http://dx.doi.org/10.1103/PhysRevB.59.R2514]

[43] Kelly BT. Physics of Graphites. London: Applied Science 1981.

[44] Benedict LX, Louie SG, Cohen ML. Heat capacity of carbon nanotubes. Solid State Commun 1996; 100: 177-80.
[http://dx.doi.org/10.1016/0038-1098(96)00386-9]

[45] Rego LG, Kirczenow G. Quantized Thermal Conductance of Dielectric Quantum Wires. Phys Rev Lett 1998; 81: 232.
[http://dx.doi.org/10.1103/PhysRevLett.81.232]

[46] Schwab K, Henriksen EA, Worlock JM, Roukes ML. Measurement of the quantum of thermal conductance. Nature 2000; 404(6781): 974-7.
[http://dx.doi.org/10.1038/35010065] [PMID: 10801121]

[47] Berber S, Kwon Y-K, Tomanek D. Unusually high thermal conductivity of carbon nanotubes. Phys Rev Lett 2000; 84(20): 4613-6.
[http://dx.doi.org/10.1103/PhysRevLett.84.4613] [PMID: 10990753]

[48] Hone J, Llaguno MC, Nemes NM, Johnson AT, Fischer JE. Electrical and thermal transport properties of magnetically aligned single wall carbon nanotube films. Appl Phys Lett 2000; 77: 666-8.
[http://dx.doi.org/10.1063/1.127079]

[49] Hone J, Llaguno MC, Biercuk MJ. Thermal properties of carbon nanotubes and nanotube-based materials. Appl Phys, A Mater Sci Process 2002; 74: 339-43.
[http://dx.doi.org/10.1007/s003390201277]

CHAPTER 4

# Preparation and Application of Carbon Nanotubes Composites

**Ranu Nayak***

*Amity Institute of Nanotechnology, Sec 125, Amity University, Noida, UP, INDIA*

**Abstract:** The outstanding properties of CNTs have instigated incorporation of CNT as filler material in various metals and polymer material. The blend of carbon material with other metals/polymers is better known as carbon nanocomposites. Significant enhancement can be brought about in the already existing properties of metals and polymers by this process. This chapter discusses the methods for synthesis of such unique nanocomposite structures. The various aspects of how and where these structures can be applied with enhanced properties are also discussed with evidences from published literatures.

**Keywords:** CNT, Filler, Metal, Metal oxide, Nanocomposites, Polymer, Synthesis.

## 4.1. INTRODUCTION

There are abundant evidences on the outstanding properties of CNTs as discussed in the previous chapter. The unique mechanical, electrical and thermal properties of CNTs and other forms of carbon make it apt as a filler material for strengthening the properties of other materials. CNTs and CNFs have been widely explored and incorporated with metal nanoparticles, metal oxide nanoparticles, and polymers. Nanomaterials blended with CNTs/CNFs is better known as CNT nanocomposites.

The CNT-nanocomposites have been widely explored for several applications like – enhancing catalytic properties using combinations of CNT with Co [1], Ni nanoparticles [2], *etc.*; CNT-polymer nanocomposites as gas sensor [3]; magnetic applications using CNT with iron oxide or Fe nanoparticles; biomedical applications; water treatment, *etc.* [4 - 6]. This book chapter discusses the different methods of synthesis reported for the CNT-nanocomposites and their uses in different applications.

* **Corresponding author Ranu Nayak:** Amity Institute of Nanotechnology, Sec 125, Amity University, Noida, Uttar Pradesh, India; Tel: +919818936883; E-mails: rnayak@amity.edu; ranunayakbose@gmail.com

**Jagriti Narang & Chandra Shekhar Pundir (Eds.)**

## 4.2. METHODS OF SYNTHESIS

This section discusses the methods of synthesis employed for obtaining CNT–nanocomposites using different types of materials. The three main categories of materials used for nanocomposite formation are polymers, metals, and metal oxides.

### 4.2.1. CNT-Polymer Nanocomposite

#### *4.2.1.1. In-situ Polymerization*

In this method, CNTs are adsorbed on the polymer surface while polymerization of the monomers. Cochet *et al.* were the first group to report this method, to prepare a MWNT-polyaniline composite [7]. Some other reported literatures on CNT-polyaniline composite are by Feng *et al.* and Abdulla *et al.* [8, 9]. In these reports, MWNT was mixed in aniline/HCl (1:1) solution at 0 °C that initiated adsorption of aniline monomer on the CNT surface. Thereafter, ammonium peroxydisulfate solution was added to carry out the polymerization process.

Several other CNT-polymer composites have been prepared using *in-situ* polymerization, such as, MWNT-polystyrene [10], SWNT- polyimide [11], *etc.*

Electrochemical polymerization is another type of *in-situ* polymerization technique. Some groups have used this process to grow porous composite films of MWNT-polypyrrole where pyrrole dissolved in the electrolyte is electrochemically deposited by applying a potential across two electrodes [12, 13]. The work electrode used is a CNT film on a conductive electrode.

#### *4.2.1.2. Solution Based Method / Polymer Graphting*

These processes are similar and a very simple and cost-effective approach where a suitable solvent needs to be used to disperse the polymer and CNTs. Strong agitation is required for proper mixing of the two materials and it also prevents aggregation of CNTs. Finally a controlled evaporation is needed for formation of the composite. CNTs are often functionalized to disperse in organic solvents and the polymer is then solubilized in the same solution.

Several literatures have reported polymer grafting process for the synthesis of CNT-polymer nancomposite. Couple of typical literatures based on this approach are referred below where both pristine and functionalized CNTs were used.

Qian *et al.* dispersed pristine MWNTs in toluene by strong ultrasonication and then mixed with a dilute solution of polystyrene in toluene and wet casted [14].

Hill *et al.* solubilized carboxilated SWNTs and MWNTs with a polystyrene copolymer [15]. There are several other polymers that have been used to form composite with CNT using a similar approach, such as, Polyamide [6, 16]; poly(*m*-phenylenevinylene-*co*-2,5-dioctyloxyp-phenylenevinylene), polyvinyl-alcohol [17 - 19]; poly(9-vinyl carbazole) (PVK) [18]; chlorinated polypropylene [19]; poly(styrene-*co*-butyl acrylate) latex [20, 21]; Polymethyl methacrylate (PMMA) [22].

Chitosan, a polysaccharide biopolymer material has been widely explored and incorporated with CNTs using a similar method. Chitosan, has been widely used in medical applications because of its unique properties like, biocompatible, biodegrdable, non-toxic, low-cost and ease of processing [23 - 26]. In one of the literature [26], chitosan was mixed with acidic deionized water and suspended MWNTs in water was mechanically agitated by long hour sonication, followed by crosslinking with glutaraldehyde for increased homogeneity.

#### *4.2.1.3. Shear Mixing*

It is an alternative process of wet mixing as mentioned in the above section. Here, thermoplastic polymers are melted and CNTs are homogeneously shear mixed in the molten polymer. The nanocomposite can be obtained thereafter by injection moulding process. Some of the literatures using the melt mixing process are discussed below.

Andrews *et al.* used shear mixing to disperse catalytically produced CNTs in different polymers such as, polystyrene, acrylonitrile butadiene styrene, and polypropylene [27]. Haggenmueller *et al.* used a combination of solvent casting and melt mixing methods to form composite of SWNT in PMMA [28]. Sennet *et al.* used melt processing technique to form composite of aligned CNTs with polycarbonate [29]. CNT-epoxy composites prepared by Ajayan *et al.*, Sandler *et al.*, Xu *et al.*, and Puglia *et al.* [30 - 33] also reported a similar method of synthesis.

### 4.2.2. Metal Oxide–CNT Nanocomposite

Metal oxides (many metal oxides are Ceramics) have high stiffness and thermal stability but quite fragile. Incorporating CNTs into a ceramic line alumina, silica matrix have produced composites with both toughness and high temperature stability. However, homogeneous dispersion and bonding of CNTs in metal oxides is much more difficult than preparing CNT-polymer nanocomposite. Several approaches have been explored to prepare CNT-metal oxide (alumina, silica, titania, magnetite) nanocomposites and are discussed in the text below.

#### *4.2.2.1. Impregnation*

Peigney *et al.* pioneered in this technique to synthesize CNT-alumina composites (Fe–$Al_2O_3$ and Fe/Co–$MgAl_2O_4$ matrices) [34 - 36]. An *et al.* also reported assembly of hot-pressed the alumina–CNT composite powder that was synthesized by catalytic decomposition of $C_2H_2$ gas over alumina powders impregnated with Fe catalysts [37]. Jang *et al.* reported magnetic carbon nanocomposite preparation using hard anodic alumina template [38]. Authors used anodic alumina template to grow CNTs inside the template and then iron oxide was added by a precursor impregnation. Use of hard template definitely results in a high quality composite; however removal of the template without damaging the composite material is a challenge.

#### *4.2.2.2. Sintering*

***Spark Plasma Sintering***

Spark plasma is a very effective and fast method of sintering, especially for synthesis of ceramic nanocomposites. It involves initial activation of powders by a high power pulsed voltage followed by a resistance sintering with a mechanical pressure. Major benefits of this technique are reduced sintering time, good grain to grain bonding, clean grain boundaries.

The initial studies on synthesis of SWNT-ceramic composite using this technique was described by Zhan *et al.* [39]. Authors prepared alumina composites containing 10 vol% SWNTs. Unlike some of the other processes used to prepare CNT-metal oxide composites, spark plasma sintering does not damage the CNTs. Metal oxide matrix are often mixed by ball milling to produce homogeneous dispersion without damaging the CNTs.

***LASER Aided Sintering***

Seeger *et al.* used Nd:YAG laser to rapidly heat a Tetraethylorthosilicate (TEOS) and CNT mixture. Partial melting of the matrix fused CNTs onto an amorphous silica matrix [40].

#### *4.2.2.3. Sol-gel Method*

Vincent *et al.* reported the production of CNT-$TiO_2$ composites using sol–gel methods [41]. Sol–gel methods were also used by Sakamoto and Dunn to prepare SWNT/$V_2O_5$ composites for use as electrodes in secondary lithium batteries [42].

#### *4.2.2.4. Hydrothermal Method*

It is modified sol-gel process where the reaction occurs under a constant pressure condition. Use of this method prevents high temperature post-calcination of the prepared powders. A recent literature by Yang *et al.* reports synthesis of CNT-$TiO_2$ nanocomposite by hydrothermal method. Dispersed CNTs (in a mixture of ethanol and water) was mixed with tetrabutyltitanate (TBOT) solution with constant agitation. Thereafter the suspension was transferred into a Teflon lined autoclave hydrothermal bomb and kept at 180 °C for 12 h [43]. Xuan *et al.* synthesized amorphous carbon-encapsulated $Fe_3O_4$ nanocomposites using $FeCl_3$ as iron precursor and glucose using a one-pot hydrothermal conditions [44]. Chen *et al.* also synthesized a Clay-CNT nanocomposite *via* a hydrothermal treatment. Authors dispersed ATP clay in glucose solution under vigorous agitation to form a homogeneous dispersion at room temperature. Then, the mixed solution was put into a hydrothermal bomb and kept at 160 $^0$C for 48 h. The as- synthesized nanocomposite was then filtered, rinsed, and dried [5].

#### *4.2.2.5. Xerogel Drop Casting*

Xerogels of magnetite/starch nanocomposites were prepared by drying a colloidal suspension of the composite material. Further, the xerogels were calcined at 500 $^0$C in $N_2$ atmosphere to form the desired composite powder. The same nanocomposite film was formed by drop coating the magnetite/starch colloidal suspensions onto stainless steel substrates followed by calcination at 500 $^0$C in $N_2$ atmosphere [45].

#### *4.2.2.6. Solvothermal Method*

$SnO_2$-MWNTs nanocomposite was synthesized by a solvothermal method as reported by Ba *et al.* and Wang *et al.* [46, 47]. Pre-treated MWNT was dispersed in $SnCl_4$ solution dissolved in benzyl alcohol. The resultant solution was refluxed and the final product was collected by centrifugation.

### 4.2.3. Metal-CNT Nanocomposite

Composite materials containing CNT in a metal matrix have outstanding property enhancement in the metal matrix by combining metal's light weight with high strength and modulus from CNTs, making them quite attractive for application aerospace industry, *etc.*

#### *4.2.3.1. Colloidal Dispersion*

Chitosan- CNT-gold nanoparticle composite solution was prepared by dispersing carboxylated MWNT in a Chitosan solution (pH 5); followed by ultrasonically mixing gold colloid to obtain the desired nanocomposite [48].

#### *4.2.3.2. Electroless Plating*

As the name suggests, this method does not use electrical power for electroplating. Thus, it is a non-galvanic plating method that involves several concurrent reactions in an aqueous solution.

Chen *et al.* prepared nanotube/nickel/phosphorus composite coatings on stainless steel substrates using an electroless plating method [49]. The coatings had very high wear resistance and low friction coefficients.

#### *4.2.3.3. Impregnation*

Lang *et al.* prepared a sulfur and mesoporous carbon nanocomposite [50]. 10 wt% Sulfur solution in carbon disulfide was prepared. The filtered was loaded to a mesoporous carbon template through liquid phase infiltration. The infiltration of sulfur was carried out by a repetitive solution impregnation/drying procedure and finally the carbon disulfide solvent was evaporated by annealing at 140 $^0$C in nitrogen atmosphere.

#### *4.2.3.4. Chemical Vapor Deposition*

Chemical vapor deposition (CVD) is a chemical process in which reactive gases of precursor materials are used to synthesize high-purity, high performance solid materials in a reactor system. In CVD gases containing the deposition material is introduced into a reaction chamber and subsequently deposited on a substrate in presence of a carrier gas. Generally CNTs are grown by CVD process. Flahaut *et al.* reported growth of small Co nanoparticles (5 nm to 15 nm) encapsulated in graphitized carbon shells by a catalytic CVD method [51]. Wang *et al.* also presented preparation of carbon-coated cobalt nanocapsules by a chemical vapor-condensation process using cobalt carbonyl ($Co_2(CO)_8$) as the precursor and carbon monoxide as the carrier gas. The composites consisted of a Co metal core and an amorphous carbon shell [52].

#### *4.2.3.5. Hydrothermal/Solvothermal Method*

He *et al.* hydrothermally synthesized carbon encapsulated magnetic nanoparticles by heating an aqueous glucose solution containing Fe-Au or Ni nanoparticles. The composite obtained by this method has –OH and –COOH functional groups on the surface, which provides active sites for further modification [53]. Caiulo *et al.* hydrothermally synthesized carbon-decorated Fe-Pt nanoparticles by adding of Fe and Pt precursors in glucose solution followed by calcination at 900 $^0$C [54].

#### *4.2.3.6. Pyrolysis*

Pyrolysis method involves heat treatment of a mixture containing soluble metal salt and an organic compound as a source of carbon, such as glucose, starch, cane sugar, cellulose, *etc.*

Tomita *et al.* synthesized carbon-coated Co nanoparticles ba solid-state reaction of a mixture of Co nanoparticles and diamond at a high temperature of 1700 $^{0}$C. This method yielded carbon-encapsulated Co nanoparticles of diameter 60 nm [55].

Ma *et al.* reported mass scale production of carbon encapsulated iron and nickel nanoparticles by an instant pyrolysis method [56]. Paraskevas *et al.* have synthesized carbon-encapsulated Fe-Ni nanoparticles *via* the thermal decomposition of methane and benzene [57].

## 4.3. APPLICATIONS

Wide variety of applications of CNT nanocomposites is being reported. Some of them are discussed in the text below.

### 4.3.1. Water Treatment

MWNT-Polyamide composite water permeable membranes were synthesized for salt removal from water [6]. Addition of MWNTs to the 10 wt% polyamide membrane could increase salt rejection up to 76%. Incorporation of MWNT created a structural compactness in the composite membranes due to the strong interaction between MWNTs and polyamide matrix. This networked structure steered to a high salt rejection.

Chitosan/MWNT composite was synthesized for heavy metal ions from water [25]. Copper, Zinc, Cadmium and Nickel ions were successfully removed from aqueous solution. Different factors were studied like, nanocomposite mass, pH, solution flow rate, and temperature for enhancement in efficiency. Chen *et al.* demonstrated removal of toxic metal ions using Clay-CNT nanocomposite [5]. This composite could serve as a promising material for low-cost of fabrication, sustainability, and effective adsorbency of toxic ions like Chromium (IV) and Lead (II). The as prepared composite showed excellent adsorption capacities of 177.74 $mg^3$ $g^{-1}$ for Cr (IV) and 263.83 $mg^3$ $g^{-1}$ for Pb (II) ions.

Magnetic separation technology is a rapid and quite effective method implemented for water purification. For instance, Zhang *et al.* reported higher adsorption rates and a superior removal capacity of organic pollutants from waste water using mesoporous carbon capsules encapsulated in $Fe_3O_4$ nanoparticles [58].

Carbon-encapsulated magnetic nanoparticles have also been efficiently used for removal of heavy metal ions from aqueous solutions [59]. MWNTs filled with $Fe_2O_3$ nanoparticles were shown to have strong adsorption capacities for water soluble dyes from water. The composite nanoparticles were found to be recyclable and the adsorption capacities for Methylene Blue and Neutral Red were in the range of 42.3 $mg^3/g$ and 77.5 $mg^3/g$ respectively [60]. In yet another study the MWNT-Fe composite, nanoparticles were utilized as sorbents for the removal of aromatic compounds [61]. Removal of Eu (III) ions from waste water was reported by Lu *et al.* [62], again by using MWNT-iron oxide nanocomposite.

### 4.3.2. Gas Sensor

MWNT-Polyanilne nanocomposite based gas sensor was fabricated for the detection of ammonia gas [9]. High sensitivity was detected with a limit of detection for ammonia gas ranging from 2-10 ppm. High sensitivity was attributed to the increased charge transfer due to the presence of conductive Polyaniline on the surface of MWNTs. Fast response and good reversibility was also observed. Pd nanoparticle modified CNT-Ni nanorod composite was developed by Lin *et al.* for hydrogen gas sensing [63]. At 200 ppm gas concentration, response of 7.3% within a small time of 312 s indicated a high and fast response for $H_2$ gas sensing using the Pd nanoparticle modified CNT-Ni composite. Another hydrogen gas sensor using CNT-polypyrrole nanocomposite was developed by Park *et al.* MWNT-$WO_3$ nanocomposite based $NO_2$ gas sensor was developed by Su *et al.* [64]. Comparative study indicated significantly higher sensitivity for the MWNT/$WO_3$ composite than that on a $WO_3$ film sensor. Only 1.7wt% MWNT in combination of $WO_3$ could reduce the LOD to 100 ppb for $NO_2$ gas.

### 4.3.3. Biomedical Applications

Composite of amino-modified $NaYF_4$ doped with Yb and Er up-conversion luminescent nanoparticles with SWNT (nanohorns) for cell labeling application [65]. NaYF4:Yb,Er/SWNT/Antibody was incubated with bovine serum albumin (BSA) blocked live cells and when imaged using a fluorescence microscope It can be seen from Fig. (**1C**) that conjugated Antibody with the composite show a bright green luminescence. When the nanocomposite without the antibody, was incubated with the BSA-blocked live cells, the cells showed weak green luminescence that may be due to the nonspecific binding of the nanocomposite (Fig. **1B**); Whereas cells alone only showed dark region without any luminescence (Fig. **1A**).

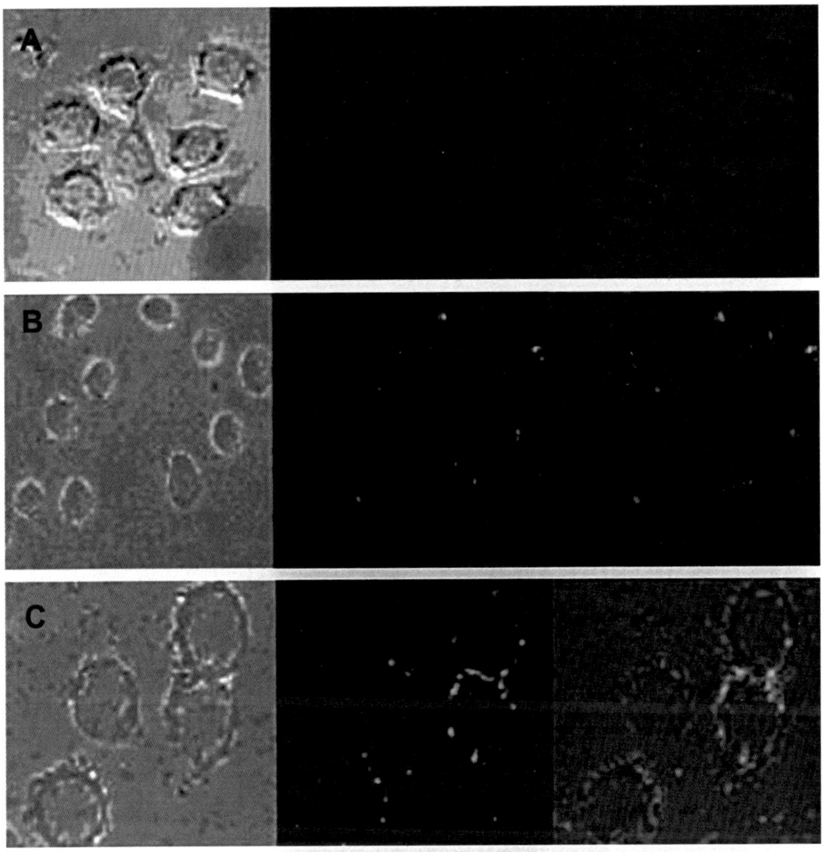


**Fig. (1).** Images of HeLa cells alone (A), HeLa cells after incubation with NaYF4:Yb,Er/SWNH (B), and HeLa cells with rabbit anti-CEA8 Ab-conjugated NaYF4:Yb,Er/SWNH (C) for 1 h obtained under the irradiation of a 980-nm laser. The left images are in bright field, the middle ones are fluorescent images in dark field, and the right ones are overlays of the left and middle images.(Source: Ref [65]).

Chitosan-Carbon-Gold nanoparticle composite was electrochemically deposited for application as a carcino-embryonic antigen immunosensor [48]. Immunoassay techniques are based on the highly specific molecular recognition of antigens by antibodies. It can be a very sensitive technique for disease detection and other fields as well. The key criteria for immunosensing is effective immobilization of biomolecules to retain their activity and stability. Chitosan is preferred bio-matrix for biomolecule immobilization because of its low cost, excellent film-forming ability, high water permeability, good adhesion, non-toxicity and biocompatibility. Gao *et al.* incorporated CNTs and gold nanoparticles inside the porous chitosan film by one step electrodeposition. Addition of CNT and gold increased the chemical stability, mechanical strength, and the conductivity. Moreover the porous structure of chitosan film provided high surface area environment for immobilization of antibodies [48]. Luo *et al.* reported electrochemical fabrication of glucose biosensor, using a similar composite structure of CNT-Chitosan [66]. The detection scheme was based Glucose oxidase catalyzed oxidation of glucose that would consume oxygen and produce hydrogen peroxide. A very similar work was reported by Cui *et al.*, where the authors designed a lactate biosensor using a novel electrochemically deposited chitosan/(polyvinylimidazole-Os)/CNT/lactate oxidase nanocomposite [67]. The as deposited biosensor exhibited significantly improved the electro-activity, stability, signal to noise ratio and sensitivity.

Hydrogels are 3D porous polymeric network that can hold 20 to 40 fold more water than its own dry weight. Thus, they can be of potential biological application to mimic the physical, chemical, electrical, and biological properties of many biological tissues. Hydrogel nanocomposites are also being explored for enhanced biomedical applications. CNT/grapheme based hydrogel nanocomposites have been explored for a variety of biomedical applications such as biosensors, tissue engineering scaffolds, drug delivery systems, and biomedical devices [68, 69]. In an initial study Shin *et al.* embedded –COOH functionalized MWNTs in methacrylated gelatin (GelMA) film to fabricate a hydrogel nanocomposite [70]. Addition of 0.5% CNTs could increase the tensile strength by three folds. Authors demonstrated 3-fold increase in the spontaneous beating frequency of cardiomyocyte cells seeded onto CNT-GelMA than cells seeded on only GelMA hydrogels (Fig. **2**).

Li and Mezzenga hybridized MWNTs with amyloid fibrils to generate fibrous hydrogels [71]. A significant reduction in the gelling concentration of the amyloid fibrils was observed due to the incorporation of CNTs. This approach was anticipated for practice as protein-CNT systems as many proteins can form amyloid fibers, thus, offering potential applications in the field tissue engineering, drug delivery, and biosensing.

Wang *et al.* detected several biomolecules electrochemically, such as, brucine, acetaminophen (ACOP), dopamine (DA), ascorbic acid (AA), norepinephrine (NE), using glassy carbon electrode modified with carbon-coated nickel nanoparticles [72 - 74]. Li *et al.* developed a magnetic nanocomposite with a combination of Pt, $Fe_3O_4$ nanoparticles and MWNTs on glassy carbon electrode. This modified electrode was further used to construct a sensitive and selective amperometric glucose biosensor [75]. Xu *et al.* have demonstrated apoptotic effects of magnetic metals (Fe, Co, and Fe/Co)/graphite nanocomposites on cancer cells [76].

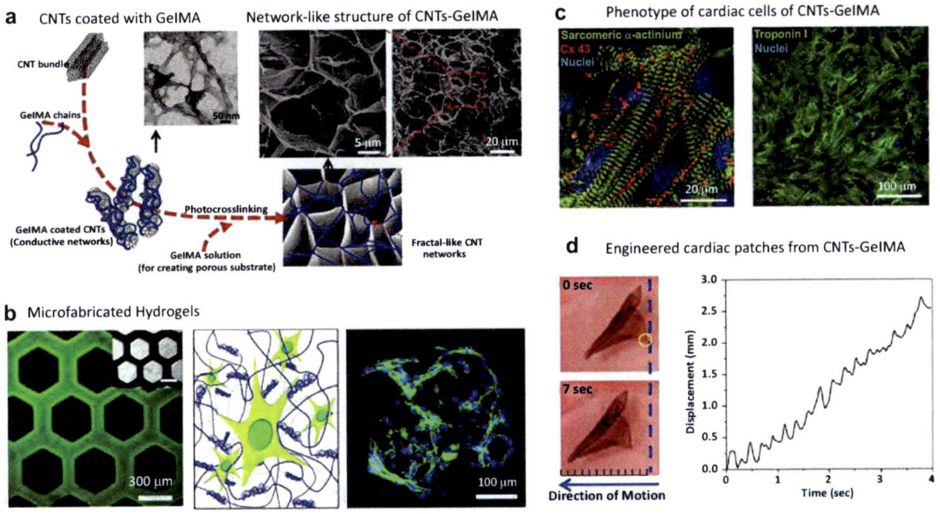


**Fig. (2).** Nanocomposite hydrogels from CNTs and GelMA. a: Schematic showing synthesis of nanocomposite network. First, CNTs are coated with GelMA and then the composite is subjected to UV radiation to obtain photocrosslinked network. b: Due to photocrosslinking ability of the nanocomposite network, microfabrication technologies can be used to control cellular interactions. c: Cardiac cells that were seeded on CNTs-GelMA nanocomposites retained their phenotype as determined by the expression of sarcomeric a-actinin and troponin I. d: The engineered cardiac patch obtained by seeding cardiac cells on CNTs-GelMA surface showed macroscopic mechanical displacement due to continuous contraction and relaxation of the patch. (Source: Ref [70]).

NiO nanoparticles and MWNT composite was synthesized to realize a novel microfluidics nanobiochip. Authors reported a rapid detection and a low-cost approach on the miniaturized microfluidics-integrated biochip using bienzyme functionalization approach on the NiO-MWNT composite [77].

### 4.3.4. Reinforcement of Polymer Foams

Filling any conventional foaming material with CNTs can significantly enhance the mechanical properties of foam, like tensile strength, modulus, resilience, *etc.* Park *et al.* prepared CNT- ethylene vinyl acetate (EVA)-MWNTs nanocomposite foams and observed significant increase in mechanical properties [78]. Zeng *et al.* prepared PMMA- MWNTs nanocomposite foams strengthen the PMMA foam [22]. MWNT-poly(3-caprolactone) [79] and MWNT-polystyrene [80] nanocomposite foams were prepared and shown to have excellent em shielding properties.

### 4.3.5. Energy Storage Devices

Wang *et al.* synthesized $SnO_2$-CNT nanocomposite and used it as the anode material in Na-ion batteries [47]. This nanocomposite exhibited a high reversible Na storage capacity and a good cycling performance. Liang *et al.* developed a sulfur-CNT hierarchical structured nanocomposite for application as a cathode material in rechargeable Li-ion batteries [50]. The first discharge capacity of the as-prepared cells was found to be as high as 1584 mAh $g^{-1}$ at a high current density of 2.5 A/g; which is better than the reports in other literature [81] (1155 mAh $g^{-1}$;current density of 0.04 A/g). Hu *et al.* synthesized Si@$SiO_x$/C core-shell nanocomposites using hydrothermal carbonization of glucose in the presence of Si nanoparticles for use as an anode material in Li-ion batteries [82]. The composite showed significantly improved Li-storage performance in terms of highly reversible Li-storage capacity and excellent cycling performance.

Peng *et al.* demonstrated an efficient super-capacitor using a hierarchical structured $MnO_2$-Carbon nanocomposite [83]. Optimization of MnO2 concentration and conductivity of the nanocomposites resulted in high specific capacitance (170 F $g^{-1}$) and excellent rate capability.

### 4.3.6. Other Applications

CNT composites are used in ice hockey sticks for enhanced durability and controllability. CNT based nanocomposites are suitable for small components such as small CNT reinforced plastic gear less than 1 mm in diameter (Fig. **3**) [84]. The plastic micro-gear has high strength, thermal and shape stability, moldability, and much less friction than conventional gears of the same size.

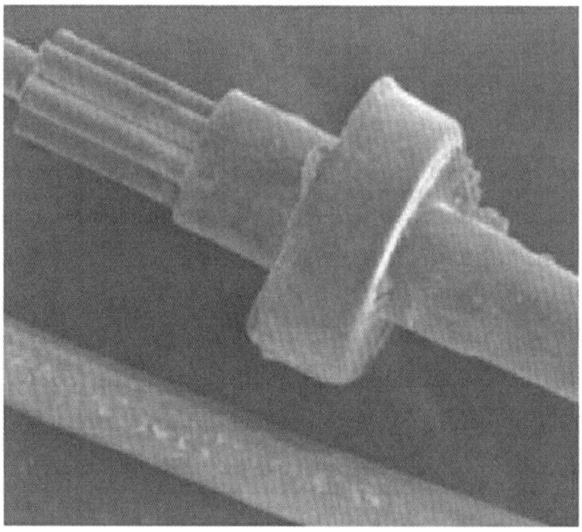

**Fig. (3).** SEM image of a microgear made of CNT nanocomposite. A human hair is on the side to show how small the gear is (Source: Ref [84]).

Endo *et al.* prepared CNT/rubber nanocomposite [85], where, CNT aided as an elastic material and good heat conductor to improve shape resiliency and thermal stability of the rubber component as can be seen from Fig. (**4**). The rubber composite was demonstrated to sustain extreme conditions such as high pressure and high temperature. These composites can be used in applications like oil mining.

**Fig. (4).** CNT/rubber composite unloaded after 10% compression at 200 $^0$C, 70 h. Permanent set rate is shown in the image (Source: Ref [85]).

## 4.4. CONCLUSION

This book chapter discusses the various forms of nanocomposites that can be prepared using carbon nanotubes as filler material. The different methods of synthesis of CNT-metal, CNT-polymer, and CNT-metal oxide nanocomposites are discussed. The enhanced features of CNT nanocomposites leads to significant improvement in its existing properties and generation of some unique properties as well. Some of the key fields application of CNT nanocomposites, such as, Water treatment, Gas sensor, Biosensor, polymer foams, Energy storage devices, *etc.* are discussed in this chapter. This book chapter is meant for readers from any field of science, who wants to explore the intervention of Nanotechnology in the futuristic world.

## CONFLICT OF INTEREST

The authors declare no conflict of interest, financial or otherwise.

## ACKNOWLEDGEMENTS

The author would like to acknowledge the scientists whose work has been referred in this chapter. Author would also like to thank Amity University for generous scientific and moral support. The author especially thanks Dr Jagriti Narang and Dr C S Pundir for their constant support and encouragement for writing this book chapter.

# REFERENCES


[1] Fu T, Li Z. Review of recent development in Co-based catalysts supported on carbon materials for Fischer-Tropsch synthesis. Chem Eng Sci 2015; 135: 3-20. [http://dx.doi.org/10.1016/j.ces.2015.03.007]

[2] Carreño NL, Garcia IT, Raubach CW, Krolow M, Santos CC, Probst LF, *et al.* Nickel–carbon nanocomposites prepared using castor oil as precursor: A novel catalyst for ethanol steam reforming. J Power Sources 2009; 188: 527-31. [http://dx.doi.org/10.1016/j.jpowsour.2008.12.004]

[3] Wei C, Srivastava D, Cho K. Thermal Expansion and Diffusion Coefficients of Carbon Nanotube-Polymer Composites. Nano Lett 2002; 2: 647-50. [http://dx.doi.org/10.1021/nl025554+]

[4] Zhu M, Diao G. Review on the progress in synthesis and application of magnetic carbon nanocomposites. Nanoscale 2011; 3(7): 2748-67. [http://dx.doi.org/10.1039/c1nr10165j] [PMID: 21611651]

[5] Chen LF, Liang HW, Lu Y, Cui CH, Yu SH. Synthesis of an attapulgite clay@carbon nanocomposite adsorbent by a hydrothermal carbonization process and their application in the removal of toxic metal ions from water. Langmuir 2011; 27(14): 8998-9004. [http://dx.doi.org/10.1021/la2017165] [PMID: 21668024]

[6] Shawky HA, Chae SR, Lin S, Wiesner MR. Synthesis and characterization of a carbon nanotube/polymer nanocomposite membrane for water treatment. Desalination 2011; 272: 46-50. [http://dx.doi.org/10.1016/j.desal.2010.12.051]

[7] Cochet M, Maser WK, Benito AM, Callejas MA, Martínez MT, Benoit J-M, *et al.* Synthesis of a new polyaniline/nanotube composite: "in-situ" polymerisation and charge transfer through site-selective interaction. Chem Commun (Camb) 2001; 1450-1. [http://dx.doi.org/10.1039/b104009j]

[8] Feng W, Bai XD, Lian YQ, Liang J, Wang XG, Yoshino K. Well-aligned polyaniline/carbon-nanotube composite films grown by in-situ aniline polymerization. Carbon N Y 2003; 41: 1551-7. [http://dx.doi.org/10.1016/S0008-6223(03)00078-2]

[9] Abdulla S, Mathew TL, Pullithadathil B. Highly sensitive, room temperature gas sensor based on polyaniline-multiwalled carbon nanotubes (PANI/MWCNTs) nanocomposite for trace-level ammonia detection. Sens Actuators B Chem 2015; 221: 1523-34. [http://dx.doi.org/10.1016/j.snb.2015.08.002]

[10] Shaffer MS, Koziol K. Polystyrene grafted multi-walled carbon nanotubes. Chem Commun (Camb) 2002; (18): 2074-5. [http://dx.doi.org/10.1039/b205806p] [PMID: 12357785]

[11] Park C, Ounaies Z, Watson KA, Crooks RE, Smith J, Lowther SE, *et al.* Dispersion of single wall carbon nanotubes by *in situ* polymerization under sonication. Chem Phys Lett 2002; 364: 303-8. [http://dx.doi.org/10.1016/S0009-2614(02)01326-X]

[12] Jurewicz K, Delpeux S, Bertagna V, Béguin F, Frackowiak E. Supercapacitors from nanotubes/polypyrrole composites. Chem Phys Lett 2001; 347: 36-40. [http://dx.doi.org/10.1016/S0009-2614(01)01037-5]

[13] Hughes M, Chen GZ, Shaffer MS, Fray DJ, Windle AH. Electrochemical capacitance of a nanoporous composite of carbon nanotubes and polypyrrole. Chem Mater 2002; 14: 1610-3. [http://dx.doi.org/10.1021/cm010744r]

[14] Qian D, Dickey EC, Andrews R, Rantell T. Load transfer and deformation mechanisms in carbon nanotube-polystyrene composites. Appl Phys Lett 2000; 76: 2868. [http://dx.doi.org/10.1063/1.126500]

[15] Hill DE, Lin Y, Rao AM, Allard LF, Sun YP. Functionalization of carbon nanotubes with polystyrene. Macromolecules 2002; 35: 9466-71.
[http://dx.doi.org/10.1021/ma020855r]

[16] Qu L, Lin Y, Hill DE, Zhou B, Wang W, Sun X, *et al.* Polyimide-functionalized carbon nanotubes: Synthesis and dispersion in nanocomposite films. Macromolecules 2004; 37: 6055-60.
[http://dx.doi.org/10.1021/ma0491006]

[17] Kilbride BE, Coleman JN, Fraysse J, Fournet P, Cadek M, Drury A, *et al.* Experimental observation of scaling laws for alternating current and direct current conductivity in polymer-carbon nanotube composite thin films. J Appl Phys 2002; 92: 4024-30.
[http://dx.doi.org/10.1063/1.1506397]

[18] Cadek M, Coleman JN, Barron V, Hedicke K, Blau WJ. Morphological and mechanical properties of carbon-nanotube-reinforced semicrystalline and amorphous polymer composites. Appl Phys Lett 2002; 81: 5123-5.
[http://dx.doi.org/10.1063/1.1533118]

[19] Coleman JN, Cadek M, Blake R, Nicolosi V, Ryan KP, Belton C, *et al.* High-performance nanotube-reinforced plastics: Understanding the mechanism of strength increase. Adv Funct Mater 2004; 14: 791-8.
[http://dx.doi.org/10.1002/adfm.200305200]

[20] Dalmas F, Chazeau L, Gauthier C, Masenelli-Varlot K, Dendievel R. Multiwalled carbon nanotube/polymer nanocomposites: processing and properties. J Polym Sci, B, Polym Phys 2005; 43: 1186-97.
[http://dx.doi.org/10.1002/polb.20409]

[21] Dufresne A, Paillet M, Putaux JL, Canet R, Carmona F, Delhaes P, *et al.* Processing and characterization of carbon nanotube/poly(styrene-co-butyl acrylate) nanocomposites. J Mater Sci 2002; 37: 3915-23.
[http://dx.doi.org/10.1023/A:1019659624567]

[22] Zeng C, Hossieny N, Zhang C, Wang B. Synthesis and processing of PMMA carbon nanotube nanocomposite foams. Polymer (Guildf) 2010; 51: 655-64.
[http://dx.doi.org/10.1016/j.polymer.2009.12.032]

[23] Liu Y, Tang J, Chen X, Xin JH. Decoration of carbon nanotubes with chitosan. Carbon N Y 2005; 43: 3178-80.
[http://dx.doi.org/10.1016/j.carbon.2005.06.020]

[24] Carson L, Hibbert K, Akindoju F, *et al.* Synthesis, characterization and stability of chitosan and poly(methyl methacrylate) grafted carbon nanotubes. Spectrochim Acta A Mol Biomol Spectrosc 2012; 96: 380-6.
[http://dx.doi.org/10.1016/j.saa.2012.05.042] [PMID: 22722071]

[25] Salam MA, Makki MS, Abdelaal MY. Preparation and characterization of multi-walled carbon nanotubes/chitosan nanocomposite and its application for the removal of heavy metals from aqueous solution. J Alloys Compd 2011; 509: 2582-7.
[http://dx.doi.org/10.1016/j.jallcom.2010.11.094]

[26] Rinaudo M. Chitin and chitosan: Properties and applications. Prog Polym Sci 2006; 31: 603-32.
[http://dx.doi.org/10.1016/j.progpolymsci.2006.06.001]

[27] Andrews R, Jacques D, Qian D, Rantell T. Multiwall carbon nanotubes: synthesis and application. Acc Chem Res 2002; 35(12): 1008-17.
[http://dx.doi.org/10.1021/ar010151m] [PMID: 12484788]

[28] Haggenmueller R, Gommans HH, Rinzler G, Fischer JE, Winey KI. Aligned single-wall carbon nanotubes in composites by melt processing methods. Chem Phys Lett 2000; 330: 219-25.
[http://dx.doi.org/10.1016/S0009-2614(00)01013-7]

[29] Sennett M, Welsh E, Wright JB, Li WZ, Wen JG, Ren ZF. Dispersion and alignment of carbon nanotubes in polycarbonate. Appl Phys, A Mater Sci Process 2003; 76: 111-3. [http://dx.doi.org/10.1007/s00339-002-1449-x]

[30] Ajayan PM, Stephan O, Colliex C, Trauth D. Aligned carbon nanotube arrays formed by cutting a polymer resin--nanotube composite. Science 1994; 265(5176): 1212-4. [http://dx.doi.org/10.1126/science.265.5176.1212] [PMID: 17787587]

[31] Sandler J, Shaffer MS, Prasse T, Bauhofer W, Schulte K. Windle a. Development of a dispersion process for carbon nanotubes in an epoxy matrix and the resulting electrical properties. Polymer (Guildf) 1999; 40: 5967-71. [http://dx.doi.org/10.1016/S0032-3861(99)00166-4]

[32] Xu X, Thwe MM, Shearwood C, Liao K. Mechanical properties and interfacial characteristics of carbon-nanotube-reinforced epoxy thin films. Appl Phys Lett 2002; 81: 2833-5. [http://dx.doi.org/10.1063/1.1511532]

[33] Puglia D, Valentini L, Kenny JM. Analysis of the cure reaction of carbon nanotubes/epoxy resin composites through thermal analysis and Raman spectroscopy. J Appl Polym Sci 2003; 88: 452-8. [http://dx.doi.org/10.1002/app.11745]

[34] Peigney A, Laurent C, Dobigeon F, Rousset A. Carbon nanotubes grown *in situ* by a novel catalytic method. J Mater Res 1997; 12: 613-5. [http://dx.doi.org/10.1557/JMR.1997.0092]

[35] Flahaut E, Peigney A, Laurent C, Marliére C, Chastel F, Rousset A. Carbon nanotube-metal-oxide nanocomposites: Microstructure, electrical conductivity and mechanical properties. Acta Mater 2000; 48: 3803-12. [http://dx.doi.org/10.1016/S1359-6454(00)00147-6]

[36] Peigney A, Flahaut E, Laurent C, Chastel F, Rousset A. Aligned carbon nanotubes in ceramic-matrix nanocomposites prepared by high-temperature extrusion. Chem Phys Lett 2002; 352: 20-5. [http://dx.doi.org/10.1016/S0009-2614(01)01441-5]

[37] An JW, You DH, Lim DS. Tribological properties of hot-pressed alumina-CNT composites. Wear 2003; 255: 677-81. [http://dx.doi.org/10.1016/S0043-1648(03)00216-3]

[38] Jang J, Lee KJ, Kim Y. Fabrication of polyimide nanotubes and carbon nanotubes containing magnetic iron oxide in confinement. Chem Commun (Camb) 2005; (30): 3847-9. [http://dx.doi.org/10.1039/b503831f] [PMID: 16041437]

[39] Zhan G-D, Kuntz JD, Wan J, Mukherjee AK. Single-wall carbon nanotubes as attractive toughening agents in alumina-based nanocomposites. Nat Mater 2003; 2(1): 38-42. [http://dx.doi.org/10.1038/nmat793] [PMID: 12652671]

[40] Seeger T, De la Fuente G, Maser WK, Benito AM, Callejas MA, Martínez MT. Evolution of multiwalled carbon-nanotube/SiO2 composites *via* laser treatment. Nanotechnology 2003; 14: 184-7. [http://dx.doi.org/10.1088/0957-4484/14/2/316]

[41] Vincent P, Brioude A, Journet C, Rabaste S, Purcell ST, Le Brusq J, *et al.* Inclusion of carbon nanotubes in a $TiO_2$ sol-gel matrix. J Non-Cryst Solids 2002; 311: 130-7. [http://dx.doi.org/10.1016/S0022-3093(02)01371-6]

[42] Sakamoto JS, Dunn B. Vanadium Oxide-Carbon Nanotube Composite Electrodes for Use in Secondary Lithium Batteries. J Electrochem Soc 2002; 149: A26. [http://dx.doi.org/10.1149/1.1425791]

[43] Yang MQ, Zhang N, Xu YJ. Synthesis of fullerene-, carbon nanotube-, and graphene-$TiO_2$ nanocomposite photocatalysts for selective oxidation: a comparative study. ACS Appl Mater Interfaces 2013; 5(3): 1156-64. [http://dx.doi.org/10.1021/am3029798] [PMID: 23324025]

[44] Xuan S, Hao L, Jiang W, Gong X, Hu Y, Chen Z. A facile method to fabricate carbon-encapsulated Fe(3)O(4) core/shell composites. Nanotechnology 2007; 18(3): 035602. [http://dx.doi.org/10.1088/0957-4484/18/3/035602] [PMID: 19636125]

[45] Pang SC, Khoh WH, Chin SF. Synthesis and Characterization of Magnetite/Carbon Nanocomposite Thin Films for Electrochemical Applications. J Mater Sci Technol 2011; 27: 873-8. [http://dx.doi.org/10.1016/S1005-0302(11)60158-8]

[46] Ba J, Polleux J, Antonietti M, Niederberger M. Non-aqueous synthesis of tin oxide nanocrystals and their assembly into ordered porous mesostructures. Adv Mater 2005; 17: 2509-12. [http://dx.doi.org/10.1002/adma.200501018]

[47] Wang Y, Su D, Wang C, Wang G. SnO2@MWCNT nanocomposite as a high capacity anode material for sodium-ion batteries. Electrochem Commun 2013; 29: 8-11. [http://dx.doi.org/10.1016/j.elecom.2013.01.001]

[48] Gao X, Zhang Y, Wu Q, Chen H, Chen Z, Lin X. One step electrochemically deposited nanocomposite film of chitosan-carbon nanotubes-gold nanoparticles for carcinoembryonic antigen immunosensor application. Talanta 2011; 85(4): 1980-5. [http://dx.doi.org/10.1016/j.talanta.2011.07.012] [PMID: 21872047]

[49] Chen WX, Tu JP, Wang LY, Gan HY, Xu ZD, Zhang XB. T ribological application of carbon nanotubes in a metal-based composite coating and composites. Carbon N Y 2003; 41: 215-22. [http://dx.doi.org/10.1016/S0008-6223(02)00265-8]

[50] Liang C, Dudney NJ, Howe JY. Hierarchically structured sulfur/carbon nanocomposite material for high-energy lithium battery. Chem Mater 2009; 21: 4724-30. [http://dx.doi.org/10.1021/cm902050j]

[51] Flahaut E, Agnoli F, Sloan J, O'Connor C, Green ML. CCVD Synthesis and characterization of cobalt-encapsulated nanoparticles. Chem Mater 2002; 14: 2553-8. [http://dx.doi.org/10.1021/cm011287h]

[52] Wang ZH, Choi CJ, Kim BK, Kim JC, Zhang ZD. Characterization and magnetic properties of carbon-coated cobalt nanocapsules synthesized by the chemical vapor-condensation process. Carbon N Y 2003; 41: 1751-8. [http://dx.doi.org/10.1016/S0008-6223(03)00127-1]

[53] He NY, Guo YF, Deng Y, Wang ZF, Li S, Liu HN. Carbon encapsulated magnetic nanoparticles produced by hydrothermal reaction. Chin Chem Lett 2007; 18: 487-90. [http://dx.doi.org/10.1016/j.cclet.2007.01.042]

[54] Caiulo N, Yu CH, Yu KM, Lo CC, Oduro W, Thiebaut B, *et al.* Carbon-decorated FePt nanoparticles. Adv Funct Mater 2007; 17: 1392-6. [http://dx.doi.org/10.1002/adfm.200601200]

[55] Tomita S, Hikita M, Fujii M, *et al.* Formation of Co filled carbon nanocapsules by metal-template graphitization of diamond nanoparticles. J Appl Phys 2000; 88: 5452-6. [http://dx.doi.org/10.1063/1.1317242]

[56] Ma C, Luo B, Song HH, Zhi LJ. Preparation of carbon-encapsulated metal magnetic nanoparticles by an instant pyrolysis method. Xinxing Tan Cailiao. N Carbon Mater 2010; 25: 199-204. [http://dx.doi.org/10.1016/S1872-5805(09)60028-7]

[57] Paraskevas I, Caps V, Tsang SC. Syntheses of carbon encapsulated magnetic FeNi nanoparticle *via* decompositions of methane and benzene. Carbon N Y 2006; 44: 820-3. [http://dx.doi.org/10.1016/j.carbon.2005.09.036]

[58] Zhang Y, Xu S, Luo Y, Pan S, Ding H, Li G. Synthesis of mesoporous carbon capsules encapsulated with magnetite nanoparticles and their application in wastewater treatment. J Mater Chem 2011; 21: 3664. [http://dx.doi.org/10.1039/c0jm03727c]

[59] Bystrzejewski M, Pyrzyńska K, Huczko A, Lange H. Carbon-encapsulated magnetic nanoparticles as separable and mobile sorbents of heavy metal ions from aqueous solutions. Carbon N Y 2009; 47: 1201-4.
[http://dx.doi.org/10.1016/j.carbon.2009.01.007]

[60] Qu S, Huang F, Yu S, Chen G, Kong J. Magnetic removal of dyes from aqueous solution using multi-walled carbon nanotubes filled with Fe2O3 particles. J Hazard Mater 2008; 160(2-3): 643-7.
[http://dx.doi.org/10.1016/j.jhazmat.2008.03.037] [PMID: 18430510]

[61] Jin J, Li R, Wang H, Chen H, Liang K, Ma J. Magnetic Fe nanoparticle functionalized water-soluble multi-walled carbon nanotubules towards the preparation of sorbent for aromatic compounds removal. Chem Commun (Camb) 2007; (4): 386-8.
[http://dx.doi.org/10.1039/B610842C] [PMID: 17220979]

[62] Lu S, Chen L, Dong Y, Chen Y. Adsorption of Eu(III) on iron oxide/multiwalled carbon nanotube magnetic composites. J Radioanal Nucl Chem 2011; 288: 587-93.
[http://dx.doi.org/10.1007/s10967-010-0973-y]

[63] Lin TC, Huang BR. Palladium nanoparticles modified carbon nanotube/nickel composite rods (Pd/CNT/Ni) for hydrogen sensing. Sens Actuators B Chem 2012; 162: 108-13.
[http://dx.doi.org/10.1016/j.snb.2011.12.044]

[64] Su PG, Pan TT. Fabrication of a room-temperature NO2 gas sensor based on WO3 films and WO3/MWCNT nanocomposite films by combining polyol process with metal organic decomposition method. Mater Chem Phys 2011; 125: 351-7.
[http://dx.doi.org/10.1016/j.matchemphys.2010.11.001]

[65] Li J, He Y, He Z, Zeng P, Xu S. Synthesis of NaYF(4):Yb,Er/single-walled carbon nanohorns nanocomposite and its application as cells label. Anal Biochem 2012; 428(1): 4-6.
[http://dx.doi.org/10.1016/j.ab.2012.05.025] [PMID: 22677625]

[66] Luo X-L, Xu J-J, Wang J-L, Chen H-Y. Electrochemically deposited nanocomposite of chitosan and carbon nanotubes for biosensor application. Chem Commun (Camb) 2005; (16): 2169-71.
[http://dx.doi.org/10.1039/b419197h] [PMID: 15846436]

[67] Cui X, Li CM, Zang J, Yu S. Highly sensitive lactate biosensor by engineering chitosan/PVI-Os/CNT/LOD network nanocomposite. Biosens Bioelectron 2007; 22(12): 3288-92.
[http://dx.doi.org/10.1016/j.bios.2007.03.004] [PMID: 17408948]

[68] Kuilla T, Bhadra S, Yao DH, Kim NH, Bose S, Lee JH. Recent advances in graphene based polymer composites. Prog Polym Sci 2010; 35: 1350-75.
[http://dx.doi.org/10.1016/j.progpolymsci.2010.07.005]

[69] Goenka S, Sant V, Sant S. Graphene-based nanomaterials for drug delivery and tissue engineering. J Control Release 2014; 173: 75-88.
[http://dx.doi.org/10.1016/j.jconrel.2013.10.017] [PMID: 24161530]

[70] Shin SR, Bae H, Cha JM, *et al.* Carbon nanotube reinforced hybrid microgels as scaffold materials for cell encapsulation. ACS Nano 2012; 6(1): 362-72.
[http://dx.doi.org/10.1021/nn203711s] [PMID: 22117858]

[71] Li C, Mezzenga R. Functionalization of multiwalled carbon nanotubes and their pH-responsive hydrogels with amyloid fibrils. Langmuir 2012; 28(27): 10142-6.
[http://dx.doi.org/10.1021/la301541d] [PMID: 22712696]

[72] Wang SF, Xie F, Hu RF. Electrochemical study of brucine on an electrode modified with magnetic carbon-coated nickel nanoparticles. Anal Bioanal Chem 2007; 387(3): 933-9.
[http://dx.doi.org/10.1007/s00216-006-0984-2] [PMID: 17180337]

[73] Wang SF, Xie F, Hu RF. Carbon-coated nickel magnetic nanoparticles modified electrodes as a sensor for determination of acetaminophen. Sens Actuators B Chem 2007; 123: 495-500.
[http://dx.doi.org/10.1016/j.snb.2006.09.031]

[74] Bian C, Zeng Q, Xiong H, Zhang X, Wang S. Electrochemistry of norepinephrine on carbon-coated nickel magnetic nanoparticles modified electrode and analytical applications. Bioelectrochemistry 2010; 79(1): 1-5.
[http://dx.doi.org/10.1016/j.bioelechem.2009.09.011] [PMID: 19846352]

[75] Li J, Yuan R, Chai Y, Che X. Fabrication of a novel glucose biosensor based on Pt nanoparticles-decorated iron oxide-multiwall carbon nanotubes magnetic composite. J Mol Catal B Enzym

[76] Xu Y, Mahmood M, Fejleh A, *et al.* Carbon-covered magnetic nanomaterials and their application for the thermolysis of cancer cells. Int J Nanomedicine 2010; 5: 167-76.
[http://dx.doi.org/10.2147/IJN.S8306] [PMID: 20463932]

[77] Ali MA, Srivastava S, Solanki PR, *et al.* Highly efficient bienzyme functionalized nanocomposite-based microfluidics biosensor platform for biomedical application. Sci Rep 2013; 3: 2661.
[http://dx.doi.org/10.1038/srep02661] [PMID: 24071971]

[78] Park K-W, Kim G-H. Ethylene vinyl acetate copolymer (EVA)/multiwalled carbon nanotube (MWCNT) nanocomposite foams. J Appl Polym Sci 2009; 112: 1845-9.
[http://dx.doi.org/10.1002/app.29736]

[79] Tomasko DL, Han X, Liu D, Gao W. Supercritical fluid applications in polymer nanocomposites. Curr Opin Solid State Mater Sci 2003; 7: 407-12.
[http://dx.doi.org/10.1016/j.cossms.2003.10.005]

[80] Yang Y, Gupta MC, Dudley KL, Lawrence RW. Novel carbon nanotube-polystyrene foam composites for electromagnetic interference shielding. Nano Lett 2005; 5(11): 2131-4.
[http://dx.doi.org/10.1021/nl051375r] [PMID: 16277439]

[81] Young-Jin C, Sang-Sik J, Ki-Won K, Hyo-Jun A, Jou-Hyeon A, Choi YJ, *et al.* Effects on the carbon matrix as conductor in sulfur electrode for lithium/sulfur battery. Eco-Materials Process Des Vii 2006; 510-511: 1082-5.

[82] Hu YS, Demir-Cakan R, Titirici MM, *et al.* Superior storage performance of a Si@SiOx/C nanocomposite as anode material for lithium-ion batteries. Angew Chem Int Ed Engl 2008; 47(9): 1645-9.
[http://dx.doi.org/10.1002/anie.200704287] [PMID: 18213561]

[83] Peng Y, Chen Z, Wen J, Xiao Q, Weng D, He S, *et al.* Hierarchical manganese oxide/carbon nanocomposites for supercapacitor electrodes. Nano Res 2011; 4: 216-25.
[http://dx.doi.org/10.1007/s12274-010-0072-y]

[84] Endo M, Hayashi T, Kim Y-A. Large-scale production of carbon nanotubes and their applications Pure Appl Chem 2006; 78: 1703-13.
[http://dx.doi.org/10.1351/pac200678091703]

[85] Endo M, Noguchi T, Ito M, Takeuchi K, Hayashi T, Kim YA, *et al.* Extreme-performance rubber nanocomposites for probing and excavating deep oil resources using multi-walled carbon nanotubes. Adv Funct Mater 2008; 18: 3403-9.
[http://dx.doi.org/10.1002/adfm.200801136]

CHAPTER 5

# Carbon Nanotubes in Electrochemical Sensing

**Shikha Wadhwa**[1,*], **Ashish Mathur**[1] and **Susanta Sinha Roy**[2]

[1] *Amity Institute of Nanotechnology, Sec 125, Amity University, Noida, UP, India*

[2] *Department of Physics, Shiv Nadar University, Greater Noida, UP, India*

**Abstract:** This chapter outlines the unique electrochemical properties of Carbon Nanotubes (CNTs) as a novel carbonaceous material. Methods of electrode preparation using CNTs to modify some commonly used electrodes and functionalization of CNTs have been considered and discussed in detail. Further, the prospective applications of CNTs in electrochemical sensors for the detection of micro-to macro-scale molecules highlighting the potential of the material in achieving ultra-sensitivity, specificity, faster response times, stability and reproducibility have been taken into consideration for the enrichment of reader's knowledge.

**Keywords:** Carbon Nanotubes, Electrochemical, Electrodes, Functionalization, Sensing.

## 5.1. INTRODUCTION

Ever since the discovery of CNTs, researchers are keen to develop CNTs-based sensors for variety of practical applications. The inherent properties of CNTs such as small size, high strength, high electrical and thermal conductivity, and high specific surface area demonstrate the potential to revolutionize the sensor industry. CNTs are an ideal material for sensing application due to the physical and catalytic properties, high electrical conductivity, chemical stability, and mechanical strength. Due to these reasons, large number of manuscripts have been published on the sensing application of CNTs in pressure, flow, thermal, gas, optical, mass, position, strain, chemical, and biological sensors [1].

A brief discussion on the electrochemical properties of CNTs is presented here with main focus on the electrochemical sensors based on CNTs.

## 5.2. ELECTROCHEMISTRY OF CNTS

The electrochemically inert nature of CNTs and the absence of voltammetric

* **Corresponding author Shikha Wadhwa:** Amity Institute of Nanotechnology, Sec 125, Amity University, Noida, UP, India; Tel: 09711202698; E-mail: swadhwa@amity.edu

**Jagriti Narang & Chandra Shekhar Pundir (Eds.)**

response in the potential window commonly used, makes it suitable for electrochemical applications. This property is analogous to other types of carbon-based materials commonly employed in electrochemistry, *e.g.*, glassy carbon (GC), graphite and diamond. The ability of CNTs to promote electron transfer in electrochemical reaction is the reason that CNTs qualify as an excellent material for sensing over other materials. There are two types of carbon atoms in CNT: those at the tube ends and those at the sidewall of nanotubes. Both these components demonstrate different electrochemistry; the tube ends behave like the edge plane of Highly Oriented Pyrolytic Graphite (HOPG) and exhibit favourable electrochemical properties, while the sidewalls are similar to the basal plane of HOPG and exhibit slow electron transfer kinetics. The electrocatalytic effect of CNTs has been attributed to the activity of edge-plane-like graphite sites at the CNTs ends [2]. The remarkable electrochemical characteristics of CNTs enable them to be appropriate for use in Faradaic processes (*e.g.*, the fast electron transfer kinetics of CNTs, owing to the presence of edge plane graphite sites within the walls and at the ends) or in non-Faradaic processes (*e.g.*, significant changes in conductance when CNTs can accept/withdraw electrical charges from or to molecules in their proximity, owing to the presence of the electronic cloud surrounding their walls). The methods used to synthesise CNTs inevitably introduces impurities into the CNTs structure, which includes nanocrystal metal catalyst, amorphous carbon and carbon nanoparticles. Therefore, the raw CNTs must be subjected to purification, such as chemical oxidation or thermal treatment prior to its use. The shortening of CNTs length and partial oxidation of CNTs producing functional oxygenated groups at the open ends and defects along the sidewall are observed as a consequence of pre-treatment processes. This functionalization modifies the electronic and structural characteristics of the CNTs radically. Some functional oxygenated groups on CNTs, produced by pre-treatment methods, are electroactive. This property is quite similar to various kinds of carbon-based materials as mentioned above [2]. The surface characteristics of CNTs direct the electrochemical attributes of carbon material-based electrodes. This is due to the fact that the formation of specific surface functional groups (especially oxygen-containing groups) can enhance the electron transfer rate (ET) significantly [3]. On the basis of their specific structures, there are two distinct surface regions in CNTs namely the sidewalls and the ends. The electrochemical features of CNTs are analogous to the Basal Planes of Pyrolytic Graphite (BPPG). In case of intact carbon nanotubes, the structure free of defects possess nearly the same features as that of BPPG except the cap regions where, due to much higher curvature strain than the sidewall, may become more reactive. The physical/chemical treatments for opening of the ends of CNTs lead to formation of a range of oxygen-containing groups, possessing comparable properties to the edge places of BPPG [4]. The critical roles of defected sites and oxygen-containing groups on the electrochemical performances of CNTs-based electrodes have been proved by Banks *et al.* [5]. They studied electrochemical oxidation of epinephrine and the electrochemical reduction of ferricyanide at CNT-modified electrodes using different types of graphite electrodes and with $C_{60}$-modified electrodes. The work suggested that the electrocatalytic nature of multi-walled carbon nanotubes (MWCNT) appear due to the electron transfer from the ends of nanotubes, which shows structural resemblance to that of an edge plane graphite (Fig. **1**).

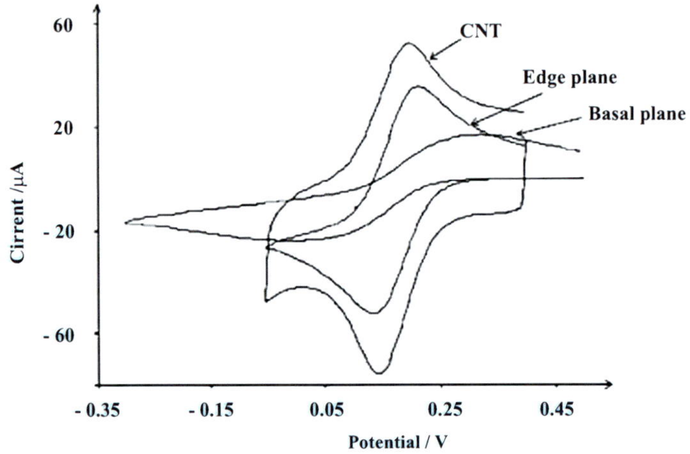


**Fig. (1).** Cyclic voltammograms for the reduction of 1mM ferricyanide for CNT- and C60 film-modified BPPG electrodes. Also shown is the response of a bare BPPG electrode and an edge plane pyrolytic graphite electrode. All scans at a rate of 100 mV/s [5].

Redox waves at purified CNTs were observed by Li *et al.* [6] and were attributed to the redox process of oxygenated groups formed at the CNTs. The reactions are shown below:

For reduction process:

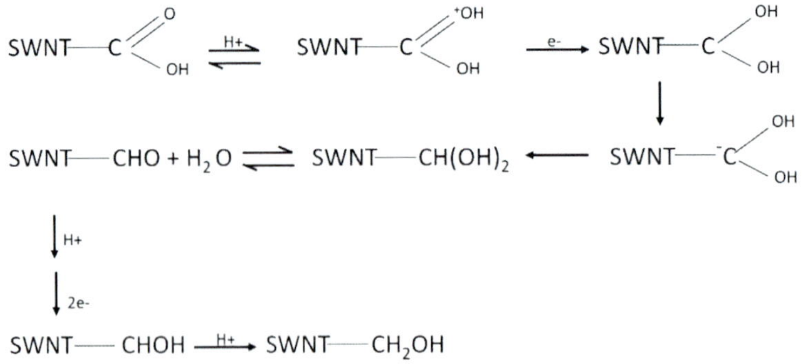

For re-oxidation process:

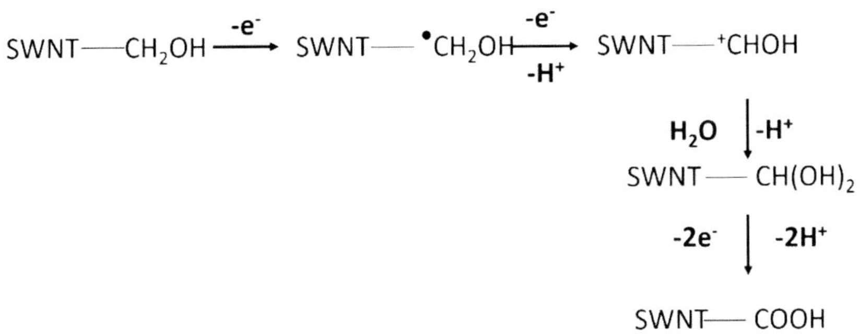

## 5.3. CNTS BASED ELECTROCHEMICAL SENSORS

A wide range of applications such as electronic and optoelectronic, biomedical, pharmaceutical, energy, catalytic, analytical, and materials sector incorporate CNTs as a novel material with unique properties. Predominantly, the small dimension in nano-regime, surface modification with active functional groups, excellent electrical conductivity, biocompatibility, modifiable sidewall, and high reactivity of CNT contribute towards an ideal material property for sensing application with high performance. For this reason, CNTs have been widely employed in creating numerous electrochemical sensors. Compared with the conventional scale materials and other types of nanomaterials, the nanostructural properties of CNTs demonstrate some overwhelming benefits for the fabrication of an electrochemical sensors. These include:

i. The large specific area leading to high sensitivity.
ii. The tubular nanostructure and the chemical stability facilitating fabrication of ultrasensitive sensors incorporating only one nanotube.
iii. The excellent biocompatibility which is appropriate for fabricating electrochemical biosensors, facilitating electron transfer of redox biomolecules such as proteins and enzymes.
iv. The ends and sidewalls of CNTs can be modified using a variety of functional groups for fabricating multifunctional electrochemical sensors.
v. The likelihood of achieving miniaturization.
vi. The probability of fabricating ultrasensitive nanoarrays. CNTs are included into sensors to decrease overpotential and to improve sensitivity [7].

The simple and well-defined responses have been observed at carbon materials and with the $Fe(CN)_6^{3-}/Fe(CN)_6^{4-}$ couple. Therefore, $Fe(CN)_6^{3-}/Fe(CN)_6^{4-}$ couple has been extensively employed to characterize the surface features of all types of carbon material based electrodes. Likewise, the electrochemical properties of CNTs-based electrodes are commonly investigated by $Fe(CN)_6^{3-}/Fe(CN)_6^{4-}$ as the probe. The electrochemical behaviours of aligned bundles of CNTs with other carbon based electrodes of analogous structures were compared by Nugent *et al.* [8]. $Fe(CN)_6^{3-}/Fe(CN)_6^{4-}$demonstrated an ideal redox peak separation (ΔEp) of 59mV at the aligned multiwalled carbon nanotubes (MWCNTs) [7]. Much larger ΔEp at single-walled carbon nanotube (SWCNT) papers (ΔEp = 96mV) and at aligned MWNTs with heat pre-treatment (to remove impurities like amorphous carbon and catalyst (ΔEp = 230mV)) were observed by Li *et al.* [4].

Mainly, for aligned MWCNTs, the correlation between the observed electron transfer rate (as determined from ΔEp) and the area of the exposed sidewalls (analogous to graphite basal-plane like characteristics) and the density of graphite edge-plane like defects was found to be good. This implies that the electrochemistry of the probe had some influence from the sidewalls of the CNTs. The small ΔEp at the SWCNT paper was attributed to the electron transfer promoted by the defect states containing oxygen which are formed during the acid purification step for formulating the papers. Papakonstantinou *et al.* recently reported the fundamental electrochemical characteristics of inherently hydrophobic MWCNT electrodes and acid-treated MWCNTs studied by Electrochemical Impedance Spectroscopy (EIS) and Cyclic Voltammetry (CV) [9]. They observed slow electron transfer kinetics with ΔEp>200mV and an electron transfer resistance of ~ 370Ω on pristine MWCNTs. The values obtained were very close to the results obtained by Li *et al.* at aligned MWCNTs [4]. Though, acid treatment significantly enhanced the electron transfer kinetics of $Fe(CN)_6^{3-}/Fe(CN)_6^{4-}$ redox couple, approaching almost reversible electron transfer (ET) kinetics. They reported that the charge carrier density at the CNT surface gets modified by acid treatment due to introduction of surface states which facilitated electron transfer. Faster ET kinetics in acid treated MWCNTs was attributed to the better wetting properties and the higher local density of states relative to untreated MWCNTs. In other words, the intact sidewalls of CNT, which resemble the basal planes of pyrolytic graphite, can be considered electrochemically inert to electroactive species. The observed enhanced electron transfer at the intact sidewalls of CNT is due to the higher strains than the basal planes. The opened caps possess electrochemical properties resembling that of edge planes of pyrolytic graphite due to the defects and oxygen-containing functional groups present at those planes. The electron transfer kinetics is improved due to the incorporation of edge-like defect sites and oxygen-containing functional groups at the end-caps and the sidewalls by chemical/physical treatments which considerably improves the electrochemical properties of CNTs by altering the electronic structures, the surface states and the wettability of the sidewall.

The well-defined structure, the stability, chemistry and the electrocatalytic properties of CNTs towards a range of substances, carrier platforms for numerous electrochemical sensors are being extensively constructed using CNTs. Detection of biomolecules such as neurotransmitters, proteins, glucose, and DNA has been made possible by CNT-based electrochemical sensors. Various types of electrochemical methods have been employed in these sensors, which are as follows:

1. Direct electrochemical detection with amperometry or voltammetry.
2. Indirect detection of an oxidation product using enzyme sensors.
3. Detection of conductivity changes using CNTs-field effect transistors (FETs).

CNTs have been utilised for detection of molecules (analytes) at lower potentials than any other electrode materials. These analytes include dopamine [9, 10], ascorbic acid [10, 11], uric acid [11, 12], norepinephrine [13], NADH [14 - 17], hydrogen peroxide [18], glucose [19], and nucleic acid [20]. The aforementioned analytes are vital in a biological context. Particularly, hydrogen peroxide and NADH are imperative for the interfacing CNT modified electrodes with enzymatic detection of bio-analytes, for example in enzyme biosensors. The enzyme reaction was monitored by the detection of hydrogen peroxide produced as exemplified by Wang *et al.* [18]. The onset of the oxidation of $H_2O_2$ at +0.2 V *versus* Ag/AgCl on CNT modified glassy carbon electrode was observed while no $H_2O_2$ oxidation was observed on bare glassy carbon electrode up to +1.0V. Wang further detected the glucose level by incorporating the CNT modified electrode into a flow injection apparatus by monitoring the $H_2O_2$ produced due to the enzymatic reaction between glucose oxidase in the carrier solution and glucose. A considerable glucose response was observed at −0.05 V with insignificant interference from acetamidophen, ascorbic acid and uric acid, which are common interferences. The strong sorption properties of CNTs facilitated their application on the working electrode surface and lead to development of various original methods of stripping voltammetry.

The immobilisation of enzymes and other redox proteins have also been carried out on the ends of aligned nanotube arrays [21, 22], on the walls of CNTs [23 - 25] and inside CNTs [26] using CNT modified electrodes. Majority of these studies are focussed on the expansion of electrochemical enzyme biosensors. Azamian *et al.* [23] demonstrated that glucose oxidase is adsorbed along the length of CNTs, which are randomly dispersed on a glassy carbon electrode. The glucose signal was found to increase by 10-fold in the presence of ferrocene as a mediator as compared to when the glucose oxidase was adsorbed onto the glassy carbon electrode alone. CNTs offered high surface area electrode to which enzyme could adsorb. Hence, greater amount of active enzyme adsorption on the electrode interface lead to greater signal [27].

### 5.3.1. CNTs Based Electrodes

Individual CNTs, with their excellent electrical, mechanical properties and flexibility may eventually become vital in their use as nanoprobes. It has been experimentally shown that SWCNTs can carry current densities up to $10^9$ A/cm$^2$, which is much higher than that observed for normal metals (~$10^5$ A/cm$^2$) [28]. Due to such large difference in the current densities (of SWCNTs and normal metals) the contact resistance turns out to be quite large, making CNTs less useful for potential sensor applications. For SWCNTs, the contact resistance was found to be ~ 6.5 kΩ [29]. The contact resistance can be reduced by contacting all layers in CNTs, however, cannot be eliminated completely [30]. Agglomeration of CNTs in dispersed form is considered one of the critical issues to be resolved. An inherent tendency of CNTs to aggregate together in almost all types of aqueous and organic solutions due to Van der Waals forces between tubes poses difficulty in forming homogenous CNTs composite and in turn affect the electrochemical properties. To overcome the agglomeration, functionalization of CNTs is strategically carried out in order to increase the degree of reactivity in CNTs and also for a homogenous dispersion. Early studies revealed the process of CNTs electrode preparation by simple dispersion on substrate with polymer coating. Such polymers include Teflon, Nafion [31], or paste [32]. Though such studies showed a highly selective and sensitive determination of biomolecules such as dopamine, but the fundamental contribution of the side and end wall of CNTs to such electrochemical studies was still unknown. Further to this issue, vertically aligned carbon nanotubes were investigated as an electrode material to study the electrochemical properties for the development of biosensors. Many new methods have been introduced for incorporation of CNTs on the electrode surface for electrochemical and electro analytical applications. Few of the CNTs electrode preparation methods are developing of CNTs paste electrodes, CNT- polymer nanocomposites, electrodes of CNTs- solgel nanocomposite, layer by layer assembly of CNT film electrodes. Various types of CNTs modified electrodes are discussed below:

#### *5.3.1.1. CNT-Polymer Nanocomposite Electrodes*

CNT- Polymer composite films have been used widely in electrochemical sensing. CNTs are integrated into the polymer matrix to create CNTs-polymer composite. CNTs integration inside the polymer matrix can be in two ways either randomly oriented or vertically aligned. Wang *et al.* prepared an electrode of CNTs/Teflon composite by simple handmixing of desirable amount of CNT in granular Teflon in a dry atmosphere, that was further firmly packed into the electrode cavity [33]. This particular study showed the strong association of CNT and Teflon binder, that gave a stable electrocatalytic properties by observing redox activity in analytes NADH and peroxide over the entire potential range from 0.0-1.0 V. In one of the other studies, a Conducting polymer/CNT- modified electrode of Poly(aniline) (PANI) and poly(flavin adenine dinucleotide) was prepared using cyclic voltammetry (CV) by application of repeating potential cycles over a range of from −0.35 to 1.0 V at 50 mV $s^{-1}$ [34]. Polypyrrole (PPy), another conducting polymer, can be used for fabricating CNT/PPy nanocomposite electrodes that depict high conductivity and a stable electrocatalysis, thus making a favourable choice for construction of electrochemical sensors [35]. CNT based microelectrodes are also possible to prepare using polypyrrole and polyaniline due to the ease of preparation of such macro electrodes by electrochemical polymerization. CNTs exhibit good electrochemical properties and are also comparably smaller in size which easily makes them available as microelectrodes at the micro level for investigating biological tissues. A voltammetric sensor consisting of Poly (3,4-ethylene-dioxythiophene)-coated MWCNTs was developed by Viswanathan *et al.* for detection of Cholera Toxin [36]. For an amperometric detection of biotin, Cataldo *et al.* prepared an amperometric immunosensor of anti-biotin antibodies (Ab) implanted into a polylysine (PLL)-SWCNTs composite layer [37]. This study revealed an enhanced amperometric detection limit of biotin(Ag)labelled with HRP in the range 10 pM, having improved thermal stability of the anti-biotin embedded into the PLL matrix.

CNTs-polymer nanocomposites of insulating polymer have also been developed. CNT/Nafion nano composite electrodes were prepared by Wang *et al.* [38] and were found to possess tremendous electrocatalytic activities for detecting carbohydrates, hydrogen peroxide ($H_2O_2$) and NADH. CNTs polymer composite electrodes made from insulating material such as Nafion were give more advantage over conducting polymer composite electrodes, because the insulating polymer electrodes of CNTs displayed a distinct faradaic response for the redox species at low current charges, when compared of the conducting polymer electrodes of CNTs. Considering, this CNTs/Nafion composite electrodes attracted tremendous research in electroanalytical and electrochemical applications. Some MWCNTs based polymer composites were fabricated too for detection and sensing of NADH were poly (acrylic acid)-wrapped MWCNTs complex [39], MWCNTs-poly (1,2- diaminobenzene) nanoporous composite [40] and quinone-amine polymer- MWCNTs nanocomposite [41]. These electrodes gave improved sensitivities for NADH detection due to their exceptional electrocatalysis.

### *5.3.1.2. CNT Paste Electrodes*

Carbon pastes have been used in electrochemical sensors from several years. Electrodes made out of carbon paste were easy to prepare, compatible with other modifiers and also had unique renewability. Therefore they were broadly used as a matrix for preparing modified electrodes, mainly comprising of CNTs. The CNT-paste electrodes exhibited a substantial growth in the electrochemical properties of other materials, in which they were used. Method preparation of CNTs paste electrode is almost similar to the conventional carbon paste electrodes. The process involves initial mixing of CNTs with a mineral oil binder and the firm packaging of the paste in the electrode cavity. This process was first developed by Valentini *et al.,* in which SWCNT paste electrodes were fabricated [42] by mixing CNTs into mineral oil binder in equal parts in a mortar. Due to high mineral oil content and comparably high surface area of CNTs, mechanically robust and stable paste electrodes were constructed. The CNTs paste electrodes exhibited enhanced electrochemical properties as compared to other conventional carbon paste electrodes. This property enhancement was not observed in conventional electrodes as it was not possible to load high content of mineral oil in these Carbon paste electrodes, due to continuous leakage of Carbon paste into the solution. The electrochemical studies for as prepared SWCNTs paste electrodes were investigated in redox couple ($Fe(CN)_6^{3-/4-}$) using cyclic voltammetry. The authors revealed through this study, a dependency on reversibility of the Ferri-Ferro redox couple to the amount of CNTs used in the preparation of paste electrodes. The results showed a good reversibility trend when the CNTs content was 3 quarters of the total quantity. These CNT paste electrodes found a possible application in the electrochemical detection of dopamine sensor. In another study MWCNT paste electrodes were fabricated by modifying the conventional carbon paste electrodes with MWCNTs. These MWCNT modified electrodes displayed a good electrochemical detection of Berginin, that gave a linear increase in oxidation peak current on increasing MWCNT content while preparing the paste electrodes. A detection limit of Berginin as low as 70 nM was observed using Differential pulse voltammetry (DPV) [43]. Recently, one more MWCNTs paste electrodes was prepared by Abbaspour *et al.* for the application in electrochemical sensing and determination of piroxicam [44], quercetin [45] and urapidil [46]. This MWCNTs paste electrode exhibited good sensitivity and quicker response time in a linear range of 0.15–5 μg/mL. A highly sensitive voltammetric sensor showing good electrochemical behaviour for determination and reduction of xanthine [47] and methlparathion was fabricated by Fan *et al.* [48]. This study prepared SWCNTs paste electrode with 1-butyl-3-methylimidazolium hexafluorophosphate, that is an ionic liquid.

### *5.3.1.3. CNTs/Sol-gel/Nanocomposite Electrodes*

The sol-gel method has been generally utilized as a part of electrochemistry. Throughout the most recent decades, sol-gel electrochemistry has drawn broad interest on account of its exceptional assistance in investigating electrochemical properties. Ceramic Carbon electrodes (CCEs) prepared from sol-gel method has attracted a lot of attention in recent years. The preparation of ceramic carbon electrodes generally involves doping of powdered carbon or graphite on a silica gel matrix. These electrodes based on sol-gel preparation demonstrate good electrochemical and electro analytical applications in bio-sensing and energy harvesting. Another form of electrodes prepared by this method is Ceramic-CNT Nano-composite Electrodes (CCNEs). The method of preparation is similar to conventional CCEs, only difference being that CNTs are doped in the gel matrix [48, 49]. A characteristic transformation was observed in the tuning of the electrode dimensions of the CCNEs from conventional to microscopic level. The CNTs composition in the sol was a prime factor behind the tuning of the electrode dimensions. Since, CNTs exhibit good electrochemical properties and sol-gel matrices are bio compatible, this combination of CNT/sol-gel nanocomposite electrodes lead to further development of bisoensors by entrapment of proteins and enzymes [50]. A MWCNT/ methyltrimethoxysilane ceramic composite electrode was prepared by Zhu *et al.* for electrochemical detection of Uric acid(UA), NADH, ascorbic acid(AA), dopamine(DA), $H_2O_2$, cysteine and acetaminophenol [51]. A CNT/horseradish peroxidase(HRP)/sol-gel nano-composite was prepared by Zhang *et al.* [52] The authors indicated that CNT/HRP/sol-Gel nanocomposite electrodes exhibited a better catalytic activity in reduction of $O_2$ due to a direct electron transfer as compared to the CNT/HRP/sol-gel modified glassy carbon(GC) electrode. All these observations were recorded by cyclic voltammetry. Gavalas *et al.* [53] demonstrated in their study, a method for amino acid detection by construction of CNT/amino acid oxidase/sol-gel bio composite electrodes.

### *5.3.1.4. Layer by Layer Assembly of CNT Film Electrodes*

Layer by layer self-assembly has been widely used for preparation of CNT multilayer films on the substrates. Layer by layer method of preparation is a template driven approach that utilizes depositing of oppositely charged species based on the electrostatic interaction taking place in between them. Zhai *et al.* [54] used the layer by layer technique to develop a MWCNT-chitosan multilayer film, in which chitosan is the positively charged species and MWCNTs are the negatively charged species. This study developed a nine layer chitosan-MWCNT assembly that was found to exhibit a stable and quicker response for NADH oxidation with a limit of detection as low as of $3\times10^{-7}$ M. Layer by layer method for assembly of CNT films have proved to be useful for electro analytical and electrochemical investigations as they give films that are uniform and homogenous. Though this method gives a way to efficiently increase the enzyme or catalyst concentration based on the sensor surface, it is also very important to optimize the layer thickness, such that it does not hamper the sensor response. Fabrication of SWCNT/ PDDA (poly(diallyldimethylammonium chloride) multilayer films was reported by Rouse *et al.* and Mamedov *et al.* [55, 56]. In both the cases, electrostatic interaction took place at the tubular ends of the CNTs. In PDDA/$GO_x$/PDDA/CNTs multilayer film was constructed for a stable amperometric determination of glucose sensing using glucose oxidase enzyme. The multilayer films were constructed in a sandwich like structure with alternating self-assemblies of PDDA layer and $GO_x$ layer onto the CNT surface. This sensor was demonstrated an outstanding sensitivity for glucose determination that was stable and reproducible, free of any hindrances caused by other electro-active species present [57, 58]. Zu *et al.* [59] also constructed a multilayer Au/MWCNT/$GO_x$ membrane by layer by layer assembly of PDDA species. It was characterized by AFM and SEM for its morphology. This study unveiled a glucose detection level of 128 μM, at a relatively low potential of -0.2V. Zhao *et al.* [60] also developed alternating $GO_x$ and Prussian blue nanoparticles using the layer-by-layer assembly technique. The Prussian blue nanoparticles were wrapped in PDDA, thus generating a net positive charge on the surface of the nanoparticles that leads to the assembly of nanoparticles electrostatically layer by layer on the electrode surface. Another glucose biosensor was fabricated by loading $GO_x$ and MWCNTs using layer by layer technique. This sensor showed a linear dependency of current response to number of layers of MWCNTs/GOx [61]. A detection limit of $9\times10^{-5}$μM was exhibited in 6 MWCNTs/$GO_x$ layers.

### 5.3.2. Nanoparticles-CNT Electrodes

Interest in metal nanoparticles synthesis for electrochemical sensing has grown in recent years. A lot of research has been carried out in modifying the electrodes with electro deposition of synthesized metal nanoparticles into CNTs. Nanoparticles modified CNT electrodes have proved to deliver a better electrochemical response, as compared to conventional electrodes modified with nanoparticles, thus making them a preferred choice for electrochemical sensing today. Various combinations of metal nanoparticles CNT modified GC electrodes were fabricated by Hrapoic *et al.* [62] for electrochemical detection of trinitrotoluene (TNT), an explosive material. This study exhibited the use of gold, copper and platinum nanoparticles with MWCNTs and SWCNTS dispersed in Nafion solution for nanocomposite formation. Out of all the combinations, the authors observed the most significant results for Cu/SWCNT/Nafion composite modified GC electrode by observing a detection limit of 1ppb and giving the highest sensitivity for TNT in analysing samples of tap water, river water and soil. A AuNPs/MWCNTs composite modified GC electrode was fabricated by Liu *et al.* [63] using electro polymerization of thionine. This fabricated sensor was designed to determine sensitivity of adenine and guanine in DNA, thus giving a detection limit of 10 nM and 8 nM respectively for both the purine nucleobases. Another modified electrode based on nanosilver (AgNPs) coating on MWCNTs was fabricated to determine the thiocyante concentration in urinary and salivary test analysis, conducted for smokers and non-smokers [64]. This was the first research to develop a novel method of fabricating an electrode from AgNPs-MWCNT composite for analyzing real time biological and environmental samples. These electrodes gave a linear dynamic range of the detection limit as low as 1 nmol/L. For chloramphenicol detection, a conventional GC electrode was modified with AuNPs/SWCNT composite. The ionic liquid used was 1-octyl-3-methylimidazolium hexafluorophosphate [65]. Lin *et al.* [66] demonstrated a MWCNT modified GC electrode based on platinum nano-clusters having a strong oxidation of number of estrogens such as estrone, estrogen, oestriol. These estrogens showed a linear response through the square wave voltammetric signals.

Apart from metal nanoparticle modified CNT electrodes, there is a significant importance of electrodes modified with oxide phase- CNT nanocomposites too. Tao *et al.* [67] revealed a simple and easy method based on low energy ball milling treatment, that was useful to prepare oxide based CNT composite electrodes using electrode that is insulating in nature. This technique used $LiFeTiO_4$ as the electrode material for all the testing and the results obtained exhibited a remarkable increase in the capacity of the nanocomposite in comparison to the $LiFeTiO_4$ nanoparticles, *i.e.* 105–120 mA h $g^{-1}$*vs.* 65–70 mA h $g^{-1}$. These lithium polymorph composites with CNT were found to be a good electrode material to be used in batteries. The studies indicated this technique as a significant approach towards enhancing the electrode characteristics in an insulating phase.

## 5.4. RATIONAL FUNCTIONALISATION OF CNTS

Raw CNTs having metallic nature are generally considered highly hydrophobic. This hinders the dispersion of CNTs into aqueous solutions. Therefore, it is often believed to functionalize or modify the surface of CNTs for future applications. CNT functionalization methods majorly revolve around any of the two categories *i.e.* Non- covalent functionalization of CNTs that tend to use $\pi$-$\pi$ interactions of hydrophobic CNTs for coating of other amphiphilic molecules. The second method is the covalent functionalization in which chemical bonds of carbon atoms are generated on the CNTs surface. The functionalization of CNTs not only offer high solubility in both organic and aqueous solutions but also yield too many new functional moieties that are helpful in linking them to other applications such as in Nano devices, recognition of drug delivery targets *etc*. The structure of CNTs consists of $sp^2$ carbon atoms. The sidewalls of the CNTs are hydrophobic in nature. The non-covalent interaction taking place on these side walls can be carried out without damaging the $sp^2$ structure of CNTs, thus maintaining their unique structure and properties. Researchers have explored different methods for non-covalent functionalization, based on the interactions between the sidewalls and the π–π stacking, CH- π or hydrophobic interactions. In Recent years of development of CNTs, non-covalent functionalization of CNTs has gained specific interests, due to its prime advantage of not destroying the conjugated structure of the sidewalls of CNT, and hence, enhancing its structural properties. Various approaches have been followed to non- covalently functionalize CNT *i.e.* by using polymers, aromatic small compounds, nanoparticles and biomolecules.

### 5.4.1. Functionalisation with Polymers

Conjugated polymers with a π–π stacking structure are also a good choice for coupling on the sidewalls of CNTs. Due to strong Van der Waal interaction taking place polymers with such conjugated structure and the nanotube surface, a notable increase was observed in the CNT dispersion. Use of block copolymers in non-covalent functionalization of CNTs showed a remarkable increase in dispersibility of CNT at the time of drug delivery [68]. SWCNTs functionalized with hexamethylenediamine and poly (diallyldimethylammonium chloride) enable non-covalent interactions to take place in intracellular delivery of biomolecules having few cytotoxic effects [69]. Another simple method of chemical functionalization of pyrrole on MWCNTs was reported by Zhao *et al.* [70]. This method was carried in absence of any oxidants at room temperature. In another study, Li *et al.* [71] reported crystallizing polyethylene and nylon-66 homopolymers for CNT modification. Using this novel method another research was carried out by Wang *et al.* [72], who fabricated block copolymers onto the CNTs surface using the controlled polymer crystallization technique. Uniform CNT dispersion and controlled alignment was observed in the microdomains of block copolymers. Many studies have displayed the use of polymers such as polyethylene glycol and polethyleglycolated phospholipids for non - covalent CNT functionalization due their unique biocompatible nature, dispersion and also such functionalized CNTs are free from any cytotoxins [73 - 75].

### 5.4.2. Functionalization with Aromatic Molecules

Another form of non-covalent functionalization of CNTs can be done by π-π interactions between aromatic compounds and sidewalls of CNT. These interactions are predicted to be much stronger as compared to planar graphitic structures, because of which they are an attractive choice applicable to construction of CNT Nano devices. The other crucial advantage of using aromatic compounds for functionalization of CNTs is that they provide novel electrochemical properties due to modified structures, which are different from the initial and comparatively much better than CNTs alone. For example, Hecht *et al.* reported fabrication of a non-covalent functionalization of CNTs by a zinc porphyrin derivative, that was used to fabricate a SWCNT/FET device, capable of direct detection of photo induced electron transfer process [76]. The results obtained depicted a relationship between the magnitude of PET as a function of Intensity and wavelength of light. Another group of researchers led by Hu, exhibited a CdSe based CNT hybrid system in which the CNT surface was functionalized by a wrapping of self-assembled layer of pyrene molecules onto the CdSe nanoparticles (pyrene/CdSe) [77].

### 5.4.3. Non-Covalent Functionalization with Biomolecules

The biomolecules used for CNT functionalization are saccharides and polysaccharides. Shim *et al.* [78] demonstrated in his work, a streptavidin/biotin system that investigated the adsorption properties of proteins on the SWCNTs side walls. The nanotubes were co-functionalized by protein resistant polymers and biotin that was used to detect the specific binding ability of streptavidin on SWCNTs surface. Many other biomaterials have been used for non-covalent functionalization of CNTs, such as chitosan [79], phospholipid dextran [80], pullulan [81]. Saccharides and polysaccharides of these types had no light absorption in the UV-Visible region. These polysaccharide functionalized CNTs have a good biocompatibility thus increasing their application in medicine.

### 5.4.4. Functionalization with Metal Nanoparticles

Linking nanoparticles with CNTs to give nanocomposites has attracted a lot of attention, due to unique structural properties of CNTs. Different approaches have been followed for nanoparticle deposition onto the CNT for surface modification and enhance dispersivity of CNTs. Liu *et al.* [82] reported coupling of AuNPs onto MWCNT surface using non covalent functionalization. In this work, a self-assembled AuNPs were linked on the CNT surface using 17-(1-pyrenyl)- 13-ox--heptadecanethiol *i.e.* bi-functionalized molecule as the interlinker. The bi-functionalized molecule consisted of a pyrenyl unit as the one end group that bound to the CNTs surface through $\pi$- $\pi$ interaction. The other thiol end group was linked to the gold nanoparticle surface, thus inducing a charge transfer interaction between the nanoparticles and the nanotubes. Sadegh *et al.* [83] described a novel and effective method for surface modification of MWCNTs using magnetic iron oxide nanoparticles. This approach led to an enhanced dispersion of CNTs into the aqueous media. Yuan *et al.* [84] reported non covalent functionalization of CNTs with platinum nanoparticles (PtNPs) that has a significant effect on the electro catalytic activities of CNTs for applications in PEM fuel cells.

## 5.5. CNT BASED ELECTROCHEMICAL SENSORS FOR ELECTROANALYSIS

Even though CNT electrode possesses electrochemical properties analogous to the edge plane pyrolytic graphite electrodes, there are some benefits for making CNT based electrochemical sensors. These include good conductivity, biocompatibility, the large specific surface area, a range of functional groups on its surface post pre-treatments, and able to form stable solutions or dispersions. The electrochemical sensing using CNTs has been focussed mainly on two different types: the non-enzymatic and enzymatic CNT based electrochemical sensors.

### 5.5.1. CNT Based Non-Enzymatic Electrochemical Sensors

#### *5.5.1.1. For Biomedical Analysis*

##### *5.5.1.1.1. Thiols*

The levels of amino-thiols in biological environment such as plasma and urine are important biomarkers in various clinical situations. The risk of cardiovascular diseases is related to increased levels of HcySH and CySH. For the development of electrochemical techniques for thiol detection, the direct oxidation of thiols at the solid electrode surface is slow and requires large over-potentials [85]. A direct estimation of L-CySH and GSH at a MWCNT modified electrode was demonstrated by Cao *et al.* [86]. CNT-modified electrode demonstrated high efficiency for electrocatalytic oxidation for L-CySH and GSH with significantly high sensitivity, stability and long-life. This electrode was employed for amperometric detection of these thiols along with the high performance liquid chromatography (HPLC). Cao *et al.* demonstrated that thiols can be successfully determined using CNT modified electrochemical sensors and lead to a number of studies based on the determination of thiols by CNT based electrochemical sensors. CNT-based electrochemical sensors can be constructed using two types of sensing, which include CNTs and Pt-CNT composites. High sensitivity for thiols was observed with Pt decorated CNTs. Similarly, Pt decorated CNT modified glassy carbon electrode (GCE) was employed for the determination of thiols, such as L-CySH and GSH. Due to high sensitivity, and good stability for thiol detection in comparison to MWCNT or Pt microparticle modified electrode, nanocomposite modified electrode was effectively employed for L-CySH and GSH detection in rat striatum.

### *5.5.1.1.2. Carbohydrates*

For applications ranging from medical applications of the blood glucose sensing to ecological approaches (*e.g.*, waste-water treatment in food industry), electrocatalytic oxidation of carbohydrates is of huge interest. Studies involve the use of $GO_x$ catalysing the oxidation of glucose to gluconolactone. However, some stability issues are associated with enzyme-based sensors owing to their intrinsic nature. Thus, great efforts are made to achieve direct determination of carbohydrates. An electrocatalytic oxidation of carbohydrates at SWCNT modified GC electrode has been reported by Deo *et al.* [87].The oxidation processes of galactose, fructose, mannose and xylose were found to start around +0.25V in 0.1M NaOH solution. The amperometric responses of the species studied at the CNT-modified electrodes were found to be highly stable, and sensitive over a constant 2-hour operation. Similarly, composites of metal particles with CNTs have also been tested to attain electro-oxidation of carbohydrates. Copper nanoparticles (CuNPs) (~4–8nm diameter), decorated CNT-based electrochemical sensor for the determination of carbohydrates has been reported by Male *et al.* [88]. Enhanced response relative to the GC electrode modified by either of the constituent particles was observed with the SWCNT/Cu nano-composite at +0.65V with a relatively low starting oxidation potential (+0.3V *versus* Ag/AgCl (3M NaCl)). Other monosaccharides, such as galactose, arabinose, xylose, fructose and mannose demonstrated similar responses as glucose.

### *5.5.1.1.3. Electroactive Microbiomolecules*

For biochemical and clinical diagnosis, the estimation of neurotransmitters is of immense significance. In general, determination of neurotransmitters is easier by electrochemical methods as they offer a simple and sensitive way for their determination. However, the selectivity at bare electrodes is difficult to be achieved due to the overlapping of the oxidation potentials of such species. Therefore, modification of bare electrodes has been attempted by many. CNTs emerged as an outstanding material as modifier of bare electrodes to achieve selectivity of neurotransmitters. Many recent works use CNTs as the sensing materials for the determination of the microbiomolecules. CNT composite films offer higher stability and enhanced uniformity of the films and takes advantage of the unique features of both the components of the composite film *e.g.* CNT-Nafion composite films can efficiently diminish interference from negatively charged amino acid on the determination of other biomolecules. This is due to the electronic repulsion between amino acids and negatively charged nafion film.

#### *5.5.1.1.4. Amino Acids*

Amino acids containing phenol moieties can be easily detected using electrochemicalmethods. Tang *et al.* [89] used electrospun carbon nanofibres modified carbon paste electrodes (CNF-CPE) for determination of L-Tryptophan (Trp), L-Tyrosine (Tyr) and L-Cysteine (Cys). The cyclic voltammetry and constant potential amperometric methods were used for the determination of these amino acids. The results demonstrated a linear ranges of Trp, Tyr and Cys to be 0.1–119, 0.2–107 and 0.15–64 μM with correlation coefficients of 0.9994, 0.9985 and 0.9996, respectively. The detection limit was 0.1 μM for all the analytes. The modified electrodes (CNF-CPE) exhibited excellent reproducibility, high sensitivity and good selectivity for the determination of the amino acids.

Wu *et al.* [90] fabricated MWNT/GCE electrode for the determination of tryptophan. The oxidation peak current of tryptophan was found to increase in comparison to a bare electrode. The detection limit for tryptophan was found to be $2.7 \times 10^{-8}$ M.

Deo *et al.* [91]. carried out an electrochemical determination of conventionally "nonelectroactive"amino acids at an electrode modified with SWNT and coated by a nickelhydroxide film The electrocatalytic oxidation of "nonelectroactive" amino acids at the $Ni(OH)_2$ layer electrodeposited on surface SWCNTs facilitated the electrochemical determination of these biomolecules. In an electrocatalytic process, the amino acids reduce the newly formed NiO(OH) species on the Ni-CNTlayer. A linear response was observed in the range of 10-80mM arginine with a detection limit of around 1 mM. Other amino acids such as histidine, lysine, asparagine, methionine and phenylalanine have also been detected with this electrode.

### *5.5.1.2. For Drug Analysis*

Various nanomaterials are employed to fabricate modified electrodes with high performance and improved detection sensitivity. CNTs have demonstrated to be one of the best materials for enhancing sensitivity. Some advantages of CNTs over other nanomaterials include good conductivity and strong electrocatalytic activity. Some water insoluble drugs get adsorbed on the hydrophobic and $\pi$-conjugated sidewalls of CNTs and allows sensitive detection of these species at the electrode surface. For a list of applications of CNT-based electrochemical sensors for drug analysis, the reader is directed to a review article by Chengguo Hu [92].

## 5.5.2. CNT-Based Enzymatic Electrochemical Sensors

### *5.5.2.1. Electrochemical Biosensors for Electroanalysis*

Enzymes incorporated into transducer facilitates fabrication of enzyme electrodes. These enzyme electrodes demonstrate not only enhanced selectivity and sensitivity but also a long term stable response to different bioanalytes. The employment of CNTs as an electrochemical transducer is beneficial owing to the exceptional electron transfer rate, the well-defined nanostructure, the excellent biocompatibility, and the possibility to modify surface which allows immobilization of biomaterial to CNTs. Some of the CNT based enzymatic sensors are discussed in this section:

1. *Glucose Oxidase:* An accurate monitoring of blood glucose level is required for the diagnosis of diabetes mellitus. For this reason, glucose biosensors are extensively studied. The amperometric glucose biosensors, most commonly, are based on the enzyme glucose oxidase ($GO_x$), which catalyses the oxidation of glucose to gluconolactone:

   Glucose + $O_2$ → ($GO_x$) → Gluconolactone + $H_2O_2$

   The quantification of glucose can be achieved *via* the electrochemical detection of the enzymatically liberated $H_2O_2$. However, the overpotential necessary for the oxidation or the reduction of $H_2O_2$ at solid electrodes is rather high. Recently, Wang *et al.* demonstrated that the overpotential of $H_2O_2$ can be greatly reduced at CNT-based electrodes [91] indicating the promising applications of CNTs in constructing glucose electrochemical biosensors. The hybrids of CNTs and some noble metal nanoparticles, such as palladium, platinum, gold, copper and iridium, have also been demonstrated to lower the $H_2O_2$oxidation/reduction overpotential efficiently. Besides the materials, the immobilization methods effectively influence the activity and the quality of enzyme electrodes. Typical $GO_x$ immobilization methods include electro-polymerization embedment, composite blending, covalent bonding, electrodeposition and physical adsorption.
2. *Other Enzymes:*
   a. *Cholesterol Oxidase:* For some health conditions such as coronary heart disease, arteriosclerosis, myocardial infarction and brain thrombosis, the clinical diagnosis involving determination of cholesterol is vital. Cholesterol oxidase ($ChO_x$)-based electrochemical biosensors which offer a simple, rapid and inexpensive diagnostic seems attractive. The oxidation of

      cholesterol to 4- cholesten-3-one by the $ChO_x$ enzyme forms the basis of most of the amperometric cholesterol biosensors:

Cholesterol + $O_2$–$ChO_x$ → 4-Cholesten-3-one + $H_2O_2$

Cholesterol can be quantified by the electrochemical detection of the enzymatically liberated $H_2O_2$.A CNT-based amperometric cholesterol biosensor fabricated using through the layer-by-layer (LBL) deposition of poly (diallyldimethylammonium chloride (PDDA) and $ChO_x$ on an MWNT-modified gold electrode is proposed by Guo *et al.* proposed [94]. The strong electrocatalytic properties of MWNTs towards $H_2O_2$lead to high sensitivity and good anti-interferent ability for the determination of cholesterol. The detection limit for cholesterol was found to be 0.2 mM. Recently, Platinum nanoparticles decorated MWCNT-chitosan film fabricated by cyclic voltammetry method demonstrated high sensitivity of 0.044 A $M^{-1}$ $cm^{-2}$ and a response time of about 8s [95].

b. *Dehydrogenases:* To fabricate an electrochemical biosensor, some dehydrogenases can also be used. These enzymes include alcohol-dehydrogenase (ADH), D-fructose dehydrogenase (FDH) and glucose dehydrogenase(GDH). The current response, in this case is due to the electro-oxidation of NADH produced in the following enzymatic reaction:

   Substrate + $NAD^+$ Dehydrogenase → Product + NADH

   $NAD^+$ acts as an enzymatic mediator in this case.

c. *Other Enzymes:* Some other enzymes that have been employed to fabricate electrochemical biosensors include acetylcholinesterase, alkaline phosphatase, organophosphorus hydrolase, urease and tyrosinase. Xu *et al* [96]. fabricated voltammetric urease and acetylcholinesterase biosensors based on immobilization of the enzymes with a sol-gel hybrid material on the surface of the SWNT modified electrode. A reversible Nernstian response is obtained from the SWCNT modified electrode as a function of analytes associated with solution pH change. The potential of the anodic peak of SWNTs shifts in response to the urea and acetylthiocholine. The fabrication methods of these biosensors is simple and reproducible, therefore, lead to the determination of a range of analytes.

### *5.5.2.2. Direct Electron Transfer of Redox Proteins and Enzymes*

The direct electron transfer of redox proteins and enzymes at the surface of electrode is of huge significance which enables the study of biological redox process. However, direct electron exchange at the electrode surface for these macrobiomolecules is difficult to attain. This is due to the fact that they normally have bigger and complex structures containing redox centres deep inside the bodies. In addition, the native configuration of these macromolecules is changed due to the adsorption of these species at the electrode surface which often causes them to denature. Previously, some mediators have been employed to enhance the electron contact between enzymes and electrodes, however, the use of mediators resulted in a slow response and made the biosensors more complicated. This lead to the research into the development of mediator-less biosensors. Since the first work of direct electron transfer of proteins at the surface of CNT electrodes in 1997 [97], several studies came up which focused on the direct electron transfer of redox proteins and enzymes at the surface of CNT electrodes. To name a few, these include haemoglobin (Hb), myoglobin (Mb), Cyt c, peroxidases (HRP), catalase (CAT) and MP-11, oxidases *e.g.*, GOx, putrescine oxidase (POx), xanthine oxidase (XOx).

(1) Direct electron transfer of redox proteins and enzymes at CNT electrodes

a. **Heme Proteins**

i. *Myoglobin:* A small sized heme protein, Myoglobin (Mb), functions in both transport and short-term storage of oxygen. Mb is an ideal model protein for the study of heme proteins or enzymes owing to its commercial availability and a well-defined structure. The tubular nature of CNTs and other unique features such as good conductivity, and the availability of functional groups allows them to act as molecular wires facilitating direct electron exchange between the electrodes and Mb. Pakapongpana *et al.* [98] developed electrochemical sensors for the detection of haemoglobin (Hb) and myoglobin (Mb) by self-assembly of methylene blue-multiwalled carbon nanotubes (MB-MWNTs) nanohybrid modified on glassy carbon electrode and without immobilizing any enzymes. Excellent electrocatalytic activity for reduction of Hb and Mb along with high stability, sensitivity and reproducibility were obtained for the constructed MB-MWNTs nanohybrid modified electrode. The detection limits for the Hb and Mb observed were 1.5 nM and 20 nM, respectively.

ii. *Haemoglobin:* A biological macromolecule responsible for the storage and transport of molecular oxygen in the blood, Haemoglobin (Hb), consists of four polypeptide subunits (two$\alpha$ and two $\beta$ subunits). Each of these subunits, similar in structure, contain iron-bearing heme. Hb has been shown to possess enzyme-like catalytic activity, however, it is not an electron carrier in biological systems. The electrochemical analysis results indicated that Cyt c could be adsorbed firmly on MWNTs electrode surface. MWNTs played a significant role in the direct electron transfer between Cyt c and electrode. The adsorption of Cyt c onto MWCNTs lead to a significant electrocatalytic activity for the reduction of hydrogen peroxide.

b. **Peroxidases:** Horseradish peroxidase (HRP), another bio-macromolecule is a well - known biocatalyst of $H_2O_2$.Zhao *et al.* [99] demonstrated the direct electron transfer from adsorbed HRP to CNT microelectrode. The transfer coefficient, α, was found to be 0.57, and the heterogeneous electron transfer rate constant was 2.48±0.03 $s^{-1}$. The adsorbed HRP still kept its electrocatalytic activity to $H_2O_2$. Cui *et al.* [100] developed a biosensor consisting of a combination of gold nanoparticles (GNPs)/carbon nanotubes (CNTs) hybrids platform with horseradish peroxidase (HRP)-functionalized gold nanoparticle label for the sensitive detection of human IgG (HIgG). Improved sensitivity, stability and excellent precision was obtained. A linear response range between 0.125 and 80 ng/mL was observed with a detection limit of 40 pg/mL.
   Other peroxidases include catalase and Microperoxidase (MP-11). Catalase is a redox enzyme which catalyses disproportionation of hydrogen peroxide into oxygen and water without generation of free radicals. It protects cells partly from the toxic effects of hydrogen peroxide. Microperoxidase (MP-11), an oligopeptide containing eleven amino acids and a covalently linked FeIII-protoporphyrin IX heme site. Direct electron transfer has been shown between MP-11 and electrodes and therefore it can be used in biosensors.

c. **Oxidases:** Glucose oxidase ($GO_x$) facilitates the electron transfer from glucose to oxygen with the formation of gluconic acid and hydrogen peroxide. A direct electron transfer between GOx and the electrode has been achieved in few studies.Cai *et al.* [101] demonstrated enhanced direct electron transfer rate between immobilized $GO_x$ onto CNT electrode. The direct electron transfer reaction of $GO_x$ was found to be a two-electron-transfer coupled with a two-proton-transfer reaction process. The bioelectrocatalytic activity of $GO_x$ for the oxidation of glucose was found to be retained after immobilization on the surface of CNTs. Liu *et al.* [102] employed an electrode modified with the composite of carbon nanotubes/chitosan containing entrapped $GO_x$ and observed enhanced electron transfer rate. The bioelectrocatalytic activity of $GO_x$ was also found to be retained. Very recently, some workers demonstrated boron doped CNT modified glassy carbon electrode [103] and CdTe (3 nm) quantum dots modified CNT electrodes [104] as novel electrochemical biosensing platforms and achieved further higher sensitivity, stability and reproducibility.

### 5.5.3. Electrochemical DNA Sensors

Electrochemical DNA biosensors, owing to their high sensitivity, portability, low-cost and rapid response, are receiving much attention. These devices convert the nucleic acid recognition processes into a useful electrical signal. The immobilization of a single-stranded (ss-) DNA probe onto the transducer surface leads to a useful electrical signal. By employing CNTs based platforms for the development of electrochemical DNA biosensors, one can attain improved oxidation of the oxidablc bases, and the enhanced current responses of the indicators.The enhanced detection of target guanine or of the product of an enzyme label, as well as to the use of CNT carrier platforms offer such improvements. CNT-coated GC electrodes have shown to be more advantageous in comparison to the commonly used carbon electrodes such as unmodified GC, carbon paste and pencil graphite electrodes. A label-free electrical detection of DNA hybridization along with the significant amplification of the guanine signal has been achieved. Similar enhanced guanine signals were reported at MWCNT paste electrodes [105] and at a SWCNT-coated glassy-carbon electrode [106]. Ultrasensitive detection of DNA hybridization has been demonstrated by an array of vertically aligned MWCNT, entrenched in $SiO_2$ [107]. By the combination of CNT nanoelectrode array with $Ru(bpy)^{2+}_3$ mediated guanine oxidation, measurement of subattmoles of DNA targets could be possible (Fig. **2**). The detection of trace (mg/L) of nucleic acids have been achieved following short accumulation times.

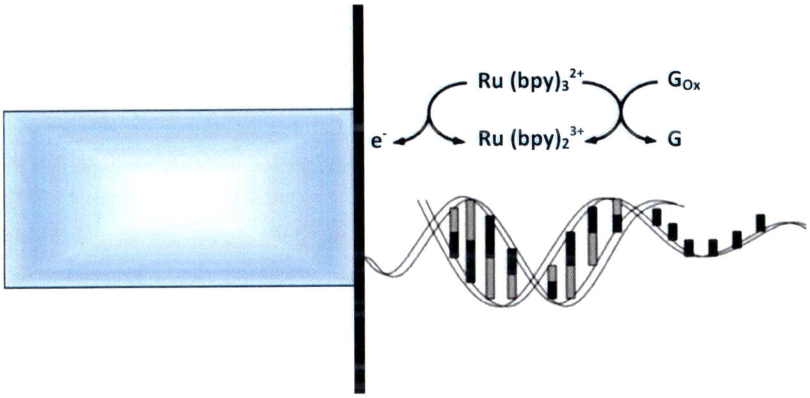


**Fig. (2).** Schematic mechanism of the $Ru(bpy)^{2+}_3$ mediated guanine oxidation for sensitive detection of DNA hybridization using CNT electrode array.

## 5.6. CONCLUSION

CNTs, a category of new carbonaceous materials, offer promising potential applications in electrochemical chemistry due to the unique properties demonstrated by the material which are significantly differentfrom other conventional scaled materials or nanoparticles. These excellent properties include superior and well-definedtubular structure, outstanding chemical stability, the flexible surface for functionalization and the strong electrocatalytic activity. Enhanced responses have been observed for electrodes modified with CNTs for molecules as small as $H_2O_2$ to as big as redox proteins. The small diameter and large aspect ratio of CNTs render them to be employed as molecular wires for allowing electron transfer between macromolecules and electrodes with ultra-sensitivity. However, to attain much higher sensitivities, stability and specificity from CNTs for the detection of macromolecules, it is very important to control the orientation of CNTs on the surface of electrode through simpler physical or chemical immobilisation methods, and rational functionalization of CNTs. The development of such methods is a matter of great urgency.

## CONFLICT OF INTEREST

The authors declare no conflict of interest, financial or otherwise.

## ACKNOWLEDGEMENTS

Authors would like to thank all referenced citations.

## REFERENCES

[1] Sinha N, Ma J, Yeow JT. Carbon nanotube-based sensors. J Nanosci Nanotechnol 2006; 6(3): 573-90. [http://dx.doi.org/10.1166/jnn.2006.121] [PMID: 16573108]

[2] Gong K, Yan Y, Zhang M, Su L, Xiong S, Mao L. Electrochemistry and electroanalytical applications of carbon nanotubes: a review. Anal Sci 2005; 21(12): 1383-93. [http://dx.doi.org/10.2116/analsci.21.1383] [PMID: 16379375]

[3] McCreery RL. Carbon electrodes: structural effects on electron transfer kinetics. Electroanal Chem 1991; 17: 221-374.

[4] Li J, Cassell A, Delzeit L, Han J, Meyyappan M. Novel three-dimensional electrodes: electrochemical properties of carbon nanotube ensembles. J Phys Chem B 2002; 106(36): 9299-305. [http://dx.doi.org/10.1021/jp021201n]

[5] Banks CE, Moore RR, Davies TJ, Compton RG. Investigation of modified basal plane pyrolytic graphite electrodes: definitive evidence for the electrocatalytic properties of the ends of carbon nanotubes. Chem Commun (Camb) 2004; (16): 1804-5. [http://dx.doi.org/10.1039/b406174h] [PMID: 15306892]

[6] Luo H, Shi Z, Li N, Gu Z, Zhuang Q. Investigation of the electrochemical and electrocatalytic behavior of single-wall carbon nanotube film on a glassy carbon electrode. Anal Chem 2001; 73(5): 915-20. [http://dx.doi.org/10.1021/ac000967l] [PMID: 11289436]

[7] Hu C, Hu S. Carbon nanotube-based electrochemical sensors: principles and applications in biomedical systems. Journal of Sensors 2009. [http://dx.doi.org/10.1155/2009/187615]

[8] Papakonstantinou P, Kern R, Robinson L, *et al.* Fundamental electrochemical properties of carbon nanotube electrodes. Fuller Nanotub Carbon Nanostruct 2005; 13(2): 91-108. [http://dx.doi.org/10.1081/FST-200050684]

[9] Britto PJ, Santhanam KS, Ajayan PM. Carbon nanotube electrode for oxidation of dopamine. Bioelectrochem Bioenerg 1996; 41(1): 121-5. [http://dx.doi.org/10.1016/0302-4598(96)05078-7]

[10] Wang Z, Liu J, Liang Q, Wang Y, Luo G. Carbon nanotube-modified electrodes for the simultaneous determination of dopamine and ascorbic acid. Analyst (Lond) 2002; 127(5): 653-8. [http://dx.doi.org/10.1039/b201060g] [PMID: 12081044]

[11] Rubianes MD, Rivas GA. Carbon nanotubes paste electrode. Electrochem Commun 2003; 5(8): 689-94.
[http://dx.doi.org/10.1016/S1388-2481(03)00168-1]

[12] Ye JS, Wen Y, De Zhang W, Gan LM, Xu GQ, Sheu FS. Selective voltammetric detection of uric acid in the presence of ascorbic acid at well□aligned carbon nanotube electrode. Electroanalysis 2003; 15(21): 1693-8.
[http://dx.doi.org/10.1002/elan.200302752]

[13] Wang J, Li M, Shi Z, Li N, Gu Z. Electrocatalytic oxidation of norepinephrine at a glassy carbon electrode modified with single wall carbon nanotubes. Electroanalysis 2002; 14(3): 225-30.
[http://dx.doi.org/10.1002/1521-4109(200202)14:3<225::AID-ELAN225>3.0.CO;2-I]

[14] Moore RR, Banks CE, Compton RG. Basal plane pyrolytic graphite modified electrodes: comparison of carbon nanotubes and graphite powder as electrocatalysts. Anal Chem 2004; 76(10): 2677-82.
[http://dx.doi.org/10.1021/ac040017q] [PMID: 15144174]

[15] Valentini F, Amine A, Orlanducci S, Terranova ML, Palleschi G. Carbon nanotube purification: preparation and characterization of carbon nanotube paste electrodes. Anal Chem 2003; 75(20): 5413-21.
[http://dx.doi.org/10.1021/ac0300237] [PMID: 14710820]

[16] Wang J, Musameh M. Carbon nanotube/teflon composite electrochemical sensors and biosensors. Anal Chem 2003; 75(9): 2075-9.
[http://dx.doi.org/10.1021/ac030007+] [PMID: 12720343]

[17] Musameh M, Wang J, Merkoci A, Lin Y. Low-potential stable NADH detection at carbon-nanotub--modified glassy carbon electrodes. Electrochem Commun 2002; 4(10): 743-6.
[http://dx.doi.org/10.1016/S1388-2481(02)00451-4]

[18] Wang J, Musameh M, Lin Y. Solubilization of carbon nanotubes by Nafion toward the preparation of amperometric biosensors. J Am Chem Soc 2003; 125(9): 2408-9.
[http://dx.doi.org/10.1021/ja028951v] [PMID: 12603125]

[19] Lin Y, Lu F, Tu Y, Ren Z. Glucose biosensors based on carbon nanotube nanoelectrode ensembles. Nano Lett 2004; 4(2): 191-5.
[http://dx.doi.org/10.1021/nl0347233]

[20] Erdem A, Karadeniz H, Caliskan A. Single-Walled Carbon Nanotubes Modified Graphite Electrodes for Electrochemical Monitoring of Nucleic Acids and Biomolecular Interactions. Electroanalysis 2009; 21(3-5): 464-71.
[http://dx.doi.org/10.1002/elan.200804422]

[21] Gao M, Dai L, Wallace GG. Biosensors based on aligned carbon nanotubes coated with inherently conducting polymers. Electroanalysis 2003; 15(13): 1089-94.
[http://dx.doi.org/10.1002/elan.200390131]

[22] Wang SG, Zhang Q, Wang R, *et al.* Multi-walled carbon nanotubes for the immobilization of enzyme in glucose biosensors. Electrochem Commun 2003; 5(9): 800-3.
[http://dx.doi.org/10.1016/j.elecom.2003.07.007]

[23] Azamian BR, Davis JJ, Coleman KS, Bagshaw CB, Green ML. Bioelectrochemical single-walled carbon nanotubes. J Am Chem Soc 2002; 124(43): 12664-5.
[http://dx.doi.org/10.1021/ja0272989] [PMID: 12392405]

[24] Xu JZ, Zhu JJ, Wu Q, Hu Z, Chen HY. An amperometric biosensor based on the coimmobilization of horseradish peroxidase and methylene blue on a carbon nanotubes modified electrode. Electroanalysis 2003; 15(3): 219-24.
[http://dx.doi.org/10.1002/elan.200390027]

[25] Callegari A, Cosnier S, Marcaccio M, *et al.* Functionalised single wall carbon nanotubes/polypyrrole composites for the preparation of amperometric glucose biosensors. J Mater Chem 2004; 14(5): 807-

10.
[http://dx.doi.org/10.1039/B316806A]

[26] Davis JJ, Green ML, Hill HA, *et al.* The immobilisation of proteins in carbon nanotubes. Inorg Chim Acta 1998; 272(1): 261-6.
[http://dx.doi.org/10.1016/S0020-1693(97)05926-4]

[27] Hall EA, Gooding JJ, Hall CE. Redox enzyme linked electrochemical sensors: Theory meets practice. MicrochimicaActa 1995; 121(1-4): 119-45.

[28] Yao Z, Kane CL, Dekker C. High-field electrical transport in single-wall carbon nanotubes. Phys Rev Lett 2000; 84(13): 2941-4.
[http://dx.doi.org/10.1103/PhysRevLett.84.2941] [PMID: 11018981]

[29] Yao Z, Dekker C, Avouris P. Electrical transport through single-wall carbon nanotubes InCarbon Nanotubes. Springer Berlin Heidelberg 2001; pp. 147-71.

[30] Baughman RH, Zakhidov AA, de Heer WA. Carbon nanotubes--the route toward applications. Science 2002; 297(5582): 787-92.

[31] Guo RX, Xu Q, Wang DY, Hu XY. Trace determination of clenbuterol with an MWCNT-Nafionnanocomposite modified electrode. MicrochimicaActa. 2008 Apr 1;161(1-2):265-72. [32] Odaci D, Telefoncu A, Timur S. Pyranose oxidase biosensor based on carbon nanotube (CNT)-modified carbon paste electrodes. Sens Actuators B Chem 2008; 132(1): 159-65.

[33] Wang J, Musameh M. Carbon nanotube/teflon composite electrochemical sensors and biosensors. Anal Chem 2003; 75(9): 2075-9.
[http://dx.doi.org/10.1021/ac030007+] [PMID: 12720343]

[33] Odaci D, Telefoncu A, Timur S. Pyranose oxidase biosensor based on carbon nanotube (CNT)-modified carbon paste electrodes. Sensors and Actuators B: Chemical 2008; 132(1): 159-65.

[34] Li Y, Umasankar Y, Chen SM. Polyaniline and poly(flavin adenine dinucleotide) doped multi-walled carbon nanotubes for p-acetamidophenol sensor. Talanta 2009; 79(2): 486-92.
[http://dx.doi.org/10.1016/j.talanta.2009.04.017] [PMID: 19559909]

[35] Guo M, Chen J, Li J, Tao B, Yao S. Fabrication of polyaniline/carbon nanotube composite modified electrode and its electrocatalytic property to the reduction of nitrite. Anal Chim Acta 2005; 532(1): 71-7.
[http://dx.doi.org/10.1016/j.aca.2004.10.045] [PMID: 19110118]

[36] Viswanathan S, Wu LC, Huang MR, Ho JA. Electrochemical immunosensor for cholera toxin using liposomes and poly(3,4-ethylenedioxythiophene)-coated carbon nanotubes. Anal Chem 2006; 78(4): 1115-21.
[http://dx.doi.org/10.1021/ac051435d] [PMID: 16478102]

[37] Cataldo V, Vaze A, Rusling JF. Improved Detection Limit and Stability of Amperometric Carbon Nanotube-Based Immunosensors by Crosslinking Antibodies with Polylysine. Electroanalysis 2008; 20(2): 115-22.
[http://dx.doi.org/10.1002/elan.200704040] [PMID: 18846266]

[38] Wang J, Musameh M, Lin Y. Solubilization of carbon nanotubes by Nafion toward the preparation of amperometric biosensors. J Am Chem Soc 2003; 125(9): 2408-9.
[http://dx.doi.org/10.1021/ja028951v] [PMID: 12603125]

[39] Liu A, Watanabe T, Honma I, Wang J, Zhou H. Effect of solution pH and ionic strength on the stability of poly(acrylic acid)-encapsulated multiwalled carbon nanotubes aqueous dispersion and its application for NADH sensor. Biosens Bioelectron 2006; 22(5): 694-9.
[http://dx.doi.org/10.1016/j.bios.2006.02.006] [PMID: 16564686]

[40] Zeng J, Gao X, Wei W, *et al.* Fabrication of carbon nanotubes/poly (1, 2-diaminobenzene) nanoporous composite *via* multipulsechronoamperometricelectropolymerization process and its electrocatalytic property toward oxidation of NADH. Sens Actuators B Chem 2007; 120(2): 595-602.

[http://dx.doi.org/10.1016/j.snb.2006.03.016]

[41] Tu X, Xie Q, Huang Z, Yang Q, Yao S. Synthesis and Characterization of Novel Quinone□Amine Polymer/Carbon Nanotubes Composite for Sensitive Electrocatalytic Detection of NADH. Electroanalysis 2007; 19(17): 1815-21.
[http://dx.doi.org/10.1002/elan.200703939]

[42] Valentini F, Amine A, Orlanducci S, Terranova ML, Palleschi G. Carbon nanotube purification: preparation and characterization of carbon nanotube paste electrodes. Anal Chem 2003; 75(20): 5413-21.
[http://dx.doi.org/10.1021/ac0300237] [PMID: 14710820]

[43] Zhuang Q, Chen J, Chen J, Lin X. Electrocatalytical properties of bergenin on a multi-wall carbon nanotubes modified carbon paste electrode and its determination in tablets. Sens Actuators B Chem 2008; 128(2): 500-6.
[http://dx.doi.org/10.1016/j.snb.2007.07.040]

[44] Abbaspour A, Mirzajani R. Electrochemical monitoring of piroxicam in different pharmaceutical forms with multi-walled carbon nanotubes paste electrode. J Pharm Biomed Anal 2007; 44(1): 41-8.
[http://dx.doi.org/10.1016/j.jpba.2007.01.027] [PMID: 17324550]

[45] Xiao P, Zhao F, Zeng B. Voltammetric determination of quercetin at a multi-walled carbon nanotubes paste electrode. Microchem J 2007; 85(2): 244-9.
[http://dx.doi.org/10.1016/j.microc.2006.06.004]

[46] Zheng L, Song J. Voltammetric behavior of urapidil and its determination at multi-wall carbon nanotube paste electrode. Talanta 2007; 73(5): 943-7.
[http://dx.doi.org/10.1016/j.talanta.2007.05.015] [PMID: 19073125]

[47] Xiao F, Ruan C, Li J, Liu L, Zhao F, Zeng B. Voltammetric Determination of Xanthine with a Single□Walled Carbon Nanotube□Ionic Liquid Paste Modified Glassy Carbon Electrode. Electroanalysis 2008; 20(4): 361-6.
[http://dx.doi.org/10.1002/elan.200704042]

[48] Fan S, Xiao F, Liu L, Zhao F, Zeng B. Sensitive voltammetric response of methylparathion on single-walled carbon nanotube paste coated electrodes using ionic liquid as binder. Sens Actuators B Chem 2008; 132(1): 34-9.
[http://dx.doi.org/10.1016/j.snb.2008.01.010]

[49] Gong K, Zhang M, Yan Y, *et al.* Sol-gel-derived ceramic-carbon nanotube nanocomposite electrodes: tunable electrode dimension and potential electrochemical applications. Anal Chem 2004; 76(21): 6500-5.
[http://dx.doi.org/10.1021/ac0492867] [PMID: 15516147]

[50] Salimi A, Compton RG, Hallaj R. Glucose biosensor prepared by glucose oxidase encapsulated sol-gel and carbon-nanotube-modified basal plane pyrolytic graphite electrode. Anal Biochem 2004; 333(1): 49-56.
[http://dx.doi.org/10.1016/j.ab.2004.06.039] [PMID: 15351279]

[51] Zhu L, Tian C, Zhai J, Yang R. Sol–gel derived carbon nanotubes ceramic composite electrodes for electrochemical sensing. Sens Actuators B Chem 2007; 125(1): 254-61.
[http://dx.doi.org/10.1016/j.snb.2007.02.012]

[52] Zhang M, Yan Y, Gong K, Mao L, Guo Z, Chen Y. Electrostatic layer-by-layer assembled carbon nanotube multilayer film and its electrocatalytic activity for O2 reduction. Langmuir 2004; 20(20): 8781-5.
[http://dx.doi.org/10.1021/la048923l] [PMID: 15379506]

[53] Gavalas VG, Law SA, Christopher Ball J, Andrews R, Bachas LG. Carbon nanotube aqueous sol-gel composites: enzyme-friendly platforms for the development of stable biosensors. Anal Biochem 2004; 329(2): 247-52.
[http://dx.doi.org/10.1016/j.ab.2004.02.025] [PMID: 15158483]

[54] Zhai X, Wei W, Zeng J, Gong S, Yin J. Layer-by-layer assembled film based on chitosan/carbon nanotubes, and its application to electrocatalytic oxidation of NADH. MicrochimicaActa 2006; 154(3-4): 315-20.

[55] Mamedov AA, Kotov NA, Prato M, Guldi DM, Wicksted JP, Hirsch A. Molecular design of strong single-wall carbon nanotube/polyelectrolyte multilayer composites. Nat Mater 2002; 1(3): 190-4. [http://dx.doi.org/10.1038/nmat747] [PMID: 12618809]

[56] Rouse JH, Lillehei PT. Electrostatic assembly of polymer/single walled carbon nanotube multilayer films. Nano Lett 2003; 3(1): 59-62. [http://dx.doi.org/10.1021/nl025780j]

[57] Liu G, Lin Y. Amperometric glucose biosensor based on self-assembling glucose oxidase on carbon nanotubes. Electrochem Commun 2006; 8(2): 251-6. [http://dx.doi.org/10.1016/j.elecom.2005.11.015]

[58] Zhao H, Ju H. Multilayer membranes for glucose biosensing *via* layer-by-layer assembly of multiwall carbon nanotubes and glucose oxidase. Anal Biochem 2006; 350(1): 138-44. [http://dx.doi.org/10.1016/j.ab.2005.11.034] [PMID: 16430853]

[59] Liu Y, Wu S, Ju H, Xu L. Amperometric glucose biosensing of gold nanoparticles and carbon nanotube multilayer membranes. Electroanalysis 2007; 19(9): 986-92. [http://dx.doi.org/10.1002/elan.200603814]

[60] Zhao W, Xu JJ, Shi CG, Chen HY. Multilayer membranes *via* layer-by-layer deposition of organic polymer protected Prussian blue nanoparticles and glucose oxidase for glucose biosensing. Langmuir 2005; 21(21): 9630-4. [http://dx.doi.org/10.1021/la051370+] [PMID: 16207046]

[61] Huang J, Yang Y, Shi H, *et al.* Multi-walled carbon nanotubes-based glucose biosensor prepared by a layer-by-layer technique. Mater Sci Eng C 2006; 26(1): 113-7. [http://dx.doi.org/10.1016/j.msec.2005.06.002]

[62] Hrapovic S, Majid E, Liu Y, Male K, Luong JH. Metallic nanoparticle-carbon nanotube composites for electrochemical determination of explosive nitroaromatic compounds. Anal Chem 2006; 78(15): 5504-12. [http://dx.doi.org/10.1021/ac060435q] [PMID: 16878889]

[63] Liu H, Wang G, Chen D, Zhang W, Li C, Fang B. Fabrication of polythionine/NPAu/MWNTs modified electrode for simultaneous determination of adenine and guanine in DNA. Sens Actuators B Chem 2008; 128(2): 414-21. [http://dx.doi.org/10.1016/j.snb.2007.06.028]

[64] Yang P, Wei W, Tao C. Determination of trace thiocyanate with nano-silver coated multi-walled carbon nanotubes modified glassy carbon electrode. Anal Chim Acta 2007; 585(2): 331-6. [http://dx.doi.org/10.1016/j.aca.2007.01.011] [PMID: 17386682]

[65] Xiao F, Zhao F, Li J, Yan R, Yu J, Zeng B. Sensitive voltammetric determination of chloramphenicol by using single-wall carbon nanotube-gold nanoparticle-ionic liquid composite film modified glassy carbon electrodes. Anal Chim Acta 2007; 596(1): 79-85. [http://dx.doi.org/10.1016/j.aca.2007.05.053] [PMID: 17616243]

[66] Lin X, Li Y. A sensitive determination of estrogens with a Pt nano-clusters/multi-walled carbon nanotubes modified glassy carbon electrode. Biosens Bioelectron 2006; 22(2): 253-9. [http://dx.doi.org/10.1016/j.bios.2006.01.005] [PMID: 16487699]

[67] Xue H, Zhao J, Wang T, Guo H, Fana X, He J. *In situ* synthesis of graphene/carbon nanotube modified ordered mesoporous carbon as protective film of stainless steel bipolar plates for proton exchange membrane fuel. RSC Advances electroanalysis 2014; 37: 39.

[68] Adams ML, Lavasanifar A, Kwon GS. Amphiphilic block copolymers for drug delivery. J Pharm Sci 2003; 92(7): 1343-55.

[http://dx.doi.org/10.1002/jps.10397] [PMID: 12820139]

[69] Krajcik R, Jung A, Hirsch A, Neuhuber W, Zolk O. Functionalization of carbon nanotubes enables non-covalent binding and intracellular delivery of small interfering RNA for efficient knock-down of genes. Biochem Biophys Res Commun 2008; 369(2): 595-602. [http://dx.doi.org/10.1016/j.bbrc.2008.02.072] [PMID: 18298946]

[70] Zhao Y, Yang X, Tian J, Wang F, Zhan L. A facile and novel approach toward synthetic polypyrrole oligomers functionalization of multi-walled carbon nanotubes as PtRu catalyst support for methanol electro-oxidation. J Power Sources 2010; 195(15): 4634-40. [http://dx.doi.org/10.1016/j.jpowsour.2010.02.023]

[71] Li L, Li CY, Ni C. Polymer crystallization-driven, periodic patterning on carbon nanotubes. J Am Chem Soc 2006; 128(5): 1692-9. [http://dx.doi.org/10.1021/ja056923h] [PMID: 16448143]

[72] Wang WR, Xie XM, Ye XY. Crystallization induced block copolymer decoration on carbon nanotubes. Carbon 2010; 48(5): 1680-3. [http://dx.doi.org/10.1016/j.carbon.2009.12.059]

[73] Bhirde AA, Patel S, Sousa AA, *et al.* Distribution and clearance of PEG-single-walled carbon nanotube cancer drug delivery vehicles in mice. Nanomedicine (Lond) 2010; 5(10): 1535-46. [http://dx.doi.org/10.2217/nnm.10.90] [PMID: 21143032]

[74] Prencipe G, Tabakman SM, Welsher K, *et al.* PEG branched polymer for functionalization of nanomaterials with ultralong blood circulation. J Am Chem Soc 2009; 131(13): 4783-7. [http://dx.doi.org/10.1021/ja809086q] [PMID: 19173646]

[75] Nakayama-Ratchford N, Bangsaruntip S, Sun X, Welsher K, Dai H. Non-covalent functionalization of carbon nanotubes by fluorescein-polyethylene glycol: supramolecular conjugates with pH-dependent absorbance and fluorescence. J Am Chem Soc 2007; 129(9): 2448-9. [http://dx.doi.org/10.1021/ja068684j] [PMID: 17284037]

[76] Hecht DS, Ramirez RJ, Briman M, *et al.* Bioinspired detection of light using a porphyrin-sensitized single-wall nanotube field effect transistor. Nano Lett 2006; 6(9): 2031-6. [http://dx.doi.org/10.1021/nl061231s] [PMID: 16968021]

[77] Hu L, Zhao YL, Ryu K, Zhou C, Stoddart JF, Grüner G. Light-Induced Charge Transfer in Pyrene/CdSe-SWNT Hybrids. Adv Mater 2008; 20(5): 939-46. [http://dx.doi.org/10.1002/adma.200701125]

[78] Shim M, Shi Kam NW, Chen RJ, Li Y, Dai H. Functionalization of carbon nanotubes for biocompatibility and biomolecular recognition. Nano Lett 2002; 2(4): 285-8. [http://dx.doi.org/10.1021/nl015692j]

[79] Yan LY, Poon YF, Chan-Park MB, Chen Y, Zhang Q. Individually dispersing single-walled carbon nanotubes with novel neutral pH water-soluble chitosan derivatives. J Phys Chem C 2008; 112(20): 7579-87. [http://dx.doi.org/10.1021/jp711039s]

[80] Goodwin AP, Tabakman SM, Welsher K, Sherlock SP, Prencipe G, Dai H. Phospholipid-dextran with a single coupling point: a useful amphiphile for functionalization of nanomaterials. J Am Chem Soc 2009; 131(1): 289-96. [http://dx.doi.org/10.1021/ja807307e] [PMID: 19061329]

[81] Ikeda A, Tanaka Y, Nobusawa K, Kikuchi J. Solubilization of single-walled carbon nanotubes by supramolecular complexes of barbituric acid and triaminopyrimidines. Langmuir 2007; 23(22): 10913-5. [http://dx.doi.org/10.1021/la702747r] [PMID: 17894514]

[82] Liu L, Wang T, Li J, *et al.* Self-assembly of gold nanoparticles to carbon nanotubes using a thiol-terminated pyrene as interlinker. Chem Phys Lett 2003; 367(5): 747-52.

[http://dx.doi.org/10.1016/S0009-2614(02)01789-X]

[83] Sadegh H, Shahryari-ghoshekandi R, Kazemi M. Study in synthesis and characterization of carbon nanotubes decorated by magnetic iron oxide nanoparticles. Int Nano Lett 2014; 4(4): 129-35. [http://dx.doi.org/10.1007/s40089-014-0128-1]

[84] Yuan W, Lu S, Xiang Y, Jiang SP. Pt-based nanoparticles on non-covalent functionalized carbon nanotubes as effective electrocatalysts for proton exchange membrane fuel cells. RSC Advances 2014; 4(86): 46265-84. [http://dx.doi.org/10.1039/C4RA05120C]

[85] Chen G, Zhang L, Wang J. Miniaturized capillary electrophoresis system with a carbon nanotube microelectrode for rapid separation and detection of thiols. Talanta 2004; 64(4): 1018-23. [http://dx.doi.org/10.1016/j.talanta.2004.04.022] [PMID: 18969705]

[86] Cao XN, Lin L, Xian YZ, Zhang W, Xie YF, Jin LT. *In vivo* Monitoring of the Thiols in Rat Striatum by Liquid Chromatography with Amperometric Detection at a Functionalized Multi□Wall Carbon Nanotubes Modified Electrode. Electroanalysis 2003; 15(10): 892-7. [http://dx.doi.org/10.1002/elan.200390111]

[87] Wang J. Electrochemical detection of carbohydrates at carbon-nanotube modified glassy-carbon electrodes. Electrochem Commun 2004; 6(3): 284-7. [http://dx.doi.org/10.1016/j.elecom.2004.01.003]

[88] Male KB, Hrapovic S, Liu Y, Wang D, Luong JH. Electrochemical detection of carbohydrates using copper nanoparticles and carbon nanotubes. Anal Chim Acta 2004; 516(1): 35-41. [http://dx.doi.org/10.1016/j.aca.2004.03.075]

[89] Tang X, Liu Y, Hou H, You T. Electrochemical determination of L-Tryptophan, L-Tyrosine and L-Cysteine using electrospun carbon nanofibers modified electrode. Talanta 2010; 80(5): 2182-6. [http://dx.doi.org/10.1016/j.talanta.2009.11.027] [PMID: 20152470]

[90] Wu FH, Zhao GC, Wei XW, Yang ZS. Electrocatalysis of tryptophan at multi-walled carbon nanotube modified electrode. MicrochimicaActa 2004; 144(4): 243-7.

[91] Deo RP, Lawrence NS, Wang J. Electrochemical detection of amino acids at carbon nanotube and nickel-carbon nanotube modified electrodes. Analyst (Lond) 2004; 129(11): 1076-81. [http://dx.doi.org/10.1039/B407418A] [PMID: 15508037]

[92] Hu C, Hu S. Carbon nanotube-based electrochemical sensors: principles and applications in biomedical systems. Journal of Sensors 2009. [http://dx.doi.org/10.1155/2009/187615]

[93] Wang J, Musameh M, Lin Y. Solubilization of carbon nanotubes by Nafion toward the preparation of amperometric biosensors. J Am Chem Soc 2003; 125(9): 2408-9. [http://dx.doi.org/10.1021/ja028951v] [PMID: 12603125]

[94] Guo M, Chen J, Li J, Nie L, Yao S. Carbon Nanotubes□Based Amperometric Cholesterol Biosensor Fabricated Through Layer□by□Layer Technique. Electroanalysis 2004; 16(23): 1992-8. [http://dx.doi.org/10.1002/elan.200403053]

[95] Tsai YC, Chen SY, Lee CA. Amperometric cholesterol biosensors based on carbon nanotube–chitosan–platinum–cholesterol oxidase nanobiocomposite. Sens Actuators B Chem 2008; 135(1): 96-101. [http://dx.doi.org/10.1016/j.snb.2008.07.025]

[96] Xu Z, Chen X, Qu X, Jia J, Dong S. Single-wall carbon nanotube-based voltammetric sensor and biosensor. Biosens Bioelectron 2004; 20(3): 579-84. [http://dx.doi.org/10.1016/j.bios.2004.03.001] [PMID: 15494242]

[97] Davis JJ, Coles RJ, Allen H, Hill O. Protein electrochemistry at carbon nanotube electrodes. J Electroanal Chem 1997; 440(1): 279-82.

[98] Pakapongpan S, Palangsuntikul R, Surareungchai W. Electrochemical sensors for hemoglobin and myoglobin detection based on methylene blue-multiwalled carbon nanotubes nanohybrid-modified glassy carbon electrode. Electrochim Acta 2011; 56(19): 6831-6. [http://dx.doi.org/10.1016/j.electacta.2011.05.089]

[99] Zhao YD, Zhang WD, Chen H, Luo QM, Li SF. Direct electrochemistry of horseradish peroxidase at carbon nanotube powder microelectrode. Sens Actuators B Chem 2002; 87(1): 168-72. [http://dx.doi.org/10.1016/S0925-4005(02)00232-0]

[100] Cui R, Huang H, Yin Z, Gao D, Zhu JJ. Horseradish peroxidase-functionalized gold nanoparticle label for amplified immunoanalysis based on gold nanoparticles/carbon nanotubes hybrids modified biosensor. Biosens Bioelectron 2008; 23(11): 1666-73. [http://dx.doi.org/10.1016/j.bios.2008.01.034] [PMID: 18359217]

[101] Cai C, Chen J. Direct electron transfer of glucose oxidase promoted by carbon nanotubes. Anal Biochem 2004; 332(1): 75-83. [http://dx.doi.org/10.1016/j.ab.2004.05.057] [PMID: 15301951]

[102] Liu Y, Wang M, Zhao F, Xu Z, Dong S. The direct electron transfer of glucose oxidase and glucose biosensor based on carbon nanotubes/chitosan matrix. Biosens Bioelectron 2005; 21(6): 984-8. [http://dx.doi.org/10.1016/j.bios.2005.03.003] [PMID: 16257668]

[103] Deng C, Chen J, Chen X, Xiao C, Nie L, Yao S. Direct electrochemistry of glucose oxidase and biosensing for glucose based on boron-doped carbon nanotubes modified electrode. Biosens Bioelectron 2008; 23(8): 1272-7. [http://dx.doi.org/10.1016/j.bios.2007.11.009] [PMID: 18178424]

[104] Liu Q, Lu X, Li J, Yao X, Li J. Direct electrochemistry of glucose oxidase and electrochemical biosensing of glucose on quantum dots/carbon nanotubes electrodes. Biosens Bioelectron 2007; 22(12): 3203-9. [http://dx.doi.org/10.1016/j.bios.2007.02.013] [PMID: 17416515]

[105] Pedano L, Rivas GA. Adsorption and electrooxidation of nucleic acids at carbon nanotubes paste electrodes. Electrochem Commun 2004; 6(1): 10-6. [http://dx.doi.org/10.1016/j.elecom.2003.10.008]

[106] Wang J, Li M, Shi Z, Li N, Gu Z. Electrochemistry of DNA at single□wall carbon nanotubes. Electroanalysis 2004; 16(1□2): 140-4. [http://dx.doi.org/10.1002/elan.200302915]

[107] Li J, Ng HT, Cassell A, *et al.* Carbon nanotube nanoelectrode array for ultrasensitive DNA detection. Nano Lett 2003; 3(5): 597-602. [http://dx.doi.org/10.1021/nl0340677]

## CHAPTER 6

# Carbon Nanotubes in Drug Delivery

**Mohammad Faiyaz Anwar**[1,*], **Neelam Yadav**[2] and **Swati Jain**[3]

[1] *National Post-Doctoral Fellow, Renal Pathology Lab,Department of Pathology,AIIMS, New Delhi-29, India*

[2] *Centre for Biotechnology, Maharshi Dayanand University Rohtak, Rohtak, Haryana, India*

[3] *Amity Institute of Nanotechnology, Amity University, Sec. 125, Noida, Uttar Pradesh, India*

**Abstract:** Carbon nanotubes are one of the most commonly used nanomaterials in research field and consumer industries. They have attained popularity in biomedical applications pertaining to novel drug delivery carriers, as building blocks for tissue regeneration as well as probes for bio-imaging and bio-sensing. Both multi-walled and single walled CNTs have been employed for therapeutic activity for the treatment of several diseases such as cancer therapy, Neurodegenerative Diseases and Alzheimer Syndrome *via* multiple approaches. Potential application of CNTs as nano-carrier is their ability to translocate through cellular membranes, facilitate their use for the delivery of therapeutically active molecules resembling cell-penetrating peptides. This chapter highlights the importance of CNTs as a vehicle for delivering wide variety of payloads including drugs, small organic molecules, oligonucleotides, proteins, siRNA, vaccines and nutrients. We have focussed on demonstrating their synthesis and functionalization as important procedural steps for creating a smart and intelligent system for delivery.

**Keywords:** Biodegradation, Biosafety, Cancer therapy, Cargoes, Drug delivery, Encapsulation, Encapsulation efficiency, Genes, MWCNTs, Nano-carrier, Neurons, Nutrients, Personalized therapeutics, Proteins, PL-PEG, Smart vehicle, Surface functionalization, SWNTs.

## 6.1. INTRODUCTION

Researchers are continuously trying to search an effective drug delivery system (DDS) for improving the pharmacological and therapeutic properties of administered drugs. Carbon nano tubes are tiny hollow cylinders of one-ato- -thick carbon sheets and exhibit exceptional applications because of their strength, elasticity, and other significant characteristics [1]. CNTs have showed higher

* **Corresponding author Mohammad Faiyaz Anwar**: National Post-Doctoral Fellow, Renal Pathology Lab, Department of Pathology, All India Institute of Medical Sciences (AIIMS), New Delhi-29, India; Tel: +918860816199; E-mail: faiyazinnanotechnology@gmail.com

mechanical, thermal and electronic properties when they are compared to several other artificial nanomaterials. They can be further activated by using functional bioactive groups like polypeptides, oligonucleotides and drugs to transport their cargos to the target cells and organs [2, 3]. Several biomolecules such as drugs [4, 5], proteins [6], peptide [7], plasmid DNA [8] and small interfering RNA siRNA) [9, 10] have been effectively delivered by SWCNTs by the process known as endocytosis [11].

## 6.2. PHYSICOCHEMICAL STRUCTURE OF CNTS

Carbon nanotubes (CNTs) are cylindrical structures composed of hexagonal arrangement of $SP^2$ hybridized carbon atoms having bond distance1.4′ ° A. The wall of CNTs may be comprised of one or more layers of graphene sheets. Therefore, this arrangement of graphene sheets helps in discrimination between single-walled carbon nanotubes (SWNT) which is formed by rolling up of a single graphene sheet and multi-walled CNTs (MWCNTs) are formed by rolling up of more than one graphene sheet [12]. Both ends of CNTs are capped with the help of SWNTs and MWCNTs this hemisphere carbon arrangement is called as fullerenes wrapped by graphene layer (Fig. **1**). The interlayer distance between the graphene sheets of MWCNTs is .34nm, each forming an individual tube and the outer diameter ranges from 2.5 to 100 nm, whereas SWNTs comprised of this value 0.6 to 2.4 nm [13].

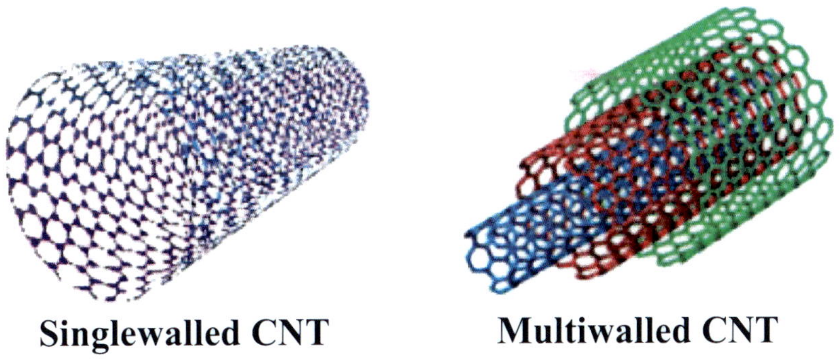


**Fig. (1).** Structure of the different types of carbon nanotube.

In past years, CNTs have attracted much more attention of scientific community due to their unique physical and chemical properties and have capability to use in various branches of science and engineering. These CNTs posses an exceptional combination of mechanical, thermal and electronic properties; hence they have been considered as an ideal nanotubes in various applications as well as a test bed for fundamental science [14]. Researchers have observed the usefulness of CNTs conductivity and large surface to volume ratio to produce conductive plastics with exceedingly low percolation thresholds. CNTs have also been used in conductive, electromagnetic and microwave absorbing and high-strength composites, fibers, sensors, inks, energy storage and energy conversion devices, radiation sources and nanometer-sized semiconductor devices, probes, and interconnect using carbon nanotubes [15]. Carbon nanotubes exhibit impressive mechanical properties with Young's modulus in the range 100-1000 GPa and strengths between 2.5 and 3.5 GPa [16].Nanotubes possess diameter ranges from 1 to 100 nm and lengths of up to millimetres (Fig. **2**). Their densities can be as low as ~1.3 g / cm3 and their Young's moduliare superior to all carbon fibres with valves [17].

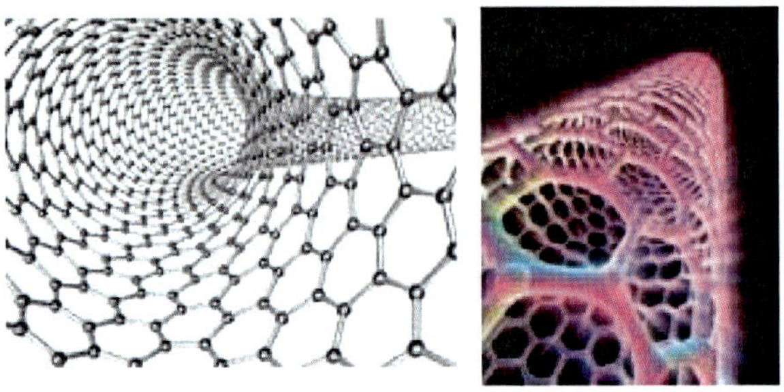

**Fig. (2).** Structure of the carbon nanotube.

## 6.3. SYNTHESIS OF CARBON NANOTUBE

Characterization of nanotubes and devices based on nanotubes has been investigated over the years. Characteristics of basic research and practical application requirements have been driving and motivating synthetic methods for improved and harmonized materials. Due to high sensitivity of carbon nanotubes has improved the associations among diverse field of sciences and in turn enhanced the applications. Hence, it is necessary to understand the mechanism of how to synthesize nanotubes.

There are three approaches to synthesize CNT such as (i) Arc-discharge method which involves evaporation of graphite electrodes in electric arcs that require very high (~4000°C) temperatures. Though, arc-shaped CNTs are well crystallized, they are impure; about 60–70% of the arc-grown product contains metal particles and amorphous carbon [18]. (ii) Laser-vaporization technique involves evaporation of high-purity graphite target by using high intensity lasers in combination with high-temperature furnaces [19]. These laser-grown CNTs are of highly pure but their production yield is very low (in milli gram order). Thus, both of these approaches are poor in terms of energy use efficiency as well as in resources. (iii) Chemical vapour deposition (CVD), is most popular approach for the synthesis of CNTs. It integrate catalyst-assisted thermal decomposition of hydrocarbons, and offers many benefits such as economic and scalable technique for bulk production of CNTs [20] Here, we will review the materials aspects of CNT synthesis by CVD and discuss the CNT growth mechanism in the light of latest progresses in the field.

Fig. (**3**) shows a schematic illustration of the experimental set-up used for CNT growth by CVD method in its simplest form. This method requires decomposition of hydrocarbon by passing a hydrocarbon vapor (typically 15–60 min) *via* a tubular reactor in which a catalyst material is present at adequate high temperature *i.e.* 600–1200°C. CNT formation occurs on the catalyst in the reactor, which are collected after cooling the system to room temperature. While in liquid hydrocarbon (benzene, alcohol, *etc.*), the liquid is heated in a flask and an inert gas is flushed out which carries the hydrocarbon vapor into the reaction zone. If a solid hydrocarbon is used for CNT synthesis which is directly kept in the low-temperature zone of the reaction tube. Volatile materials including camphor, naphthalene, ferrocence *etc.* directly turn from solid to vapor, and perform CVD while passing over the catalyst kept in the high-temperature zone. Similar to the CNT precursors, catalyst precursors in CVD may also be employed in any form such as solid, liquid or gas, which are placed inside the reactor or fed from outside. Pyrolysis of the catalyst vapor at an appropriate temperature produces metal nanoparticles in-situ this method is called as floating catalyst method. Instead, catalyst-coated substrates could also be placed in the hot zone of the furnace to catalyze the CNT growth.

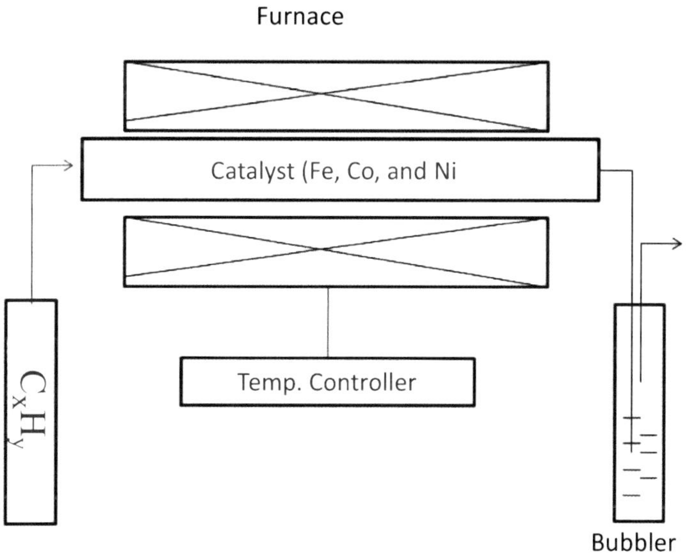


**Fig. (3).** Schematic diagram of a Chemical vapour deposition method for synthesis of carbon nanotube.

## 6.4. GROWTH MECHANISM

Different growth methods use nano tubes for growth but the phenomenon of growth is similar to almost all the growth methods. However, the precise methods by which nanotubes are formed are still not well understood. The mechanism includes three steps of growth of nanotubes which has been depicted in Fig. (**4**). First of all there is a formation of round or pear-shaped precursor on the surface of the metal catalyst. In step two involves the diffusion of carbon from the sides of precursor however, the top of precursor part is free/empty this is the reason why CNTs are hollow. Upward growth of hollow rod like nanotubes starts from the base of metal particle which is attached to the substrate which is followed by the detachment of tip of particle and persists on the growing nanotubes. Depending on the size of the catalyst particles, single walled or multi-walled nanotubes are grown [21].

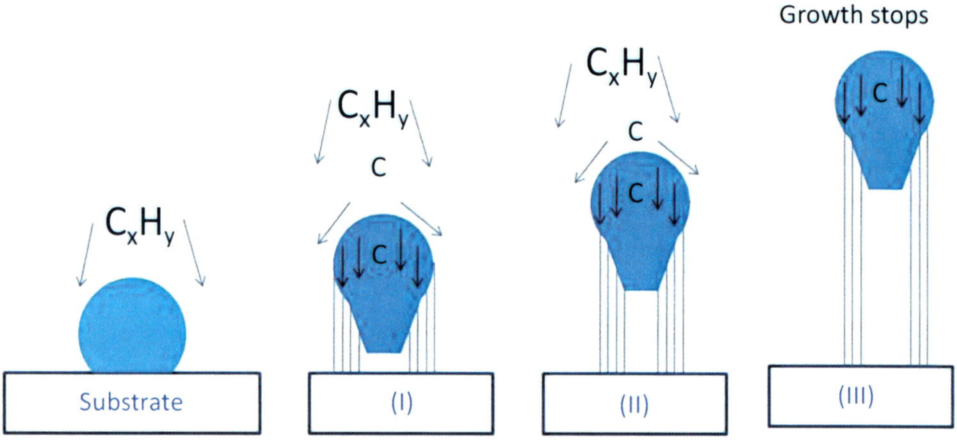


**Fig. (4).** Diagrammatic illustration of widely-accepted growth mechanisms for CNTs: tip-growth model.

## 6.5. FUNCTIONALIZATION OF CNTS

CNTs have been used as drug carrier because they are soluble in aqueous environment which is significant for gastrointestinal absorption, blood transportation, secretion, and biocompatibility and many more. Likewise, it is necessary that dispersion of such CNT dispersions should be homogeneous and stable at a suitable temperature to obtain exact concentration data. However, solubilisation of pristine CNTs in aqueous solvents is one of the major obstacles for them to use as practical drug carriers rather hydrophobic character of the graphene side walls, coupled with the strong π-π interactions between the individual tubes. These characteristics features promote the conjugation of CNTs into bundles.

Proper dispersion of CNTs requires the medium that should be capable of both wetting the hydrophobic tube surfaces and changing the tube surfaces to reduce tube's bundle formation. This has been achieved by Foldvari *et al.* four basic approaches [22] *i.e.* surfactant-assisted dispersion, solvent dispersion, functionalization of side walls, and biomolecular dispersion. Among these approaches, functionalization is the most effective approach. Furthermore, functionalization has exhibited decrease cytotoxicity, improved biocompatibility, and giving opportunity to appendage molecules of drugs, proteins, or genes for developing efficient delivery systems [23].

## 6.6. DRUG DELIVERY

Nanotubes technology has played an important role in drug delivery. An effective drug delivery system enhances the physiological and therapeutic properties of a cargo drug molecule. Carbon nanotubes have played an imperative role medical research, in effective delivery of drug and have also been used in bio-sensing methods for the treatment of disease and health monitoring. Delivery of drug has been accompanied by using nanomaterials like dendrimers, polymers, and liposomes, but carbon nanotubes offer opportunity to work with efficient structures that are having high drug loading capacities and good cell penetration qualities [24]. Recently, carbon nano tubes have attracted researchers in the field of medicine as CNT technology has potential to modify drug delivery and bio-sensing methods for getting the better results. Various methods have been used for transport of molecules with the help of nanotubes [25]. Drugs attached on the external surface of functionalized CNTs *via* covalent mode or non-covalent using hydrophobic and electrostatic interactions or by pi-pi stacking or encapsulated inside their hollow core which allows encapsulation of both low and high molecular weight molecules. [26]. It also allows encapsulation of both hydrophilic as well as lipophilic drugs. In multi drug therapy more than one drug loaded in CNTs. Ligands and diagnostic materials complex to surface of CNTs by functionalization to target the drugs to precise action. CNT facilitate the controlled release drug system to release the loaded drugs at the precise time [27].

## 6.7. *IN VIVO* BEHAVIOUR OF CNTS

Application of CNT's in drug delivery and bio-sensing technology has the potential to transform medicine. Functionalization of SWNT's has enhanced solubility and facilitate efficient tumor targeting/drug delivery [28] that prevents SWNT's cytotoxicity and changes the function of immune cells. Cancer is a collection of diseases in which cells grow and divide infinitely. Present, cancer therapy mainly involves surgery, radiation therapy, and chemotherapy. However, these methods of treatment are painful and kill normal cells in addition to producing adverse side effects. CNTs as drug delivery vehicles have shown benefits in targeting specific cancer cells with a dosage lower than traditional drugs used that is just as effective in killing the cells; however dosages harm healthy cells and potentially decreases the side effects. Existing blood glucose monitoring methods by patients suffering from diabetics are normally invasive and often painful. For instance, one method involves a continuous glucose sensor incorporated into small needle which is inserted under the skin to monitor glucose levels every few days. Blood glucose level has also been determined by glucose. However, these methods are not only invasive but also produce inaccurate results. It was shown that 70% of glucose readings obtained by continuous glucose sensors differed by 10% more and 7% differed by over 50%. Single-walled nanotubes and multi-walled nanotubes have high electrochemically accessible surface area, high electrical conductivity and useful structural properties, highly sensitive and hence used in non-invasive glucose detectors.

## 6.8. SMALL MOLECULES LOADED IN CARBON NANOTUBES SYSTEM

The majority of the anticancer drugs are tiny molecules that can be encapsulated into or onto CNTs by physical adsorption *via* p-p stacking interactions between pseudo-aromatic double bonds of the graphene sheet and the drug molecules. These anticancer drugs can also be encapsulated with CNTs through covalent immobilization of the specific drug molecules onto the reactive functional groups confined to sidewalls of CNTs [29 - 31].

Recently, Borowiak-Palen *et al.* reported that cisplatin, a tiny molecule, loaded into SWCNTs having diameter of 1.3 to 1.6 nm [32]. The cisplatin integrated with tubes was confirmed by Raman spectroscopy, infrared spectroscopy, and high-resolution transmission electron microscopy (TEM). Release of drug using dialysis membrane method revealed that cisplatin was frequently released for about a week, with maximum release during 72 h and up to 1 week. Thermo gravimetric analysis was used for the encapsulated drug with SWCNTs (21 μg of per 100 μg). Cytotoxicity investigations were carried out on DU145 and PC3 human prostate cancer cell lines using 3-(4,5)-dimethylthiahiazo (-z-y1)-3,5-di-phenytetrazoliumromide (MTT) cell proliferation assay demonstrated that the cell viability is inversely proportional to increased concentration of the CNT-based nano-vector, whereas blank CNTs did not show significant results. Computational methods showed the possible interactions between CNTs and drug molecules [33]. For cisplatin acceptance or incorporation, CNTs must have a radius of at least 4.785 Å (0.4785 nm) while generally experimentally used CNTs have diameters larger than 4.785 Å. As a result, it is inferred that cisplatin is expected to be encapsulated inside the nanotubes [34].

Doxorubicin load on CNT to form supra-molecular complexes based on p-p stacking interactions by homogenising the drug with an aqueous dispersion of CNTs stabilized by Pluronic F127 (nonionic surfactant). Loading of doxorubicin with MWCNTs was observed by measuring the emission spectrum of doxorubicin through fluorescence spectrophotometry. The fluorescence intensity of doxorubicin significantly declined with the increase in the concentration of MWCNTs from 5 to 20 μg/ml, having final concentration in the suspension remain constant (10 μg/ml), which is required for quenching of fluorescence. It was found that a mass ratio of 1:2 is optimum for maximum interaction/quenching ratio. TEM structural characterization showed that CNTs present as well-individualized, dispersed nanotubes, confirming the polymer molecules' ability to solubilise the CNTs efficiently. On MCF-7 human breast cancer cell line, it was found that the doxorubicin-MWCNT conjugate shows improved cytotoxicity as compared to both doxorubicin only and doxorubicin-Pluronic conjugates. This improved cytotoxicity obtained with the doxorubicin-MWCNT complex reveals that MWCNTs can efficiently enhance the transport of doxorubicin and thus, improve the cellular uptake of the drug [35].

Antioxidants have been considered to play a significant role in cancer therapy due to their ability to combat oxidative stress. However, their poor dispersion alleviates the reaping of the advantages from these compounds. Covalently PEGylated ultra-short SWCNTs can be conjugated to the antioxidant, amino butylated hydroxy toluene *via*, electrostatic interactions by simple thrilling of the mixture. Remaining carboxylic acid groups on the oxidized CNT permit the ionic interactions with the amine group of the butylated hydroxy toluene. The formulation was estimated by oxygen radical absorbance capacity assay, in which a fluorescent probe's loss of fluorescent intensity is monitored in the presence of oxygen radicals. Oxygen-radical scavengers hold the fluorescence of the probes. There was no change in the fluorescence intensity until the free radical scavenger is consumed when oxygen-radical scavengers are provided to the system. The assay readout compared to the radical scavenging ability of a known radical scavenger, Trolox, a vitamin E derivative. The radical scavenging ability of the composite was found to be as high as 1,240 times that of Trolox. However, when the butylated hydroxy toluene functionalized by covalently to the sidewall, the antioxidant activity of the system was decreased [36], which signify that not all functionalization are useful for antioxidant activity.

This approach is also true for CNTs.

Novel class of drug delivery system include polymeric drug complex have been envisaged for transport of tumor tissue-specific anticancer drugs [38]. Accumulation within tumor tissues have been achieved by the macromolecular size of polymeric drug complexes, which makes them specific for cancerous tissues because of better permeability and retention (EPR) effects. The pathophysiologic factors of cancerous cells include EPR, poor venous and lymph drainage, acidic pH, and relatively high temperature, enhance the pharmacological performance of polymeric-based systems. CNT complexes are being used for improving cancer therapy, though; it is a serious challenge. Methotrexate (MTX) is a drug has been widely used against cancer; however, it suffers from low cellular uptake. Combination of MTX to CNTs improves its internalization through functionalized CNTs, which is a promising approach to overcome its minimal cellular uptake. Two orthogonally protected amino groups were complexed onto the side walls of CNTs and then derivatized with FITC and MTX using the 1,3-dipolar cyclo-addition of azomethineylides. Epifluorescence and confocal microscopy studies exhibited that MTX was rapidly internalized by CNTs and drug efficiency was improved [39]. Magnetic CNTs conjugated with a layer of magnetite ($Fe_3O_4$) nanoparticles on the internal surface of the nanotubes have been used for targeting lymphatic tumor. Through nanoprecipitation, PL-PEG-FA functionalized magnetic CNTs could be impregnated with chemotherapeutic agents, such as 5-fluorouracil and cisplatin. Such a system directed by an exterior placed magnet to target regional lymphatic nodes [40].

Antibiotic amphotericin B (AmB) was attached on the surface of CNTs for the treatment of chronic fungal infections [41]. This functionalized NTs were rapidly internalized by mammalian cells showing reduced toxicity in comparison to the free drug administration [42]. More than 40% of cells died by effect of free AmB whereas all mammalian cells remained a live upon treatment with AmB covalently linked to CNT. The assessment of the antifungal activity against various microorganisms showed that the effect of the AmB is enhanced when conjugated to the tubes. In another work antimicrobial agent Pazufloxacinmesilate was linked with amino-MWCNT with high adsorption and was assayed for infection treatment [43].Vaccines can also be delivered using CNTs functionalized with appropriate linkers. It is necessary to keep the orientation and structure of antigen in intact conformation to allow maximum responses against antibody and right specificity. CNTs permit Functionalized CNTs can also act as vaccine delivery procedures [44, 45]. The association of a bacterial or viral antigen with CNTs keeps intact antigen conformation, thus, inducing antibody response with the target specificity [46]. The fixation of functionalized CNTs with B and T cell peptide epitopes evolve a multivalent system capable of persuading a strong immune response, thus becoming a superior candidate for vaccine delivery [47, 48]. Thus, functionalized CNTs act as a good cargo system for the delivery of candidate vaccine antigens. Further, CNTs have showed antimicrobial activity because bacteria adsorbed onto the surfaces of CNTs, such as *E. coli*. The antibacterial effect attributed due to carbon nanotube-induced oxidation of the intracellular antioxidant glutathione, consequently, increased oxidative stress on the bacterial cells and ultimately leads to cell death [46].

## 6.9. PROTEINS LOADED IN CARBON NANOTUBE SYSTEM

CNTs have been used for transport of small molecules as well as large size proteins. MWCNTs are used as cellular carriers of recombined ricin A chain protein toxin (RAT) for targeting tumor. Efficiency of breast cancer targeting and therapy, anti-HER2 IgY antibodies covalently bound to the side walls of SWCNTs using EDC chemistry have also been improved. Illustration obtained from single-cell level Raman spectroscopic revealed that signals collected from the SK-BR-3 cells treated with the targeted nano conjugates were significantly higher than that from the control cells [49]. Tumor lysate proteins (TLP) are a combination of a number of tumour proteins which are produced in large number of tumors and patients, irrespective of their genetic origins. Tumors cells are comprised of lack of specific tumor markers, like target antigens, which hamper the antibody responses. These cancerous cells also lack immunogens, the effectiveness of therapy involving such proteins has been estimated by tumor rejection, retard growth of tumor or prolonged life span of the immunized mice instead by antibody production. CNTs have also enhanced the anticancer immune response of TLP against multiple tumors [50].

## 6.10. DNA, RNA, OR GENESLOADED IN CARBON NANOTUBES

CNTs have been employed for the transport of genes and nucleic acids. In gene therapy, functional genetic material is introduced inside the cancerous cells as well as patients suffering from infectious diseases to recover them. Formation of tumor is accompanied by gene alterations and these alterations are corrected by gene therapy. All the gene-based therapeutic strategies require efficient and safe gene delivery systems. There have been two subcategories of gene vectors including many viral and non-viral vectors. Viral vectors are aim to modify to reduce their toxicity and maintain their high gene transfer capability. However, due to their restricted capability for transgenic organisms and safety, particularly immunogenicity, has attracted researchers to more shift towards non-viral vectors as an alternative. Various non-viral vectors such as dendrimers, polymers, liposomes, peptides have been used and are mainly based on cationic moieties present in these polymeric entities. Very recently scientific community have progressed to design CNT based delivery systems for DNA and exhibits exceptional physical, chemical, and biological properties. Investigative analysis revealed positively charged CNTs can complex with nucleic acids for the effective delivery. Pure NTs are hydrophobic; hence, it is essential to functionalize them for their aqueous solubility. Efficient delivery of pDNA,

SiRNA and miRNA in mammalian cells has been reported in literature using CNTs, which are functionalized by several alteration steps involving both covalent and non-covalent strategies [51 - 57].

In gene delivery applications, both multi and single walled CNTs have been used and due to better encapsulation of MWCNTs have showed favourable results. Brando *et al.* have illustrated that MWCNTs can penetrate pellucide zone (PZ) present as protective layer in embryonic cells and also working as barrier against chemicals and biological present in the environment. The report illustrates effective delivery of pDNA into *in vitro*-fertilized bovine embryos without obstructing the growth of embryo [58]. Real-time PCR assay results demonstrate that MWCNT does not change the gene expression levels in the embryos.

## 6.11. CARBON NANOTUBES IN CANCER THERAPY

Cancer diagnosis and treatment are key challenges due to very high incidences of the disease in developed and developing countries similarly, more death rate, recurrence of the problem and appearance of several other complications. Cancer refers to abnormal proliferation of cells growing in undifferentiated manner leads to formation of tumours in one or multiple organs. Globally, data of the disease are incredible because patients die due to primary cancer growth as well as due to systemic effects of metastases on other organ away from the original position of tumour. It is a big challenge for scientists and medical professionals working in cancer biology, research and patient management to diagnose cancer in early stages and understanding the conditions of recurrence of disease. Present armoury against these threatened diseases mainly includes surgical resection, treatment *via* chemicals or drugs called chemotherapy, *via* radiations or radiotherapy, or their combination of these three modalities.

However, these approaches have many limitations. Conventionally used chemotherapeutic formulations in form of tablets, injectables or capsules have been restrained by systemic toxicity, narrow therapeutic window and destructive effect to nearby tissues. They exhibit drug resistance and restricted cellular penetration beside, neurotoxicity, nephrotoxicity and vascular toxic as well as pose risks of infertility, and thromboembolic complications. Both radiotherapy and chemotherapy are associated with serious side effects of nausea, myocardial infarction and hair loss. Hence, researchers are trying to search nanotechnological ways for finding solutions for cancer therapy and diagnosis. CNTs function as drug container because of large inner volume, large aspect ratios for various functionalization attachments, and the ability to penetrate inside the cell. It is expected that anticancer drugs administer with CNTs deliver the therapeutic compound at tumour site in intact form as compared to conventional system as

CNTs can easily break cytoplasmic and nuclear membranes.

Many drugs have been complexed with CNTs and effectively tested *in vitro* and *in-vivovia* linkers. These are particularly useful for targeted delivery of anticancer drugs directly in tumours only. Antibody tagged CNTs loaded with chemotherapeutic agents are specific for overexpressed antigen honing in target site only where drug release is facilitated *via* bond cleavage in slow and sustained manner. *In vivo* introducing paclitaxel-CNT complex in murine breast cancer model has been observed to show high efficiency in tumour suppression with low toxicity to surrounding normal cells. The group has reported experimentation for the first time on animal model, where CNT based DDS showed higher efficacy than free taxol drug [59]. In another work it was observed that SWCNTs functionalized with p-glycoprotein antibodies were loaded with doxorubicin anticancer drug [60]. Lippard *et al.* [61] used SWCNT coated with PEG *via* non-covalent interactions for the intracellular transportation of a Platinum (IV) complex. This complex on endocytosis was converted to native Pt (II) for cancer cell destruction. This DDS showed higher cytotoxicity against K562R leukaemia cells as compared to free drug molecules. Anticancer drugs can be effectively loaded inside the cavity of CNTs *via* different methods. According to Arsawang *et al.* [62], CNT having diameter of 80 nm can hold up to 5 million drug molecules. In addition to passive delivery of drugs, active or targeting module has also been approached for treating cancer. Smart or intelligent nanomaterials tagged with cancer specific biomolecules have been demonstrated in many experiments. The results indicated higher tumour uptake as compared with non-targeted nano-systems. Fadden and his co-workers [63] have worked on multi-model approach for drug delivery by loading two anticancer molecules-doxorubicin and mitoxantrone using PEGylated SWCNTs after fine tuning this functionalization steps in terms of influence of temperature, pH and light. They showed the release of drugs was significantly influence by pH of the solution on internalization of nano-system. Some of the CNT based formulations are summarized in Table **1**.

## 6.12. HYPERTHERMIA THERAPY

It has been proved that SWCNTs exhibit strong absorption in NIR window of light (700-1100 nm) owing to their inherent photo-physical properties as compared to MWCNTs. This has propelled their use in diminishing tumour growth. Photo thermal applications of nanostructures for the treatment of non-superficial cancerous lesions *in vivo* have been accompanied by penetrating the light *via* tissues. The benefits of such a minimally invasive approach are manifolds such as minimal postoperative pain, rare hospitalization, fast return to social activities, and finally, superior cosmetics and reduced wound problems. CNTs have been regarded as very effective candidates for hyperthermia therapy because they produce huge amount of heat on excitation in NIR light. This localized heating destructs multiple cellular functions *via* induced ablation of tumour cells. The broad electromagnetic absorbance spectrum of CNTs has provided considerable usefulness than plasmonically heated nanomaterials (such as gold nanoshells and nanorods). In these nanomaterials the excitation spectra are highly dependent upon the size and shape of the particles, however they can be synthesized in uniform dimensions opposed to CNTs which are not mono-disperse in organization. Singh *et al.* [64] and Iancuetal [65] have reviewed recent activity on therapeutic action of CNTs on various kinds of cancers in-vitro and in-vivo as well. CNTs show great promise in this filed since they can convert NIR light energy to heat energy effectively as compared to many other nano-materials and their high surface to volume ratio enables multi-model approach of therapeutic as well as imaging characteristics. They can easily cross cell membranes and have been demonstrated to have low toxicity and be degraded in the cell structure.

**Table 1. Assessment of CNT based preclinical evaluation on human generated with cancer models.**

| Cancer Type | Experimental model | Material (effective dose) | Laser (wavelength; radiant exposure) | Reference |
|---|---|---|---|---|
| Brain | *in vitro*; adherent primary glioblastoma cells | SWNT (2.5 μg/well of 24 well plate) | 808 nm; 600 J/cm$^2$ | [69] |
| | *in vivo* murine flank tumor; primary glioblastoma cells | MWCNT (10 μg - *ex vivo*) | 808 nm; 600 J/cm$^2$ | [69] |
| | *in vitro*; non-adherent U251 glioblastoma cells | SWNT (10 μg/ml) | 808 nm; 600 J/cm$^2$ | [70] |
| | *in vitro*; non-adherent MDA-MB-231 cells | gold coated SWNTs (estimated 0.5-1×10$^4$ SWNTs per cell) | 850 nm; 0.5 J/cm$^2$ | [71] |
| Breast | *in vitro*; transformed human mammary epithelial cells | MWCNT (50 μg/mL) | 1064 nm; 90 J/cm$^2$ | [72] |
| | *in vivo* murine flank tumor; transformed human mammary epithelial cells | MWCNT (100 μg IT) | 1064 nm; 90 J/cm$^2$ | [72] |
| | *in vitro*; adherent BT-474 cells | SWNT (5-10 μg/ml) | 808 nm; 5130 J/cm$^2$ | [73] |
| | *in vitro*; adherent SK-BR-3 cells | SWNT (4 μg/ml) | 808 nm; 900 J/cm$^2$ | [74] |
| | *in vitro*; non-adherent BT474 cells | SWNT (100 μg/ml) | 800 nm;12 J/cm$^2$ | [75] |
| | *in vitro*; non-adherent MCF-7 cells | SWNT (not reported) | 808 nm; 144 J/cm$^2$ | [76] |

*(Table 1) contd.....*

| Cancer Type | Experimental model | Material (effective dose) | Laser (wavelength; radiant exposure) | Reference |
|---|---|---|---|---|
| Cervical | *in vitro*; adherent Hela cells | MWCNT (10 μg/ml) | 1064 nm; 128 J/cm$^2$ | [77] |
| | *in vitro*; non-adherent HeLa cells | SWNT (25 μg/ml) | 808 nm; 168 J/cm$^2$ | [78] |
| Kidney | *in vitro*; adherent CRL 1932 cells | Nitrogen-doped MWCNT (approx. 83 μg/ml) | 1064 nm; 720 J/cm$^2$ | [79] |
| Liver | *in vitro*; adherent HepG2 or CRL 4020 cells | MWCNT (1-50 μg/ml) | 808 nm; 7680 J/cm$^2$ | [65] |
| | *in vitro*; adherent SK-BR-3 cells | SWNT (20 μg/ml) | 1064 nm; 4 J/cm$^2$ | [80] |
| Lymph | *in vitro*; non-adherent DaudiBurkitt's lymphoma cells | SWNT (90 μg/ml) | 808 nm; 2100 J/cm$^2$ | [81] |
| Mouth | *in vivo* murine flank tumor; KB epidermoid mouth carcinoma cells | SWNT (12 μg IT) | 808 nm; 684 J/cm$^2$ | [67] |
| Neuroendocrine | *in vitro*; adherent stNB-V1 neuroblastoma cells | MWCNT (5-10 μg/ml) | 808 nm; 3000 J/cm$^2$ | [82] |
| Prostate | *in vitro*; non-adherent PC-3 cells | MWCNT (100 μg/ml) | 1064 nm; 4590J/cm$^2$ | [83] |
| | *in vivo* murine flank tumor; PC-3 cells | MWCNT (50 μg IT) | 1064 nm; 175 J/cm$^2$ | [66] |

Ghosh *et al.* [66] showed complete elimination of PC3 xenograft tumours in 8/8 nude mice taken for PTT experimentation. They injected DNA coated MWCNTs once following irradiation with 1064 nm laser at a power of 2.5W/cm2 inducing heating of localized environment of CNTs killing all cancer cells while the controls which were given either laser or MWCNTs were not affected. Many other reported work have demonstrated successful ablation of cancer cells using SWCNTs when injected directly into the local tumour sites in mice models with subsequent exposure to laser light[67]. Dai group in consequence have shown heat killing of cancer cells by systemic administration of functionalized SWCNTs injected intravenously in blood circulation system. They optimized non-covalent PEGylation modification of CNTs rendering them with higher half-life propelling long circulation time in body fluids. It was demonstrated that nanotubes were not accumulated in RES organs and their uptake was high in tumour only Liu *et al.* [68].

CNTs drug delivery systems have facing many problems such as lack of solubility, clumping occurrences, and half-life. However, these issues have been presently being addressed and altered for more advancement in the field of Carbon nanotubes. Use of specific use of carbon nanotubes as nano-vectors for drug delivery system offer many benefits such as efficient uptake of these structures have showed better results, reduced side-effects. Drug encapsulation has improved water dispersion, bioavailability, and reduced toxicity. Encapsulation of molecules has been used for material storage, protection and precise discharge of loaded molecules. Hence, these encapsulated molecules act as basis of effective transport of drug and furthermore, helps to understand research which improve further advancements including increased water dispersion, reduced toxicity, sustained half-life, better cell penetration and their uptake.

## 6.13. CARBON NANOTUBE TARGETED TO CENTRAL NERVOUS SYSTEM (CNS)

CNTs have found phenomenal usage in nervous system. They are utilized as substrates for adhesion and growth of neurons, promoters for stem cell differentiation into neurons and stimulators for the propagation of dendritic back currents in isolated neurons. Their unique electrical conductivity and nanoscale texturing induce directed growth in primary motor neuritis along the length of horizontally aligned nanotubes. They provide viable interface for neurons working as grafts for stimulating physical and electrical regeneration of axons. These results indicate the potential application of functionalized CNTs as therapeutic systems for neurodegenerative pathologies such as Alzheimer's and Parkinson's disease and spinal cord malfunctions [84].

Transport of drugs to any part of central nervous system which is confined for tumor is serious challenge because of the blood-brain barrier (BBB). Brain require essential amounts of fluids and gases for proper working, however BBB a vasculature of CNS acts as a physical barrier and imposing many problems. It prevents the delivery of drugs to the CNS restricting transport of several therapeutic agents such as antibiotics, anti-neoplastic agents, and neuropeptides. These therapeutic components are not permitted to pass by endothelial capillaries to brain.

Transport of non specific BBB against drugs have been facilitated by means of diffusion or active transport and generates obstacles for effective CNS drug development. However, several approaches have been used to transport injected therapeutic agents to brain or cerebrospinal fluid or intranasal delivery. These techniques are highly risky, invasive local, and short lasting. In contrast to this, there are some safe methods are also available that transport the therapeutic agents through vascular route that infuse and reach in larger portion of the brain. Neurodegenerative diseases occur when the basic units of the nervous system, *i.e.*

neurons in the brain and spinal cord, start to deteriorate or do not work properly. Modifications in these nerve cells result in abnormal function and ultimately cell death.

Zhang *et al.* [85] have explained the electrical conductive capacity and unique mechanical properties of CNTs, and exhibited the morphological characteristics of CNTs which are analogous to those of neurons. The structure and dimensions of CNTs are similar to certain neuronal structures, like ion channels, signalling proteins, and elements of the neuronal cytoskeleton. The similarity of CNTs to these neuronal components have some additional benefit which enhances neuronal interaction at the molecular level and accordingly recover the physiological activity of neurons and processing of neuronal information. Carbon nanotubes could improve neuronal performance by favouring electrical shortcuts. SWNTs have been used to stimulate brain circuit activity directly [86].

In the treatment of CNS disorders CNTs have been used as nano carrier (delivery of drugs) due to their better biocompatibility with neurons, easy functionalization, large surface to volume ratio and modification with drugs improves dispersion in physiological environment. Zhang *et al.* [87] have reported that modified SWNTs have been used in delivery of neurotransmitter acetylcholine. They showed SWNT dose as high as 300mg/kg enabled delivery of drug into lysosomes of neurons implementing treatment modality. Many kinds of neurodegenerative pathologies and malignancies can be treated using CNTs. It was observed that amine functionalized SWCNTs by amide reaction improves tolerance of neurons to ischemic damage. Therefore, amine modified SWCNTs can protect and recuperated neuronal functions without using any drug molecule or therapeutic moiety. The mechanism of amine modified SWCNTs is still unknown [88].

## 6.14. BIOSAFETY

Even though, there are exciting prospects for the application of CNTs as drug carriers in medicine still one of the key obstacles in the application of carbon nanotubes as drug carriers is their biosafety. This difficulty has been quite controversial. There are several reports that believe CNT is safe, while many studies have reported the harmful effects on cells *in vitro* and on tissues *in vivo* [89, 90]. Providentially, excitement about the fascinating physicochemical properties of CNTs has been tempered by concerns regarding their probable health vulnerability, despite some toxic effects of CNTs have been observed, which has raised issue about the biosafety of CNTs. The reported toxicity is mainly due to the structural similarity between CNTs and asbestos fibers in the high aspect ratio [91], large surface area [92], fibrosis [93], induction of inflammation [94], and malignant mesothelioma on inhalation [94], and, more importantly, their bio-persistence [95]. A recent *in vitro* study has suggested that there are similarity between the bio-durability of CNTs and the chemistry of CNT surface functionalization. However, it was revealed by advance studies that the toxicity of SWCNTs *in vivo* is due to aggregation but not due to their high aspect ratio of individual CNTs [96]. The experiment has been illustrated that granuloma-like structures with mild fibrosis observed in mice treated with aggregated SWCNTs but not observed in mice treated with nanoscale dispersed SWCNTs, even after 30 days of intra-tracheal penetration. Histo-pathological slices from the lung of the mice injected with nano-scale dispersed SWCNTs revealed uptake of the SWCNTs by macrophages and confirmed the gradual clearance over time, supporting the biosafety and biocompatibility of CNTs in the form of nanoscale dispersions in *in vivo* applications. Researchers have put more efforts to evaluate the toxicity of SWCNTs in Swiss mice on the basis of dose, length, and surface chemistry. Some researchers have reported that oral administration of CNTs, even at very high concentration (1,000 mg/kg bodyweight), has not induces death, growth or any behavioural dilemma. However, when they are intraperitoneally administered, SWCNTs causes coalesce inside the body, form fiber-like structures. Rafeeqi and Kaul have investigated the links between multi-walled carbon nanotubes and cell culture medium by spectroscopy, and the results supported the biocompatibility of these nanotubes [97]. From various studies, both positive and negative biological effects of SWCNTs have been seen, which are attributed to the attempts used in the experiments. For instance, Mittal *et al.* Have reported that interference of CNTs with traditional *in vitro* cytotoxicity assays which was confirmed by cellular system, but structural modifications, and flow cytometry exhibited the characteristics of cytotoxicity [98]. Brown *et al.* have reported that the cytotoxicity of CNTs is associated to their lengths [99]. *In vitro*, SWCNT has induced the reactive oxygen species in lysosomes though there was no effect on the ROS level in mitochondria. *In vivo*, a dose of SWCNT 300 mg/kg, was completely distributed in lysosomes but there was no ultra-structural damage revealed by transmission electron microscope. SWCNT started to enter mitochondria when the doses above 400 mg/kg and as the SWCNT entered; the pathological ultra-structural changes appeared in mitochondria. The pathological changes of mitochondria include dilation of mitochondria, reduced or disappearance of crestae and even the whole mitochondria became a vacuole. Once the mitochondria damaged developed, the lysosomes also appeared as damaged. The membrane of lysosomes destroyed, the hydrolytic enzymes of lysosomes decreased or excreted, and even the entire lysosomes became a huge vacuole. All of these results revealed that mitochondria are the principle organelles for SWCNT damage and the lysosomal damage is resultant to mitochondrial damage. Thus, mitochondria are the damaging target organelles and lysosomes are the pharmacological target organelles of SWCNT. Furthermore, experiments illustrated that the adverse effects of SWCNT on mitochondria was due to SWCNT influence on the mitochondrial membrane potentials (MMP). Decrease in MMP is the result of seepage of free electrons, which in turn increases the ROS, leading to the ultrastructural damage of the cells including lysosomes.

Recent advance progress in various aspects of the safety studies on SWCNT is of important for its realistic application in drug delivery for cancer treatment. Firstly, the modification or functionalization has remarkably reduced the toxicity of CNTs, making it possible to choose secure CNT derivatives as drug carriers. Secondly, it has been found that CNTs metabolized in liver and excreted *via.* kidneys and hap-bile systems, making less concern about the persistence residence of them in bodies. Thirdly, it has been demonstrated that the pharmacological and the toxicological effects of CNTs are arbitrated by diverse target organelles and the allocation of CNTs in organelles has been regulated by some chemicals, which makes it possible to use the advantage and decrease or avoid the disadvantage of them. Fourthly, pharmacological and toxicological effects of CNTs, depends on the dose hence, the toxicological effects can be circumvent by controlling the doses.

## 6.15. CONCLUSION

In the field of Carbon nanotubes technology as a drug delivery system is work beyond our prospect and their simple mechanism with long lasting life makes it more reliable to use. Carbon nanotube surfaces can be functionalized with various molecules by Covalent and non-covalent interactions. Functionalized CNTs exhibits the ability to deliver desired molecules across the cytoplasmic membrane and nuclear membrane without producing hazardous effect. The invention of this nanotechnology has opened novel scopes which are more efficient than the earlier drug delivery system because they can enter through cell membranes, carrying drugs, biomolecules, genes, vaccines, and into the target cells or organs which are difficult to approach. Besides, another advance is the use of short half life polymer like collagen CNTs materials as biodegradable scaffolds in tissue generation and artificial implants because CNTs resist biodegradation and are a controlling engineering applicant over other existing materials used to repair defective organs. Moreover, CNTs conjugated with biosensors or other materials have proven excellent for the therapeutic monitoring and the diagnosis of diseases as well as for the analysis of drugs in diverse areas. It is also advisable to use functionalized CNTs to develop the free radical scavenger. Use of modified CNTs in cancer treatment guarantee 70% of the cure and 100% site targeted with body-friendly in nature. The unique properties of CNTs have widen the scopes in almost all fields, particularly in drug delivery system in medicine which is useful for the mankind, however due to lack of technology for large scale production, costs of the production and rounding CNT toxicity are key issues, because there are various contradictory studies showing both toxic and non-toxic behaviour, despite of a number of drugs which have been transported with the help of carbon nanotubes. There is no real milestone for comparing results. A number of harmful effects described in the text which have shown to affect CNTs toxicity, this is an area which will entail the sustained attention of toxicologists in the future. No doubt, a lot of study is required in this area before CNT technology can be implement in cancer treatment. This is very right to say that it is a tremendously promising application of nanotechnology and certainly worth further research, since, the existing approaches for cancer treatment are arbitrarily harmful and only partially effective.

## CONFLICT OF INTEREST

The authors declare no conflict of interest, financial or otherwise.

## ACKNOWLEDGEMENTS

Thanks to all scientists referenced throughout the chapter whose valuable work has guided the way through to this research work.

## REFERENCES

[1] Abarrategi A, Gutiérrez MC, Moreno-Vicente C, *et al.* Multiwall carbon nanotube scaffolds for tissue engineering purposes. Biomaterials 2008; 29(1): 94-102.

[2] Atala A. Tissue engineering and regenerative medicine: concepts for clinical application. Rejuvenation Res 2004; 7(1): 15-31.

[3] McDevitt MR, Chattopadhyay D, Kappel BJ, *et al.* Tumor targeting with antibody-functionalized, radiolabeled carbon nanotubes. J Nucl Med 2007; 48(7): 1180-9.

[4] Feazell RP, Nakayama-Ratchford N, Dai H, *et al.* Soluble Single-Walled Carbon Nanotubes as Longboat Delivery Systems for Platinum(IV) Anticancer Drug Design. J Am Chem Soc 2007; 129: 8438-9.

[5] Liu Z, Sun X, Nakayama N, *et al.* Supramolecular Chemistry on Water-Soluble Carbon Nanotubes for Drug Loading and Delivery. ACS Nano 2007; 1: 50-6.

[6] Kam NW, Dai H. Carbon nanotubes as intracellular protein transporters: Generality and biological functionality. J Am Chem Soc 2005; 127: 6021-6.

[7] Pantarotto D, Briand JP, Prato M. Translocation of bioactive peptides across cell membranes by carbon nanotubes. Chem Commun 2004; 1: 16-7.

[8] Liu Y, Wu DC, Zhang WD. Polyethylenimine-grafted multiwalled carbon nanotubes for secure non-covalent immobilization and efficient delivery of DNA. Angew Chem Int Ed 2005; 44: 4782.

[9] Kam NW, Liu Z, Dai H. Functionalization of carbon nanotubes *via* cleavable disulfide bonds for efficient intracellular delivery of siRNA and potent gene silencing. J Am Chem Soc 2005; 36: 12492-3.

[10] Liu Z, Winters M, Holodniy M, *et al.* siRNA delivery into human T cells and primary cells with carbon-nanotube transporters. Angew Chem Int Ed 2007; 46: 2023-7.

[11] Kam NW, Liu Z, Dai H. Carbon nanotubes as intracellular transporters for proteins and DNA: An investigation of the uptake mechanism and pathway. Angew Chem Int Ed 2006; 45: 577-81.

[12] Zhang W, Zhang Z, Zhang Y. The application of carbon nanotubes in target drug delivery systems for cancer therapies. Nanoscale Res Lett 2011; 6(1): 555.

[13] Thess A, Nikolaev P. Crystalline ropes of metallic carbon nanotubes. Science 1996; 5274: 483-7.

[14] Baughman R, Zakhidov A, de Heer W. Carbon Nanotubes, The Route Toward Applications. Science 2002; 297(5582): 787-92.

[15] Dresselhaus M, Dresselhaus G, Charlier J, *et al.* Electronic, thermal and mechanical properties of carbon nanotubes. Philos T R Soc A 2004; 15(362): 2065-98.

[16] Tibbetts GG, Beetz CP. Mechanicalproperties of vapor-grown carbonfibers. J Phys D Appl Phys 1987; 20(3): 292-7.

[17] Wong EW, Sheehan PE, Lieber CM. Nanobeam mechanics: elasticity, strength, and toughness of nanorods and nanotubes. Science 1997; 277: 1971-5.

[18] Iijima S. Helical microtubules of graphitic carbon. Nature 1991; 354: 56-8.

[19] Thess AR, Lee P, Nilcolaeve H, *et al.* Crystalline of metallic nanotubes. Science 1996; 273: 483.

[20] Cassell AM, Raymakers JA, Kong J, *et al.* Large Scale CVD Synthesis of Single-Walled Carbon Nanotubes. J Phys Chem B 1999; 103: 6484-92.

[21] MacKenzie KJ, Dunens OM, Harris AT. An updated review of synthesisparameters and growth mechanismsfor carbon nanotubes in fluidizedbeds. Ind Eng Chem Res 2010; 49: 5323-38.

[22] Foldvari M, Bagonluri M. Carbon nanotubes as functional excipients fornanomedicines: I. Pharmaceutical properties. Nanomedicine (Lond) 2008; 3: 173-82.

[23] Foldvari M, Bagonluri M. Carbon nanotubes as functional excipients for nanomedicines: II. Drug delivery and biocompatibility issues. Nanomedicine (Lond) 2008; 3: 183-200.

[24] Hilder TA, Hill JM. Modeling the loading and unloading of drugs into nanotubes. Small 2009; 5: 300-8.

[25] Martincic M, Tobias G. Filled carbon nanotubes in biomedical imaging and drug delivery. Expert Opin Drug Deliv 2015; 12(4): 563-81.

[26] Modi CD, Patel SJ, Desai AB, *et al.* Functionalization and evaluation of PEGylated carbon nanotubes as novel drug delivery for methotrexate. J Appl Pharm Sci 2011; (5): 103-8.

[27] Kushwaha SK, Ghoshal S, Rai AK. Carbon nanotubes as a novel drug delivery system for anticancer therapy: a review. B. J Pharm Sci 2013; 49: 629-43.

[28] Liu Z, Davis C, Cai W, *et al.* Circulation and long-term fate of functionalized, biocompatible single-walled carbon nanotubes in mice probed by Raman spectroscopy. Proc Natl Acad Sci USA 2008; 105: 1410-5.

[29] Zhang Y, Bai Y, Yan B. Functionalized carbon nanotubes for potential medicinal applications. Drug Discov Today 2010; 15(11-12): 28-435.

[30] Kateb B, Yamamoto V, Alizadeh D. Multi-walled carbon nanotube (MWCNT) synthesis, preparation, labeling, and functionalization. Methods Mol Biol 2010; 651: 307-17.

[31] Xiao D, Dramou P, He H. Magnetic carbon nanotubes: synthesis by a simple solvothermal process and application in magnetic targeted drug delivery system. J Nanopart Res 2012; 14: 984-95.

[32] Tripisciano C, Kraemer K, Taylor A, *et al.* Single-wall carbon nanotubes based anticancer drug delivery system. Chem Phys Lett 2009; 478: 200-5.

[33] de Leon A, Jalbout AF, Basiuk VA. SWNT-amino acid interactions: a theoretical study. Chem Phys Lett 2008; 1–3: 185-90.

[34] Hilder TA, Hill JM. Carbon nanotubes as drug delivery nanocapsules. Curr Appl Phys 2008; 3–4: 258-61.

[35] Ali-Boucetta H, Al-Jamal KT, McCarthy D, *et al.* Multiwalled carbon nanotube-doxorubicin supramolecular complexes for cancer therapeutics. Chem Commun (Camb) 2008; 4: 459-61.

[36] Lucente-Schultz RM, Moore VC, Leonard AD, *et al.* Antioxidant single-walled carbon nanotubes. J Am Chem Soc 2009; 11: 3934-41.

[37] Feazell RP, Nakayama-Ratchford N, Dai H, *et al.* Soluble single-walled carbon nanotubes as longboat delivery systems for platinum (IV) anticancer drug design. J Am Chem Soc 2007; 27: 8438-9.

[38] Tripisciano C, Kraemer K, Taylor A, *et al.* Single-wall carbon nanotubes based anticancer drug delivery system. Chem Phys Lett 2009; 4–6: 200-5.

[39] Liu Z, Sun X, Nakayama-Ratchford N, *et al.* Supramolecular chemistry on water-soluble carbon nanotubes for drug loading and delivery. ACS Nano 2007; 1: 50-6.

[40] McDevitt MR, Chattopadhyay D, Kappel BJ, *et al.* Tumor targeting with antibody-functionalized, radiolabeled carbon nanotubes. J Nucl Med 2007; 48: 1180-9.

[41] Rosen Y, Elman NM. Carbon nanotubes in drug delivery: focus on infectious diseases. Expt. Opin Drug Del 2009; 6(5): 517-30.

[42] Wu W, Wieckowski S, Pastorin G, *et al.* Targeted delivery ofamphotericin B to cells by using functionalized carbon nanotubes. Angew Chem Int Ed 2005; 44: 6358-62.

[43] Jiang L, Liu T, He H, *et al.* Adsorption behavior of pazufloxacinmesilate on amino-functionalized carbon nanotubes. J Nanosci Nanotechnol 2012; 12: 1-9.

[44] Liao H, Paratala B, Sitharaman B, *et al.* Applications of carbon nanotubes in biomedical studies. Methods Mol Biol 2011; 726: 223-41.

[45] Rosen Y, Mattix B, Rao A, *et al.* Carbon nanotubes and infectious diseases Nanomed Heal Dis. 249-67.

[46] Digge MS, Moon RS, Gattani SG. Applications of carbon nanotubes in drug delivery: a review. Inter. J PharmTech Res 2012; 4(2): 839-47.

[47] Usui Y, Haniu H, Tsuruoka S, *et al.* Carbon nanotubes innovate on medical technology. Med Chem 2012; 2(1): 1-6.

[48] Yang W, Thordarson P, Gooding JJ, *et al.* Carbon nanotubes for biological and biomedical applications. Nanotech. Article ID 2007; 412001: 12.

[49] Weng X, Wang M, Ge J, *et al.* Carbon nanotubes as a protein toxin transporter for selective HER2-positive breast cancer cell destruction. Mol Biosyst 2009; 5: 1224-31.

[50] Meng J, Jie M, Duan J, *et al.* Carbon nanotubes conjugated to tumor lysate protein enhance the efficacy of an antitumor immunotherapy. Small 2008; 9: 1364-70.

[51] Qin W, Yang K, Tang H, *et al.* Improved GFP gene transfection mediated by polyamidoaminedendrimer-functionalized multi-walled carbon nanotubes with high biocompatibility. Colloids Surf B Biointerfaces 2011; 84(1): 206-13.

[52] Inoue Y, Fujimoto H, Ogino T, *et al.* Site-specific gene transfer with high efficiency onto a carbon nanotube-loaded electrode. R Soc Interface 2008; 5(25): 909-18.

[53] Al-Jamal KT, Gherardini L, Bardi G, *et al.* Functional motor recovery from brain ischemic insult by carbon nanotube-mediated siRNA silencing. Proc Natl Acad Sci USA 2011; 108(27): 10952-7.

[54] Dong H, Ding L, Yan F, *et al.* The use of polyethylenimine-grafted graphene nanoribbon for cellular

delivery of locked nucleic acid modified molecular beacon for recognition of microRNA. Biomaterials 2011; 32(15): 3875-82.

[55] Hao Y, Xu P, He C, *et al.* Impact of carbodiimidecrosslinker used for magnetic carbon nanotube mediated GFP plasmid delivery. Nanotechnology 2011; 22(28): 285103.

[56] Ladeira M, Andrade V, Gomes E, *et al.* Highly efficient siRNA delivery system into human and murine cells using single-wall carbon nanotubes. Nanotechnology 2010; 21(38): 385101.

[57] Gao L, Nie L, Wang T, *et al.* Carbon nanotube delivery of the GFP gene into mammalian cells. ChemBioChem 2006; 7(2): 239-42.

[58] Munk M, Ladeira LO, Carvalho BC, *et al.* Efficient delivery of DNA into bovine preimplantation embryos by multiwall carbon nanotubes. BrandãoSci Rep 2016; 6: 33588-91.

[59] Liu Z, Chen K, Davis C, *et al.* Drug Delivery with Carbon Nanotubes for *in vivo* Cancer Treatment. Cancer Res 2008; 68(16): 6652-60.

[60] Zhang W, Zhang Z, Zhang Y. Long boat delivery systems for platinum (IV) anticancer drug design. J Am Chem Soc 2007; 129: 8438-9.

[61] Feazell RP, Nakayama-Ratchford N, Dai H, Lippard SJ. The application of carbon nanotubes in target drug delivery systems for cancer therapies. Nanoscale Res Lett 2011; 6: 555-77.

[62] Arsawang U, Saengsawang O, Rungrotmongkol T, *et al.* How do carbon nanotubes serve as carriers for gemcitabine transport in a drug delivery system. J Mol Graph Model 2011; 29: 591-6.

[63] Sharma P, Mehra NK, Jain K, *et al.* Biomedical Applications of Carbon Nanotubes: A Critical Review. Curr Drug Deli 2015; 12: 1-20.

[64] Singh R, Torti SV. Carbon nanotubes in hyperthermia therapy. Adv Drug Deliv Rev 2013; 65(15): 2045-60.

[65] Iancu C, Mocan L. Advances in cancer therapy through the use of carbon nanotube-mediated targeted hyperthermia. Int J Nanomedicine 2011; 6: 1675-84.

[66] Ghosh S, Dutta S, Gomes E, *et al.* Increased heating efficiency and selective thermal ablation of malignant tissue with DNA-encased multiwalled carbon nanotubes. ACS Nano 2009; 3: 2667-73.

[67] Chatterjee DV, Diagaradjane P, Krishnan S, *et al.* Nanoparticle-mediated hyperthermia in cancer therapy. TherDeliv 2011; 2(8): 1001-14.

[68] Liu Z, Cai W, He L, *et al.* *In vivo* biodistribution and highly efficient tumour targeting of carbon nanotubes in mice. Nat Nanotechnol 2007; 2: 47-52.

[69] Wang CH, Chiou SH, Chou CP, *et al.* Photothermolysis of glioblastoma stem-like cells targeted by carbon nanotubes conjugated with CD133 monoclonal antibody. Nanomedicine 2011; 7: 69-79.

[70] Markovic ZM, Harhaji-Trajkovic LM, Todorovic-Markovic BM, *et al.* *In vitro* comparison of the photothermal anticancer activity of graphene nanoparticles and carbon nanotubes. Biomaterials 2011; 32: 1121-9.

[71] Galanzha EI, Kim JW, Zharov VP. Nanotechnology-based molecular photoacoustic and photothermal flow cytometry platform for in-vivo detection and killing of circulating cancer stem cells. J Biophotonics 2009; 2: 725-35.

[72] Burke AR, Singh RN, Carroll DL, *et al.* The resistance of breast cancer stem cells to conventional hyperthermia and their sensitivity to nanoparticle-mediatedphotothermal therapy. Biomaterials 2012; 33: 2961-70.

[73] Marches R, Mikoryak C, Wang RH, *et al.* The importance of cellular internalization of antibody-targeted carbon nanotubes in the photothermal ablation of breast cancer cells. Nanotechnology 2011; 22: 095101.

[74] Xiao Y, Gao X, Taratula O, *et al.* Anti-HER2 IgYantibody-functionalized single-walled carbon

nanotubes for detection and selective destruction of breast cancer cells. BMC Cancer 2009; 9: 351.

[75] Panchapakesan B, Lu S, Sivakumar K, *et al.* Single-wall carbon nanotube nanobomb agents for killing breast cancer cells. NanoBiotechnology 2005; 1: 133-9.

[76] Ning S, Lu S, Wickstrom E, *et al.* Integrated molecular targeting of IGF1R and HER2 surface receptors and destruction of breast cancer cells using single wall carbon nanotubes. Nanotechnology 2007; 18: 315101.

[77] Biris AS, Boldor D, Palmer J, *et al.* Nanophotothermolysis of multiple scattered cancer cells with carbon nanotubes guided by time-resolved infrared thermal imaging. J Biomed Opt 2009; 14: 021007.

[78] Kam NW, O'Connell M, Wisdom JA, *et al.* Carbon nanotubes as multifunctional biological transporters and near-infrared agents for selective cancer cell destruction. Proc Natl Acad Sci USA 2005; 102: 11600-5.

[79] Torti SV, Byrne F, Whelan O, *et al.* Thermal ablation therapeutics based on CNx multi-walled nanotubes. Int J Nanomedicine 2007; 2: 707-14.

[80] Kang B, Yu D, Dai Y, *et al.* Cancer-Cell Targeting and Photoacoustic Therapy Using Carbon Nanotubes as "Bomb" Agents. Small 2009; 5: 1292-301.

[81] Chakravarty P, Qian W, El-Sayed MA, *et al.* Delivery of molecules into cells using carbon nanoparticles activated by femtosecond laser pulses. Nat Nanotechnol 2010; 5: 607-11.

[82] Wang CH, Huang YJ, Chang CW, *et al.* Hsu WM, Peng CA. *In vitro* photothermal destruction of neuroblastoma cells using carbon nanotubes conjugated with GD2 monoclonal antibody. Nanotechnology 2009; 20: 315101.

[83] Fisher JW, Sarkar S, Buchanan CF, *et al.* Photothermal response of human and murine cancer cells to multiwalled carbon nanotubes after laser irradiation. Cancer Res 2010; 70: 9855-64.

[84] Roberts MJ, Leach MK, Bedewy M, *et al.* Directed growth ofprimary motor neurons on horizontally aligned carbon nanotube films and striped patterns. J Neural Eng 2014; 11(3): 036013.

[85] Zhang L, Webster TJ. Nanotechnology and nanomaterials: promises for improved tissue regeneration. Nano Today 2009; 4: 66-80.

[86] Cellot G, Cilia E, Cipollone S, *et al.* Carbon nanotubes might improve neuronal performance by favouring electrical shortcuts. Nat Nanotechnol 2009; 4: 126-33.

[87] Zhang YG, Yang Z, Yang YL, *et al.* Pharmacological and toxicological target organelles and safe use of single-walled carbon nanotubes as drug carriers in treating Alzheimer disease. Nanomedicine (Lond) 2010; 6: 427-41.

[88] Lee HJ, Park J, Yoon OJ, *et al.* Amine-modified single walled carbon nanotubes protect neurons from injury in a rat stroke model. Nat Nanotechnol 2011; 6: 121-5.

[89] Yang S, Luo J, Zhou Q, *et al.* Pharmacokinetics, metabolism and toxicity of carbon nanotubes for bio-medical purposes. Theranostics 2012; 2(3): 271-82.

[90] Shvedova AA, Kisin ER, Porter D, *et al.* Mechanisms of pulmonary toxicity and medical applications of carbon nanotubes: two faces of Janus. Pharmacology and Therapeutics 2009; 121(2): 192-204.

[91] Roda E, Coccini T, Acerbi D, *et al.* Comparative pulmonary toxicity assessment of pristine and functionalized multi-walled carbon nanotubes intratracheally instilled in rats: morphohistochemical evaluations. Histol Histopathol 2011; 26(3): 357-67.

[92] Murphy FA, Poland CA, Duffin R, *et al.* Length-dependent retention of carbon nanotubes in the pleural space of mice initiates sustained inflammation and progressive fibrosis on the parietal pleura. Am J Pathol 2011; 178(6): 2587-600.

[93] Rothen-Rutishauser B, Brown DM, Piallier-Boyles M, *et al.* Relating the physicochemical characteristics and dispersion of multiwalled carbon nanotubes in different suspension media to their oxidative reactivity *in vitro* and inflammation *in vivo*. Nanotoxicology 2010; 4(3): 331-42.

[94] Teeguarden JG, Webb-Robertson BJ, Waters KM, *et al.* Comparative proteomics and pulmonary toxicity of instilled single-walled carbon nanotubes, crocidolite asbestos, and ultrafine carbon black in mice. Toxicol Sci 2011; 120(1): 123-35.

[95] Bonanni A, Esplandiu MJ, del Valle M. Impedimetricgenosensing of DNA polymorphism correlated to cystic fibrosis: a comparison among different protocols and electrode surfaces. Biosens Bioelectron 2010; 26(4): 1245-51.

[96] Demming A. Nanotechnology under the skin. Nanotechnology 2011; 22(26): 260201.

[97] Rafeeqi T, Kaul G. Elucidation of interaction between multi-walled carbon nanotubes and cell culture medium by spectroscopy supports biocompatibility of these nanotubes. Adv Sci Lett 2011; 4: 536-40.

[98] Mittal S, Sharma V, Vallabani NV, *et al.* Toxicity evaluation of carbon nanotubes in normal human bronchial epithelial cells. J Biomed Nanotechnol 2011; 7: 108-9.

[99] David M, Brown KD, Stone V. Nuclear translocation of nrf2 and expression of antioxidant defence genes in thp-1 cells exposed to carbon nanotubes. J Biomed Nanotechnol 2010; 6: 224-33.

**CHAPTER 7**

# Intervention of Carbon Nanotubes in Tissue Engineering Applications

**Jagriti Narang**[1,*], **Tulika Dahiya**[2] and **C.S. Pundir**[3]

[1] *Amity Institute of Nanotechnology, Amity University, NOIDA, UP, India*

[2] *Shoolni University, Solan, HP, India*

[3] *Department of Biochemistry, M.D. University, Rohtak, Haryana, India*

**Abstract:** Tissue Engineering is an interdisciplinary subject which deciphers to produce functional three-dimensional (3D) tissues combining scaffolds, cells and/or bioactive molecules. Carbon nanotubes (CNTs) have received the attention of researchers for the past 25 years. It is one of the most exploratory materials in the biomedical applications. This chapter describes the properties of CNT which helps in mimicking the extracellular matrix (ECM) of various tissue materials. The chapter also discusses the role of CNT based scaffold materials in regeneration of various tissues such as neurons, cartilage and muscles.

**Keywords:** CNT properties, ECM, Musculo-skeletal tissues, Nano-scaffolds, Neurons, Tissue engineering.

## 7.1. INTRODUCTION

Tissue Engineering is an interdisciplinary subject, which deciphers to produce functional three-dimensional (3D) tissues combining scaffolds, cells and/or bioactive molecules [1]. This subject entails convergence of many research fields related to biotechnology, molecular biology, cell biology, nano-science, engineering and medicine. Tissue engineering aims to replace damaged biological parts with healthy tissues by regenerating through the combination of life science and engineering principles. Tissues used for regeneration are stem cells, which have significant potential to regenerate into many different cell types in the body. Stem cells act as a type of repair system for the body and hence can divide endlessly to regenerate other cells for treating damaged cells. Stem cell divides into cells, which have the capability to either remain a stem cell or become another type of cell with a more specialized function, such as a muscle cell, a red

* **Corresponding author Jagriti Narang**: Amity Institute of Nanotechnology, Amity University, Noida, India; Tel: 09811792572; E-mail: jags_biotech@yahoo.co.in

**Jagriti Narang & Chandra Shekhar Pundir (Eds.)**

blood cell, or a brain cell. Human embryonic stem cells (hESCs) derived from the inner cell mass (ICM) of the blastocysts have properties of self-regeneration and pluripotency [2, 3].

Tissue engineering involves injection or transplantation of healthy stem cells to a damaged site, tissue-inducing biomolecules to injured tissue, and regeneration of specific cell types in environment, which mimic the biological extracellular matrix (ECM) or three-dimensional (3D) scaffolds [4]. The main aim of tissue engineering is the regeneration of healthy tissues that sustain, develop or repair tissue function [5]. Tissue engineering in convergence with nanotechnology is a wide-ranging interdisciplinary research that attempts to repair lost, damaged, or aging cells and tissues to a state as close as possible to its native architecture and function. Nano-scaffolds (n-scaffolds) mimic to natural ECM in terms of chemical and physical conditions and its nano-scale fibrous structure.

Tissue engineering approaches, which deal with the replacement of old or damaged tissues, can be classified as cell-based approach, using biomaterials and biomaterials combining with biological engineering strategies. Tissue engineering in convergence with nanotechnology has been a remarkable progress, which provides ever-increasing potential to accomplish the objective of therapeutic treatment of myriad pathologies. Stem cells have the ability to remain either stem cells or differentiate into well specialized cell types [1]. These properties of stem cells are being exploited in tissue engineering for therapeutics strategy for damaged cells.

However, conventional methods for regeneration are injection of healthy stem cells or organ transplant and mechanical devices or surgical reconstruction. All these therapeutic approaches are involved in treating patients but these approaches are also linked to problems such as transplant rejection and lack of donor. Mechanical devices are not capable of accomplishing all the functions associated with the tissue and cannot prevent progressive patient deterioration. Hence, there is a requirement to afford further solutions to tissue repair in clinics and aims to achieve this goal by the development of *in vitro* devices that would repair, *in vivo*, the damaged tissue.

## 7.2. NANO-SCAFFOLD-BASED TISSUE ENGINEERING

Nano-scaffold-based tissue engineering has become the most popular, due to the potential of current scaffold fabrication technologies to incorporate chemical, physical, and biological stimuli at different scales that can direct cell activity. ECM usually provides structural support to the cells in addition to performing various other important functions.

Many researchers have attempted to control cell fate determination by producing a niche that mimics ECM in terms of mechanical properties, surface topography, and chemical microenvironment. ECM plays significant role in differentiation and proliferation of cells during development and wound healing. There is an exchange of proteins and hormones required for growth between cells and surrounding matrix environment. Various biomaterials have been exploited to develop nano-fibrous structures in tissue-engineering application such as collagen [6], chitosan [7], chitin [8], and silk fibroin [9], poly (DL-lactide-co-glycolide) [10], poly-caprolactone [11] and polystyrene [12]. These nano-fibers are being employed for growth and differentiation of cartilage [13], bone [14], arterial blood vessel [15], heart [16] and nerve [17 - 19]. Electrospun nano-fibrous scaffold of multiwalled carbon nanotubes (MWNT) fascinated many researchers as a model that had a physical structure of nano-fibrous network and a unique chemical component of carbon nanotubes. Integration of MWCNT in nano-scaffold material affects cell adhesion, proliferation, secretion, and cellular morphology. It has been widely established that cell–cell binding and cell-ECM adhesion takes place with cell-adhesive proteins. Some cells require some solid substratum for its growth and proliferation. Cells secrete adhesive proteins and many kinds of growth factors whose activity must be highly spatially regulated [20]. Consequently, a protein surface is a favorable and key factor for the scaffolds to cell adhesion and proliferation (Fig. **1**).

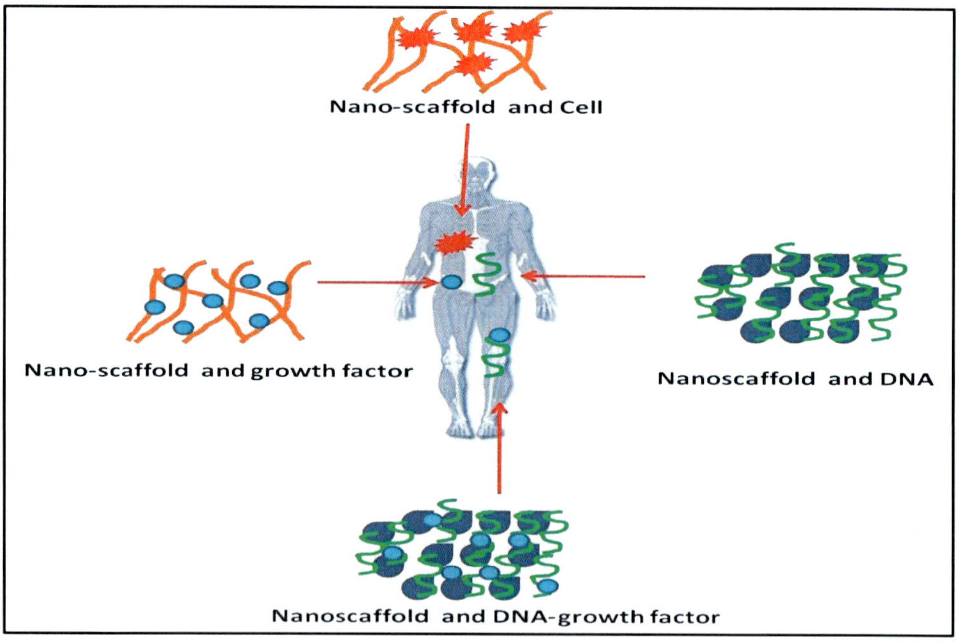


**Fig. (1).** Schematic drawing of nanoscaffold-based tissue engineering.

## 7.3. STRUCTURE AND COMPOSITION OF ECM

ECM is composed of different proteins and proteins arranged in a linear fashion incorporating macromolecules with nanoscale features. ECM proteins mainly consist of following groups *i.e.* collagens, proteoglycans and glycoproteins. Variation in ECM is mainly due to difference in composition of proteins, their composition and assembly [21 - 23]. Collagen is the most abundant protein in the human body and constitutes around 30% of all proteins by mass. Collagen provides structural framework to the ECM. Proteoglycans consist of carbohydrates and proteins, which further bind with other acidic groups. Proteoglycans with attached functional groups, permit it to provide binding sites throughout the ECM. Glycoproteins are also composed of proteins and carbohydrates which help in performing other biochemical reactions (Fig. **2**).

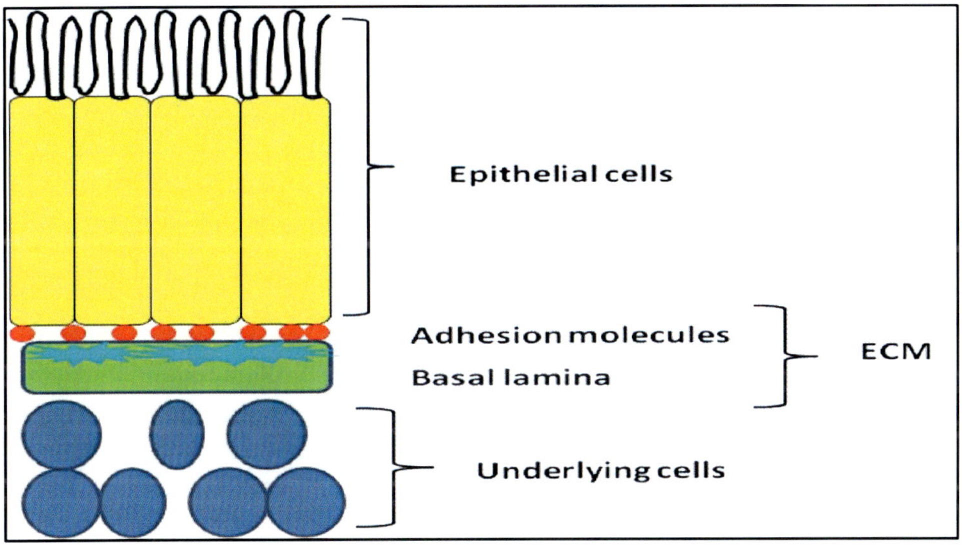


**Fig. (2).** Schematic representation of structure and composition of the extracellular matrix.

### 7.3.1. Functions of ECM

ECM proteins are mainly involved in the cell adhesion, proliferation and differentiation of stem cells. Different composition of these proteins makes the ECM distinctive in terms of mechanical, chemical and architectural diversity. ECM uniqueness is according to its cell type. Functional groups present in the active sites of ECM induce interaction which further increases the cell adhesion [24]. Morphology and alignment of cells in ECM matrix also affects cell differentiation and regeneration [25, 26]. Many ECM proteins show nano-scale size and size of the proteins influences cell adhesion. Furthermore signaling molecules also affect cell-ECM interactions. Small nano-scale biomolecules helps in signaling in between two adjacent cells. Cell –cell interactions is important for development of new blood vessels, growth and formation of new tissue [27]. The structure and functions of cell-cell interactions are reviewed elsewhere (Fig. **3**) [28, 29].

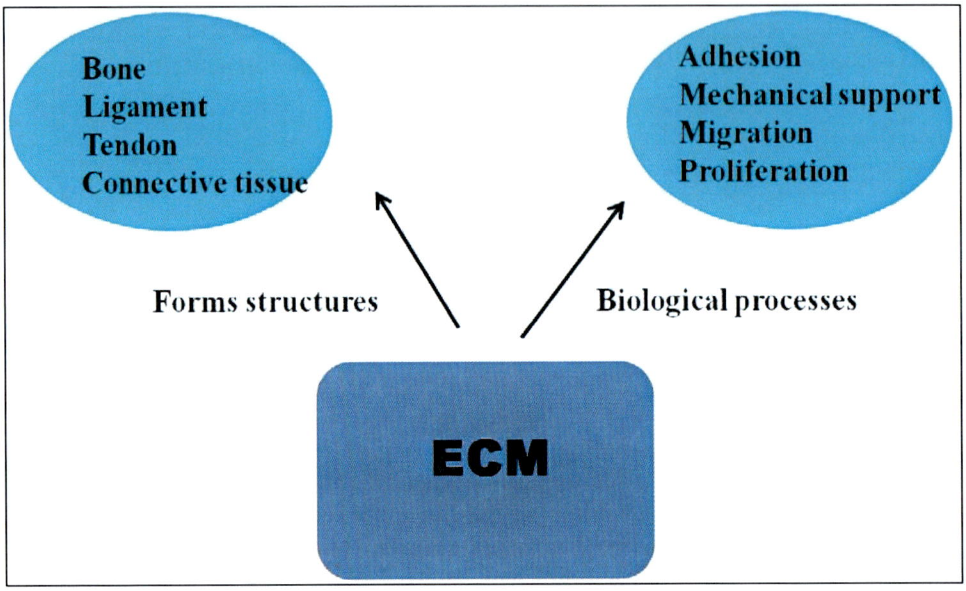


**Fig. (3).** Schematic representation showing function of extracellular matrix.

## 7.4. EXTRACELLULAR MATRIX OF BONE AND ITS NANOSCALE FEATURES

ECM of bone contains composite structure with two major constituents such as organic collagen and inorganic bone mineral. The composition of all the components should be optimized such that bone should be enough stiffer and remain stable under strains [30]. Signalling molecules in the form of growth factors, cytokines and enzymes are also embedded in bone ECM. Spongy tissue of mature adult bone is attracting to researchers as it is porous in nature which allows transport of nutrients and minerals and vascularization. At the nano-scale, mature adult bone is composed of aligned fibrillar type I collagen over which apatite as particles are arranged [31 - 34]. The diameter of collagen molecule present in the bone is of nanometer diameter. CNTs are of nanometer range and can be tuned up to dimensional similarity to collagen fibers. Longitudinal distributions in electric charge along the tropo-collagen molecule direct the axial staggering of tropo-collagen macromolecules within a collagen fiber. CNT has also capability to incorporate nano-scale bone like carbonate apatite (Fig. **4**)[35 - 41].

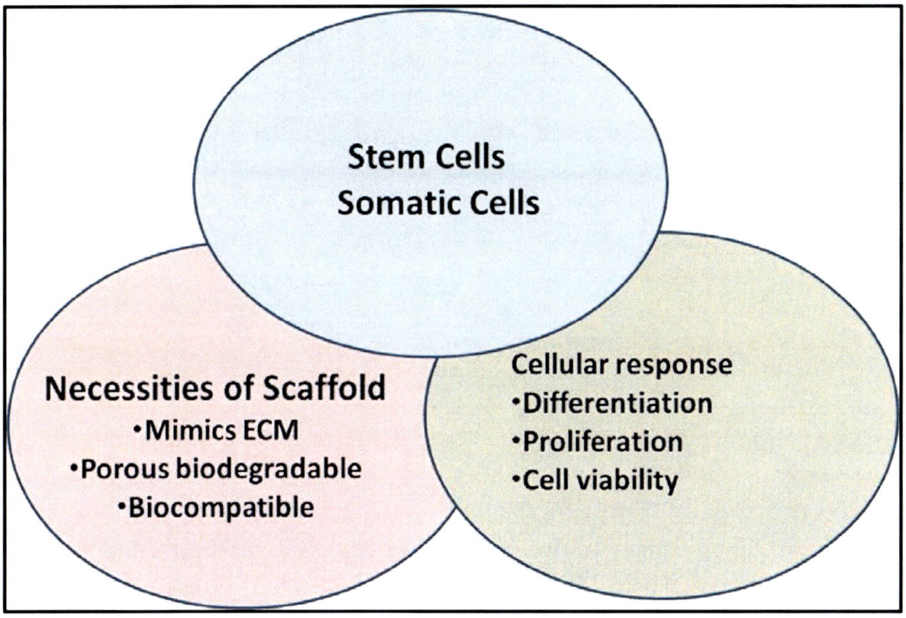


**Fig. (4).** Schematic representation of the three mainstay of tissue engineering.

### 7.4.1. Physical Properties of Carbon Nanotubes

CNTs provide high mechanical strength to the substratum. CNT also shows high tensile strength even greater than bone [42]. This makes them ideal candidates for use as reinforcement in composite materials. CNT can prove to be best structural substratum with comparable mechanical strength of bone. CNT-polymer composites have established stiffness of nearly 700% as well as increases to strength and toughness of 1200. Introduction of polymer into CNT matrix can improve both the mechanical properties and regenerative potential of a biomaterial [43 - 49]. In 2012, Ogihara *et al.* used CNTs to reinforce an alumina composite for bone tissue replacement. CNT alone shows less proliferative capacity for stem cells as compared to CNT based composites. Carbon nanotubes can display either metallic or semiconducting electronic properties according to the chirality of the nanotube. CNT act as conducting substrates which can be employed to electrically stimulates formation and repair of bone tissue [50]. Furthermore CNT also increases ECM protein synthesis and other signaling molecules [51, 52]. Some metal nano-scaffold has high conductivity but they are not biocompatible and even integrate with the cells [53]. Thus, CNTs and CNT composites show promise for use as electrically active biomaterials with high conductivity, biocompatibility and a potential to be distributed in a network throughout a composite thus inducing uniform bone growth.

CNT in combination with polymers shows enhanced effects as some researchers demonstrated that use of electro-spun PLA/MWNT nano-fibers showed greater extensions, directional outgrowth and proliferation over fibers without giving electrical stimulation [54]. Alignment of CNT on the nano-scaffold also affects tissue regeneration. Aligned CNT alone improved cell elongation, proliferation and electrical stimulation. Similar results were shown by Vila *et al.* in 2013 [55] where it was shown that electrical stimulation can have positive synergistic effects to cell proliferation when combined with the drug Zoledronic acid for treating osteoporosis.

### 7.4.2. *In vivo* Response of Carbon Nanotubes

Researchers investigated the role of a CNTs and chitosan as a nano-scaffold for bone tissue engineering [56]. Bone tissue developed on this scaffold are also transplanted in the in the back tissue of Balc type mice for 3 weeks. Results proved that there is increase in production of collagen, no inflammation and removal of the scaffold. Sitharaman *et al.* [57] evaluated CNT composites as porous biodegradable polymer *in vivo* for use as a synthetic bone substrate. Authors also depicted that CNT incorporation in the polymers showed drastically increase in bone tissue regeneration as compared to polymer alone. The histological scoring at 12 weeks was also significantly higher, indicating higher tissue quality, greater tissue ingrowth, reduced inflammatory cell density, increased connective tissue organization and a greater bone volume [14]. Scientists also proved that with decrease in size of CNT induces inflammation and its toxicological potential [56]. This is because nano-materials such as CNTs have relatively high surface area to volume ratios over which more chemical interactions can occur. Researchers depicted that CNTs showed both toxicological [57] and non-toxicological effects [58]. The true mechanism behind the toxicity is still to be understood [59]. The potential toxicological effects also depends on dosage and method of uptake [60]. CNTs in many studies showed proliferation and differentiation, induces no inflammation and toxicological aspects [60].

### 7.4.3. Necessities of Scaffolds

Tissue engineering aims to regenerate and maintain the function of human tissues by incorporating cell biology, materials science and engineering principles. Important components required for tissue engineering are harvested cells, recombinant signaling molecules, and 3D matrices. Cells and signalling molecules such as growth factors are seeded into highly porous biodegradable scaffolds, cultured *in vitro*, and subsequently the scaffolds are implanted into bone defects to induce and direct the growth of new bone. Signalling molecules can be coated onto the scaffolds or directly incorporated into them. Hence, the first and foremost function of a scaffold is its role as the substratum that allows cells to attach, proliferate, differentiate (*i.e.*, transform from a non-specific or primitive state into cells exhibiting the bone specific functions), and organize into normal, healthy bone as the scaffold degrades. Fig. (**1**) (4) summarizes most important factors involved in the optimized design of tissue engineering scaffolds.

### 7.4.4. Carbon Nanotubes as Scaffold Material

CNTs are the most exceptional candidate having advantageous features such as high surface area, high conductivity and biocompatible which can be employed as biomaterials or drug delivery system or carriers for molecules. Single-walled CNTs (SWCNTs) are quasi one-dimensional (1D) material and have been investigated *in vitro* as novel drug delivery vehicles [61 - 63]. SWCNTs can easily migrate and is permeable to all cell membranes and can transport various biomolecules into cells such as drugs, proteins, and small interfering RNA (siRNA) [64 - 67]. Biomaterials with exceptional mechanical properties are being exploited for importance in the development of implants used for tissue regeneration. CNTs have been characterized to possess excellent mechanical properties [68]. CNTs are employed as a significant material for various applications such as bio-sensing, transferring various agents and scaffolding for incorporating within the host's body [69]. Facile introduction of some functional group on the CNT incorporated scaffold enhanced the biocompatibility property which enhances the organ regeneration [69]. CNTs and polymer based composites are also used as scaffold material for musculoskeletal regeneration and are promising for orthopaedic application [68]. CNTs have potential to increase the conductivities of polymer based matrices which is required in some tissue where electrical signals are propagated, namely, neural tissue and cardiac muscle (Fig. **5**)[69].

Carbon is known for absorbing protein molecules efficiently through electrostatic interaction, which lead to a bland surface. In contrast with carbon, MWNTs have high surface area which provides microenvironment for surface immobilization of proteins. Some research groups also analyzed that carbon nanotubes have strong interactions with many kinds of proteins [70, 71]. The experimental results proved that MWNTs integrated in the nano-fibrous scaffold were favorable for the cell differentiation, enhancing absorption ability and proliferation. Roughness of the MWCNT integrated scaffold promotes better absorption of protein molecules [72]. MWCNT promotes cell growth, as MWCNT provides substratum for cells adhesion and proliferation. Researchers also reported that cells cultured on the MWNT incorporated nano-fibrous scaffold released the largest amount of proteins. The absorbed proteins formed a bioactive layer that was favorable to further cell attachment, proliferation, ECM secretion, and reconstruction.

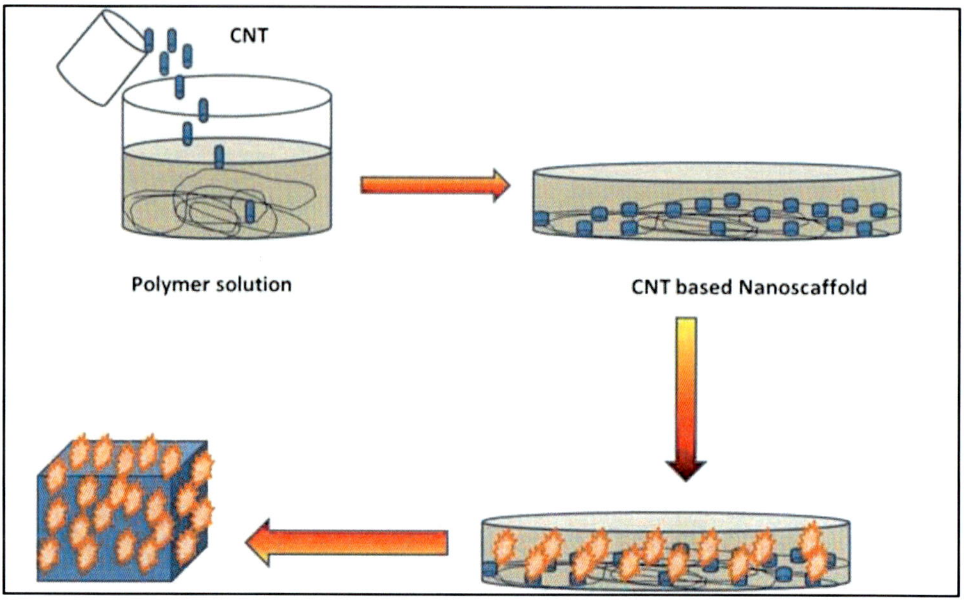


**Fig. (5).** Diagrammatic representation of carbon nanotubes as scaffold material.

### 7.4.5. Cell Secretion

Fibroblasts cells are directly related with tissue regeneration as they produce collagen during the process of wound healing and organization of new connective tissues. It was analyzed from earlier reported studies that MWCNTs incorporation in nanoscaffold approach play increment in cell secretion. The nano-fibrous architecture and carbon nanotubes component enhanced the cell proliferation and growth through stimulating cell secretion. The increased protein secretion provided favorable microenvironment for cell growth. Bioactive substances released act as signaling molecules and send signals to other cells of the same type as well as to themselves.

### 7.4.6. Cellular Morphology

Cells cultured on MWCNT incorporated substratum also shows migration between the cells. Migration in between the cells causes rapid assembly and disassembly, which depend upon complex cellular regulation that would be highly affected by the microenvironment around provided around by the engineered scaffolds for cell culture. Cell migration is likely to be an important phenomenon in tissue engineering, because the ability of cells to move will be an essential part of new tissue formation or regeneration. Researchers reported that the above-mentioned facts revealed that the nano-fibrous architecture incorporated with MWNTs provided a suitable microenvironment for cells migrating to aggregate and contact to communicate each other, which would encourage cells proliferation and tissue regeneration.

Carbon nanotubes (CNTs) have many distinct features such as high chemical stability, and biocompatible properties such as strong cell adhesion [73], protein absorption [74] and cell differentiation [75]. CNTs have exceptional mechanical strength while being flexible. consequently, they are able to maintain scaffolds' structural integrity during cell growth. In many studies, CNTs as nano-sheets in combination with natural or artificial polymers are being exploited as cell culture scaffold [76] or polymers [77]. In regeneration of many tissues such as bone, a three-dimensional (3D) scaffold is necessary. Studies suggested that CNT act as functional substrate for 3D scaffold. Currently, collagen sponges are widely used as 3D scaffolds because of their high biocompatibility. Therefore, CNT coating on the collagen sponge surface could improve its utility as a 3D scaffold.

Carbon nano-materials employed in tissue engineering are mostly carbon fibers and nanotubes (CNT), both single walled (SWCNT) and multiwalled (MWCNT) [78]. Scaffolds seeded with CNT have been reported to be biocompatible platforms for neuronal growth and differentiation [79 - 81]. Conversely, some scientist reported that CNT are cytotoxic [81 - 83] whereas some studies suggested that CNT act as excellent substrates for cellular growth [84 - 88].

CNTs have size analogous to ECM components such as collagens, which helps in growth and differentiation of neuron [89, 90]. CNT are conductive and stable. It is likely that a scaffold integrated with CNTs can facilitate electrical stimulation. All these exceptional features of CNTs make it suitable for neuron related medical applications. Furthermore, CNT also provides biocompatible microenvironment which can support neuron growth and differentiation [91 - 93]. Researchers have investigated that CNT promotes signal transmission when neurons cells seeded on a CNT network [94], as CNTs form tight contacts with neuron membranes which leads to electrical shortcuts [95]. Due to its conductivity and high mechanical strength, CNTs may act as a best scaffold material for differentiation of hESCs toward neuronal lineage.

Overall carbon nanotubes can be chosen as scaffold material because carbon nanotubes could be taken as an inorganic macromolecular material with pure carbon composition that is distinguished from the carbon element binding with hydrogen, nitrogen, or oxygen in the polymeric materials usually investigated. Carbon nanotubes have been attracted intense interests in biomedical fields in the recent years because of its excellent conductive property, mechanical strength, and binding capacity to biomolecules [95].

SWCNT–PPF nano-composites have also been explored to fabricate porous scaffolds with better mechanical properties for bone tissue engineering [96]. Scientist discovered that incorporating SWCNT into PPF matrix increases mechanical strength. However, there were large variations in the mechanical properties depending on the porosity of the scaffold. Scaffolds porosities have been shown to play a major role in determining the mechanical properties of the scaffolds in accordance with power-law relationships. In another report, MWCNTs was integrated with collagen sponge to improve the mechanical strength and surface properties of sponge. The prepared scaffold was employed to regenerate mouse osteoblast cells lines [97]. Researchers reported 2D thin film scaffolds with carbon nanotubes (CNTs), can selectively differentiate human embryonic stem cells into neuron cells while maintaining excellent cell viability. Results suggested that neuron differentiation efficiency on poly (acrylic acid) grafted CNT thin films is significant greater than that on poly (acrylic acid) thin films. Poly-L-ornithine surfaces, this new type thin film scaffold shows enhanced neuron differentiation. No noticeable cytotoxic effect difference has been detected between these two surfaces. The surface analysis and cell adhesion study have suggested that CNT-based surfaces can enhance protein adsorption and cell attachment [97].

MWCNT/CHI scaffolds were investigated for *in vitro* assays of cell adhesion, viability and proliferation onto the external surface of myoblastic mouse cell. In this approach we have studied the evolution of the cell line towards an osteoblastic lineage [98]. The study has been done both *in vitro* and *in vivo*, by implantation of MWCNT/CHI scaffolds adsorbed with rhBMP-2 in muscle tissue and evaluation of the ectopic formation of bone tissue after cofactor release [99]. Under these circumstances, scaffolds could direct cellular behavior and function, repopulating and re-synthesizing new natural matrixes and hence, opening promising perspectives for 3D construction of different tissues [99]. Porous SWCNT-polycarbonate membrane is also used for proliferation of osteoblast-like cells [100]. Honeycomb containing a MWCNT film was employed for cellular proliferation and adherence [101]. Polyurethane MWCNT nano-composites are also employed for culturing fibroblast cells [27, 101]. Cells cultured on these scaffold materials were capable of proliferating and secreting proteins. Researchers demonstrated that alginate SWCNT composite scaffold illustrated superior cellular adhesion in contrast to the alginate scaffold alone. Analysis results depicted that SWCNTs increased endothelial adhesion, and proliferation onto the alginate scaffold. Increase in cellular adhesion and proliferation is due to the fact that the increase in surface roughness property of the scaffold material (Fig. **6**).

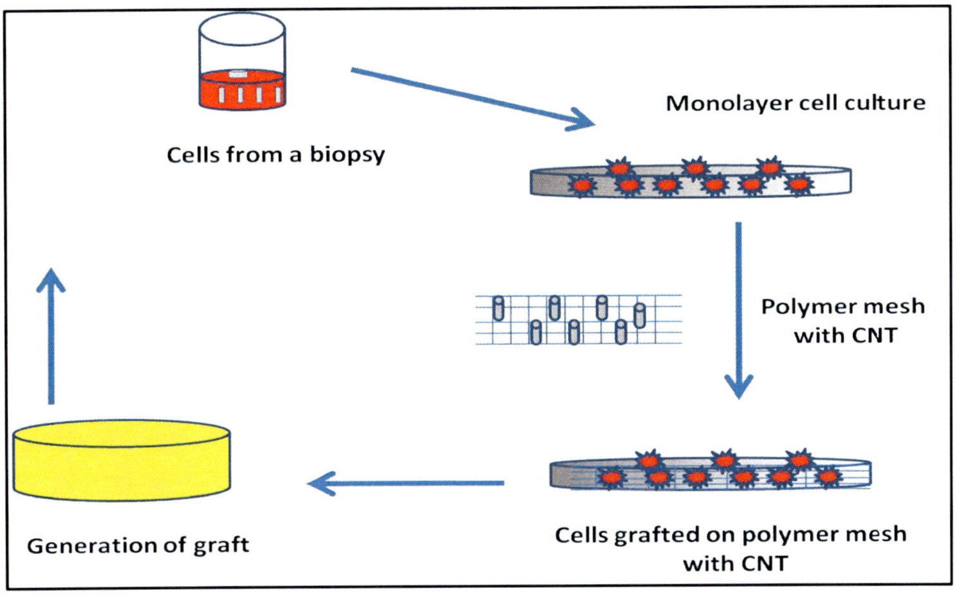


**Fig. (6).** Diagrammatic representation of culturing cells onto CNT grafted onto polymer mesh.

## 7.5. CNT FOR NEURON REGENERATION

Embryonic stem cells seeded on growth culture substratum formed large numbers of neural tube–like structures in the presence of growth factors. Neural precursors are differentiated into neurons, astrocytes, and oligo-dendrocytes. Differentiated neural are transplanted into the neonatal mouse brain where they differentiated into both neurons and astrocytes. Carbon nanotubes have been employed in nerve tissue engineering to investigate the enhanced cell behavior and growth. Researchers reported that nanotubes promote neuronal electrical activity but the mechanism behind the conductivity is inadequately understood. Carbon nanotubes might affect the collective electrical activity of cultured neuronal networks. These considerations offer a perspective that would allow us to predict or engineer interactions between neurons and carbon nanotubes.

CNT-based materials are exceptional promising nanomaterials for hESCs' neuron differentiation and implanting scaffolds that contains stem cells into injury sites and can promote axonal growth [102]. Scaffolds play an important role in injecting neural precursors into host tissues for differentiation and growth of nerve cells [103]. CNT-based scaffolds offer a new podium for co-transplanted cells to support, survive, and differentiate. Results also depicted that thin film scaffolds grafted onto CNTs, show considerable augmentation in neuron differentiation. Nanoscale fiber morphology created by CNTs surface enhances both protein adsorption and cell adhesion. Neural cells precursors can be differentiated using this type of CNT-based thin film, it will provide cell sources for different neuron lineages, thus enabling improved neuron injury repair compared to single lineage cell transplantation. Scaffolds integrated with CNT can act as compatible podium for transplanted cells to adhere, survive and differentiate in the antagonist environment. Surface modifications of CNTs can advent in tuning the hydrophobicity, surface charge and protein adsorption. It is expected that combining the topological cues offered by CNTs with the desirable biological cues which can be achieved by polymer grafting, will lead to further enhancement in selective differentiation of hESCs.

## 7.6. CNT FOR MUSCULOSKELETAL TISSUE

Musculoskeletal tissues such as bone, cartilage, meniscus, ligament, tendon, and muscles have specific architecture, composition, and functions. ECM of musculoskeletal tissues is highly specific and modified to perform the tissue specific function. Native ECM of musculoskeletal tissue is mainly composed of collagens, proteoglycans, elastin, and glycoproteins. All parameters associated with regeneration of musculo-skeletal tissue on nano-fiber scaffold should be taken into consideration.

The bone tissue is a mineralized organic matrix mostly formed from collagenous fibers and calcium phosphate in the form of hydroxyapatite (HA), with embedded osteoblasts, osteocytes and osteoclasts as cell components. Bone cells require hardness, porosity and mechanical strength for cell growth and differentiation. The majority of the prepared nano-scaffolds mimic the structural, compositional and biological characteristics of natural bone. They are based primarily on nano-HA, collagen, electro-spun silk, anodized titanium and nano-structured titanium surfaces [104 - 107]. Nanoscaffolds are known to induce osteogenesis and biomineralization with primary osteoblasts [108 - 110]; Primary growth of bone cells are restricted by many factors such as limited proliferative capacity; age or it may lose its property of differentiation through *in vitro* culture. Alternatively, for regeneration of bone, a three-dimensional (3D) scaffold is required. Bone regeneration requires biocompatibility, 3D cell proliferation and cell differentiation. Therefore CNTs would be functional as the substrate of a 3D scaffold. Researchers incorporated CNT in the collagen sponge surface which acts as 3D scaffold for regeneration of bone. MWCNTs were employed to improve its surface properties, the cell attachment, proliferation and conductivity [109]. Researchers concluded that cells sense and react to the MWNTs matrices by responding with excellent adhesion, proliferation and metabolic activity. The results presented here indicate that carbon nanotubes can mimic the nanofeatures of the native extracellular matrix and therefore, have an immense potential for application in tissue engineering. The realization of the potential of carbon nanotubes in bioapplications requires their chemical functionalization, which in turn impart the solubility of CNTs in various solvents. Surface groups of CNT permit its incorporation into polymer matrices and helped to explore the influence of CNTs surface chemistry on cellular bahavior.

Tutak *et al.*, 2010 [111] investigated the role of SWNT based substrates to support the growth of MC3T3-E1 bone osteoblasts. Authors investigated various properties such as physical and chemical properties of single-walled carbon nanotubes (SWNT) films. They investigated that the roughness of the SWNT surface promotes cellular processes, adhesion and proliferation. Result proved that wettable SWNT-COOH films exhibit the best growing conditions for MC3T3-E1 osteoblasts by promoting attachment and the development of cytoplasmic extensions.

Firkowska *et al.*, 2008 [112] investigated the interaction between human osteoblast cells and inorganic two-dimensional scaffolds based on multiwalled carbon nanotubes. In this approach, cross-linked carbon nanotubes are used as a cell growing scaffold. Cell substratum was evaluated for its biocompabitity and cell surface interactions. Osteoblast cells adhesion and morphology are investigated by means of scanning electron microscopy and cell viability assay. Furthermore, AFM is also employed to investigate the biocompatibility and adhesion of carbon nanotube (CNT)-based scaffolds. Topology observed indicates presence of tight junctions and a high-quality adhesion mechanism, which in turn indicates good biocompatibility of the matrices.

Tay *et al.* 2010 [113] described cellular behavior of human mesenchymal stem cells cultured on SWCNT. In this study, hMSCs were cultured on a thin mesh like layer of SWCNTs. SWCNT film provides large surface area which is better for cell growth as compared to cover slip. Results proved higher occurrence of filopodia at the cell boundaries when grown on SWCNT. Neurogenic markers and microtubule associated protein 2 genes were up-regulated while genes indicative of osteo-genesis remained at its nominal level. These results suggest that nano roughness alone is sufficient to modulate cellular behavior and early stage of stem cell lineage commitment without the aid of an induction medium.

Kalbacova *et al.* 2007 [114] investigated the influence of films of differently prepared single-walled carbon nanotubes (SWCNTs) and Ti6Al4V alloy on human osteoblasts was investigated. We focused on a metabolic activity of human osteoblasts as well as on their spreading and adherence ability on tested substrates. It was found that the metabolic activity of osteoblasts cultured on SWCNTs films was decreased of about 15% independently on the SWCNTs preparation technique. This result is comparable with that of osteoblasts cultured on Ti6Al4V alloy. It indicates that SWCNTs films are not toxic for human osteoblasts and could be used for biomedical applications.

Li *et al.* 2009 [115] investigated the first event occurring at the boundary between a metal implant and living tissue is the attachment of cells onto the metal surface of the implant. The attachment characteristics of the metal in this situation are critical in determining its biocompatibility and usefulness as artificial bone and tooth implants. Using the human osteosarcoma cell line Saos-2, we attempted to establish simple and reliable methods for evaluating the attachment of cultured osteo-blastic cells onto titanium samples that had been subjected to various surface treatments. Fluorescence actin imaging showed that cells cultured on titanium with hydrofluoric acid etching (HF-Ti) exhibited delayed spreading of their cytoplasm, as compared to cells cultured for the same length of time on nitrided titanium or physically polished titanium. The HF-Ti-cultured cells also exhibited poor assembly of focal contacts, as visualized by vinculin immunofluorescence. Furthermore, in motility assays based on an *in vitro* wound model, cells cultured on HF-Ti migrated more slowly than cells cultured on other titanium surfaces. These data suggest that Saos-2 cells attach less effectively to the HF-Ti surface. The methods described in this study should be useful for assessing the initial interactions of cultured cells with various materials, including metals.

Shia *et al.* 2007 [116] investigated the role of highly porous scaffolds consisting poly (propylene fumarate) (PPF) polymer, an ultra-short single-walled carbon nanotube (US-tube) nano-composite. Different characterization techniques were employed to analyze the pore structures of scaffolds. F-US-tube nano-composite provides the compressive modulus, offset yield strength, and compressive strength. Results concluded that functionalized ultra-short single-walled carbon nanotube nano-composite scaffolds with tunable porosity and mechanical properties hold great promise for bone tissue engineering applications.

### 7.6.1. Cellular Response to Carbon Nanotube-Based Matrices

Main properties of CNT *i.e.* mechanical reinforcement and electrical conductivity, which fascinated many researchers to employ CNTs in tissue engineering applications. Numerous studies have been conducted which analyze the interaction between CNTs or nano-composites and mammalian cells. If the collagen is incorporated with SWNTs, then cell shows improved cell viability and regeneration of smooth muscle cells [117]. Different functionalized CNTs are also exploited for the growth and regeneration of osteoblast cells [117]. These studies confirmed that bone cells sustained and proliferate itself on the cell surface.

Researchers have also exploited CNT for nerve cell growth [118]. Results proved that neurons grown on the CNTs functionalized with a bioactive molecule, 4-hydroxynonenal, developed multiple neurites and extensive branching. Scientist also concluded that the ability to control the neurite outgrowth also became possible by manipulating the charge carried by the functionalized carbon nanotubes. As shown by scientists [119], neurons plated on positively charged CNTs exhibited more numerous growth cones, longer neurite outgrowth, 2 n and more neurite branching in comparison with the neurons grown on negatively-charged nanotubes. Carbon nanotubes have also been used to create electrically-conductive polymers and tissue matrices with the capacity to provide controlled electrical stimulation. It has been reported that current-conducting CNT/polymer composites promote various osteoblast cell functions. By applying alternating current to these nano-composites, an increase in osteoblast proliferation by 46%, and calcium deposition by 307% has been observed [120]. This result suggests that CNT based composites may be used to stimulate bone formation. Other studies have been directed toward exploiting the electrical properties of CNTs for the purpose of healing neurological and brain-related injuries. Scientists [121] have used electrically-conductive LbL-assembled, modified SWNT films to stimulate the neurophysiological activity of neural cells.

Electrophysiological measurements showed that SWNT films can be used to electrically stimulate significant ion conductance in neuronal cells. This indicates a good electrical coupling between the LbL film and the neuronal cells in the lateral electrical configuration [122].

Although the above-mentioned studies mostly investigated cell adhesion and proliferation on randomly oriented nanotubes and CNT/polymers, carbon nanotubes can also be arranged into 3D matrices with well-defined periodic architectures. Correa-Duarte *et al.* [123], for example, created unique 3D structures from aligned CNTs grown on a silicon substrate, which subsequently formed honeycomb-like polygons in an oxidation process. This 3D CNT-based structure showed a good biocompatibility, proven by the excellent proliferation of mouse fibroblasts. In a related study, Giannona *et al.* [124] investigated the influence of periodic arrays of vertically-aligned carbon nanotubes (VACNTs) on the behavior of osteoblast cells. The authors observed a significant enhancement in cell attachment and proliferation as well as an influence on the cells' shape and orientation due to the periodicity of the CNTs. As pointed out previously, carbon nanotubes can be chemically functionalized in many different ways. This not only allows combining but even to replace other types of materials with CNTs. For instance, Scientists have demonstrated that adequately-functionalized carbon nanotubes can be used as a substitute for collagen to promote new bone growth [125]. In this study, chemically-treated carbon nanotubes served as nucleation sites for the deposition of hydroxyapatite. The negatively charged groups on the CNTs led to a self-assembly of HA crystals reaching a thickness of 3 μm after 14 days of mineralization. Similar observations for hydroxyapatite mineralization onto carbon nanotubes have been reported by other researchers [126].

## 7.7. CARBON NANOTUBES AS THERAPEUTICS AGENT

The regenerative medicine involves in developing techniques which can be useful to generate functional tissues, replace or repair tissues/organs lost due to disease [126]. In this approach, scientist employs various nanoparticles for delivery of tissue-inducing macromolecules or growth and differentiation of stem cells (Fig. 7). Among nanoparticles, CNTs are exploited much for therapeutic and drug-delivery applications in regenerative medicine owing to its exceptional physical and chemical properties.

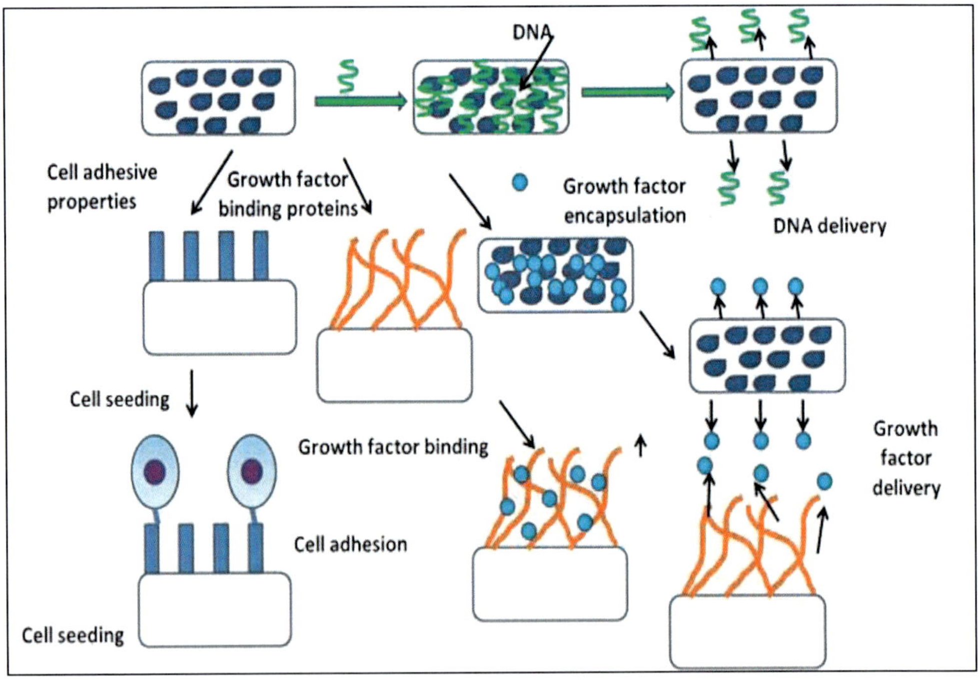


**Fig. (7).** Schematic representation of Growth factors and DNA releasing Nano scaffolds.

Several innovative drug delivery systems have been introduced to increase the therapeutic efficacy of many classes of drugs. Drug delivery should be efficient and targeted. CNTs are exploited for efficient drug delivery due to exceptional features mechanical, electrical, thermal, optical, and structural properties [127].

SWNTs have been broadly studied for their use as carriers for drugs and gene delivery *in vivo* [128, 129].

Furthermore, carbon nanotubes can be functionalized with various group entities which help in forming strong bond with biological molecules for targeted drug delivery. Functionalized CNTs have been able to cross cell membranes. For targeted drug delivery, drugs are covalently or non-covalently attached to the functionalized CNT. Drug loaded CNT are directed to the targeted cells as CNT shows internalization as it enters the cells. Cells are permeable to cell membranes and are good candidates to serve as drug delivery carriers to cells with high efficacy [130]. There are two promising means of CNT internalization: (a) through the endocytosis pathway and (b) through the endocytosis -independent pathway. CNTs can enter cells *via* the endocytosis process which is energy dependent process. CNTs can cross cell membranes *via* the endocytosis - independent pathways. CNTs can enter the cells *via* insertion and diffusion pathway. Venkatesan [131] studied the adsorption of erythropoietin (EPO) in a variety of porous carbon nano-materials, including CNTs. *In vivo* results indicated that CNTs show the highest performance and biological effect in the presence of an appropriate surfactant.

In addition, CNTs can act as therapeutic agent for instance, the strong optical absorption properties of SWCNTs provide them competent of generating acoustic waves upon irradiation. These waves have shown to induce differentiation of cells [132]. Due to nanosize of SWCNTs, it can improve retention and permeability in damaged tissues [124]. Additionally, functionalization of carbon nanotubes through various groups such as carboxylic group can form strong covalent bond with other bio-molecules [133]. Carboxylic acid groups can be introduced *via* oxidation by creating defects using acid treatments. CNTs can also be functionalized using non-covalent interactions [134 - 136]. CNT conjugated with functional group are employed for translocation of biological compound into cells and act as transfection agents.

## 7.8. CONCLUSIONS

Tissue Engineering is an interdisciplinary subject, which deciphers to produce functional three-dimensional (3D) tissues combining scaffolds, cells and/or bioactive molecules. Nowadays researchers are integrating nanotechnology in tissue engineering application for better proliferation and differentiation of stem cells. CNTs are most exceptional candidate having advantageous features such as high surface area, high conductivity and biocompatibility which can be employed as biomaterials or drug delivery system or carriers for molecules. Results proved that CNTs are mimicking the native ECM for proliferation and differentiation of stem cells. CNT along with other nanoparticles are also showing promising results in stem cells differentiation. CNT are also involved in the regenerative medicine which involves in developing techniques which can be useful to generate functional tissues, replace or repair tissues/organs lost due to disease. In this approach, scientist employs various nanoparticles for delivery of tissue-inducing macromolecules or growth and differentiation of stem cells. In nutshells, carbon nanotubes are the most exploited nano-materials in various biomedical application.

## CONFLICT OF INTEREST

The authors declare no conflict of interest, financial or otherwise.

## ACKNOWLEDGEMENTS

Thanks to all scientists referenced throughout the chapter whose valuable work has guided the way through to this research work.

## REFERENCES

[1] Griffith LG, Swartz MA. Capturing complex 3D tissue physiology *in vitro*. Nat Rev Mol Cell Biol 2006; 7(3): 211-24.
[http://dx.doi.org/10.1038/nrm1858] [PMID: 16496023]

[2] Teo WE, Ramakrishna S. A review on electrospinning design and nanofibre assemblies. Nanotechnology 2006; 17(14): R89-R106.
[http://dx.doi.org/10.1088/0957-4484/17/14/R01] [PMID: 19661572]

[3] Pham QP, Sharma U, Mikos AG. Electrospinning of polymeric nanofibers for tissue engineering applications: a review. Tissue Eng 2006; 12(5): 1197-211.
[http://dx.doi.org/10.1089/ten.2006.12.1197] [PMID: 16771634]

[4] Subbiah T, Bhat G, Tock R, Parameswaran S, Ramkumar S. Electrospinning of nanofibers. J Appl Polym Sci 2005; 96: 557-69.
[http://dx.doi.org/10.1002/app.21481]

[5] Langer R, Vacanti JP. Tissue engineering. Science 1993; 260(5110): 920-6.
[http://dx.doi.org/10.1126/science.8493529] [PMID: 8493529]

[6] Zhong S, Teo WE, Zhu X, Beuerman RW, Ramakrishna S, Yung LY. An aligned nanofibrous collagen scaffold by electrospinning and its effects on *in vitro* fibroblast culture. J Biomed Mater Res A 2006; 79(3): 456-63.
[http://dx.doi.org/10.1002/jbm.a.30870] [PMID: 16752400]

[7] Bhattarai N, Edmondson D, Veiseh O, Matsen FA, Zhang M. Electrospun chitosan-based nanofibers and their cellular compatibility. Biomaterials 2005; 26(31): 6176-84.
[http://dx.doi.org/10.1016/j.biomaterials.2005.03.027] [PMID: 15885770]

[8] Noh HK, Lee SW, Kim JM, *et al.* Electrospinning of chitin nanofibers: degradation behavior and cellular response to normal human keratinocytes and fibroblasts. Biomaterials 2006; 27(21): 3934-44.
[http://dx.doi.org/10.1016/j.biomaterials.2006.03.016] [PMID: 16574218]

[9] Min BM, Jeong L, Nam YS, Kim JM, Kim JY, Park WH. Formation of silk fibroin matrices with different texture and its cellular response to normal human keratinocytes. Int J Biol Macromol 2004; 34(5): 281-8.

[http://dx.doi.org/10.1016/j.ijbiomac.2004.08.004] [PMID: 15556229]

[10] Xin X, Hussain M, Mao JJ. Continuing differentiation of human mesenchymal stem cells and induced chondrogenic and osteogenic lineages in electrospun PLGA nanofiber scaffold. Biomaterials 2007; 28(2): 316-25.
[http://dx.doi.org/10.1016/j.biomaterials.2006.08.042] [PMID: 17010425]

[11] Jeong SI, Kim SY, Cho SK, *et al.* Tissue-engineered vascular grafts composed of marine collagen and PLGA fibers using pulsatile perfusion bioreactors. Biomaterials 2007; 28(6): 1115-22.
[http://dx.doi.org/10.1016/j.biomaterials.2006.10.025] [PMID: 17112581]

[12] Thomas V, Jose MV, Chowdhury S, Sullivan JF, Dean DR, Vohra YK. Mechano-morphological studies of aligned nanofibrous scaffolds of polycaprolactone fabricated by electrospinning. J Biomater Sci Polym Ed 2006; 17(9): 969-84.
[http://dx.doi.org/10.1163/156856206778366022] [PMID: 17094636]

[13] Baker SC, Atkin N, Gunning PA, *et al.* Characterisation of electrospun polystyrene scaffolds for three-dimensional *in vitro* biological studies. Biomaterials 2006; 27(16): 3136-46.
[http://dx.doi.org/10.1016/j.biomaterials.2006.01.026] [PMID: 16473404]

[14] Li WJ, Laurencin CT, Caterson EJ, Tuan RS, Ko FK. Electrospun nanofibrous structure: a novel scaffold for tissue engineering. J Biomed Mater Res 2002; 60(4): 613-21.
[http://dx.doi.org/10.1002/jbm.10167] [PMID: 11948520]

[15] Yoshimoto H, Shin YM, Terai H, Vacanti JP. A biodegradable nanofiber scaffold by electrospinning and its potential for bone tissue engineering. Biomaterials 2003; 24(12): 2077-82.
[http://dx.doi.org/10.1016/S0142-9612(02)00635-X] [PMID: 12628828]

[16] Xu CY, Inai R, Kotaki M, Ramakrishna S. Aligned biodegradable nanofibrous structure: a potential scaffold for blood vessel engineering. Biomaterials 2004; 25(5): 877-86.
[http://dx.doi.org/10.1016/S0142-9612(03)00593-3] [PMID: 14609676]

[17] Zong X, Bien H, Chung C, *et al.* Electrospun non-woven membranes as scaffolds for heart tissue constructs. Polymer Prepr 2003; 44: 96-7.

[18] Yang F, Murugan R, Ramakrishna S, Wang X, Ma YX, Wang S. Fabrication of nano-structured porous PLLA scaffold intended for nerve tissue engineering. Biomaterials 2004; 25(10): 1891-900.
[http://dx.doi.org/10.1016/j.biomaterials.2003.08.062] [PMID: 14738853]

[19] Silva GA, Czeisler C, Niece KL, *et al.* Selective differentiation of neural progenitor cells by high-epitope density nanofibers. Science 2004; 303(5662): 1352-5.
[http://dx.doi.org/10.1126/science.1093783] [PMID: 14739465]

[20] Shapiro SD. Matrix metalloproteinase degradation of extracellular matrix: biological consequences. Curr Opin Cell Biol 1998; 10(5): 602-8.
[http://dx.doi.org/10.1016/S0955-0674(98)80035-5] [PMID: 9818170]

[21] Hynes R O, Naba A. Overview of the matrisome – an inventory of extracellular matrix constituents and functions. Cold Spring Harb 1231 Perspect Biol 2012; 4: a004903.

[22] Naba A, Hoersch S, Hynes RO, Clauser KR, Carr SA. The Matrisome Project, The Hynes Lab, pp The Hynes Lab page of the matrisome 1234 project, complete with resources, tables and data on the matrisome 1235. 2011.

[23] Ricard-Blum S. The collagen family. Csh Perspect Biol 2011; p. 3.

[24] Chaudhuri O, Mooney DJ. Stem-cell differentiation: Anchoring cell-fate cues. Nat Mater 2012; 11(7): 568-9.
[http://dx.doi.org/10.1038/nmat3366] [PMID: 22717486]

[25] Kilian KA, Bugarija B, Lahn BT, Mrksich M. Geometric cues for directing the differentiation of mesenchymal stem cells. Proc Natl Acad Sci USA 2010; 107(11): 4872-7.
[http://dx.doi.org/10.1073/pnas.0903269107] [PMID: 20194780]

[26] Abdullah CA, Asanithi P, Brunner EW, *et al.* Aligned, isotropic and patterned carbon nanotube substrates that control the growth and alignment of Chinese hamster ovary cells. Nanotechnology 2011; 22(20): 205102.
[http://dx.doi.org/10.1088/0957-4484/22/20/205102] [PMID: 21444962]

[27] Meng WX, Takeichi M. Adherens junction: molecular architecture and regulation. Csh Perspect Biol 2009; p. 1.

[28] Humtsoe JO, Feng S, Thakker GD, Yang J, Hong J, Wary KK. Regulation of cell-cell interactions by phosphatidic acid phosphatase 2b/VCIP. EMBO J 2003; 22(7): 1539-54.
[http://dx.doi.org/10.1093/emboj/cdg165] [PMID: 12660161]

[29] Franke WW. Discovering the molecular components of intercellular junctions-a historical view. Csh Perspect Biol 2009; p. 1.

[30] Green KJ, Getsios S, Troyanovsky S, Godsel LM. Intercellular junction assembly, dynamics, and homeostasis. Csh Perspect Biol 2010; p. 2.

[31] Launey ME, Buehler MJ, Ritchie RO. On the mechanistic origins of toughness in bone. Annu Rev Mater Res 2010; 40: 25-53.
[http://dx.doi.org/10.1146/annurev-matsci-070909-104427]

[32] Fratzl P, Gupta HS, Paschalis EP, Roschger P. Structure and mechanical quality of the collagen-mineral nano-composite in bone. J Mater Chem 2004; 14: 2115-23.
[http://dx.doi.org/10.1039/B402005G]

[33] McNally EA, Schwarcz HP, Botton GA, Arsenault AL. A model for the ultrastructure of bone based on electron microscopy of ion-milled sections. PLoS One 2012; 7(1): e29258.
[http://dx.doi.org/10.1371/journal.pone.0029258] [PMID: 22272230]

[34] Ritchie RO. The conflicts between strength and toughness. Nat Mater 2011; 10(11): 817-22.
[http://dx.doi.org/10.1038/nmat3115] [PMID: 22020005]

[35] Aryal S, Bhattarai SR, Bahadur R, Khil MS, Lee DR, Kim HY. Carbon nanotubes assisted biomimetic synthesis of hydroxyapatite from simulated body fluid. Mat Sci. Eng Struct 2006; 426: 202-7.
[http://dx.doi.org/10.1016/j.msea.2006.04.004]

[36] Zhao B, Hu H, Mandal SK, Haddon RC. A bone mimic based on the self-assembly of hydroxyapatite on chemically functionalized single-walled carbon nanotubes. Chem Mater 2005; 17: 3235-41.
[http://dx.doi.org/10.1021/cm0500399]

[37] Liao S, Xu G, Wang W, *et al.* Self-assembly of nano-hydroxyapatite on multi-walled carbon nanotubes. Acta Biomater 2007; 3(5): 669-75.
[http://dx.doi.org/10.1016/j.actbio.2007.03.007] [PMID: 17512807]

[38] Lobo AO, Marciano FR, Regiani I, Ramos SC, Matsushima JT, Corat EJ. Proposed model for growth preference of plate-like nanohydroxyapatite crystals on superhydrophilic vertically aligned carbon nanotubes by electrodeposition. Theor Chem Acc 2011; 130: 1071-82.
[http://dx.doi.org/10.1007/s00214-011-0993-x]

[39] White AA, Best SM, Kinloch IA. Hydroxyapatite-carbon nanotube composites for biomedical applications: a review. Int J Appl Ceram Technol 2007; 4: 1-13.
[http://dx.doi.org/10.1111/j.1744-7402.2007.02113.x]

[40] White AA, Windle AH, Kinloch IA, Best SM. Preparation and properties of carbon nanotube-reinforced hydroxyapatite. Key Eng Mater 2008; 361–363: 419-22.
[http://dx.doi.org/10.4028/www.scientific.net/KEM.361-363.419]

[41] Beuvelot J, Bergeret C, Mallet R, *et al.* *In vitro* calcification of chemically functionalized carbon nanotubes. Acta Biomater 2010; 6(10): 4110-7.
[http://dx.doi.org/10.1016/j.actbio.2010.05.011] [PMID: 20493282]

[42] Fratzl P, Gupta HS, Paschalis EP, Roschger P. Structure and mechanical quality of the collagen-

mineral nano-composite in bone. 1285. J Mater Chem 2004; 14: 2115-23. [http://dx.doi.org/10.1039/B402005G]

[43] Ogihara N, Usui Y, Aoki K, *et al.* Biocompatibility and bone tissue compatibility of alumina ceramics reinforced with carbon nanotubes. Nanomedicine (Lond) 2012; 7(7): 981-93. [http://dx.doi.org/10.2217/nnm.12.1] [PMID: 22401267]

[44] Cheng Q, Rutledge K, Jabbarzadeh E. Carbon nanotube-poly(lactide- co-glycolide) composite scaffolds for bone tissue engineering applica tions. Ann Biomed Eng 2013; 1454.

[45] Jell G, Verdejo R, Safinia L, Shaffer MS, Stevens MM, Bismarck A. Carbon nanotube-enhanced polyurethane scaffolds fabricated by thermally induced phase separation. J Mater Chem 2008; 18: 1865-72. [http://dx.doi.org/10.1039/b716109c]

[46] Lin C, Wang Y, Lai Y, *et al.* Incorporation of carboxylation multiwalled carbon nanotubes into biodegradable poly(lactic-co-glycolic acid) for bone tissue engineering. Colloids Surf B Biointerfaces 2011; 83(2): 367-75. [http://dx.doi.org/10.1016/j.colsurfb.2010.12.011] [PMID: 21208787]

[47] Im O, Li J, Wang M, Zhang LG, Keidar M. Biomimetic three-dimensional nanocrystalline hydroxyapatite and magnetically synthesized single-walled carbon nanotube chitosan nanocomposite for bone regeneration. Int J Nanomedicine 2012; 7: 2087-99. [PMID: 22619545]

[48] Pan L, Pei X, He R, Wan Q, Wang J. Multiwall carbon nanotubes/polycaprolactone composites for bone tissue engineering application. Colloids Surf B Biointerfaces 2012; 93: 226-34. [http://dx.doi.org/10.1016/j.colsurfb.2012.01.011] [PMID: 22305638]

[49] Ormsby R, McNally T, O'Hare P, Burke G, Mitchell C, Dunne N. Fatigue and biocompatibility properties of a poly(methyl methacrylate) bone cement with multi-walled carbon nanotubes. Acta Biomater 2012; 8(3): 1201-12. [http://dx.doi.org/10.1016/j.actbio.2011.10.010] [PMID: 22023747]

[50] Yonemori K, Matsunaga S, Ishidou Y, Maeda S, Yoshida H. Early effects of electrical stimulation on osteogenesis. Bone 1996; 19(2): 173-80. [http://dx.doi.org/10.1016/8756-3282(96)00169-X] [PMID: 8853862]

[51] Hronik-Tupaj M, Rice WL, Cronin-Golomb M, Kaplan DL, Georgakoudi I. Osteoblastic differentiation and stress response of human mesenchymal stem cells exposed to alternating current electric fields. Biomed Eng Online 2011; 10: 9. [http://dx.doi.org/10.1186/1475-925X-10-9] [PMID: 21269490]

[52] Aaron RK, Boyan BD, Ciombor DM, Schwartz Z, Simon BJ. Stimulation of growth factor synthesis by electric and electromagnetic fields. Clin Orthop Relat Res 2004; (419): 30-7. [http://dx.doi.org/10.1097/00003086-200402000-00006] [PMID: 15021128]

[53] Supronowicz PR, Ajayan PM, Ullmann KR, Arulanandam BP, Metzger DW, Bizios R. Novel current-conducting composite substrates for exposing osteoblasts to alternating current stimulation. J Biomed Mater Res 2002; 59(3): 499-506. [http://dx.doi.org/10.1002/jbm.10015] [PMID: 11774308]

[54] Shao S, Zhou S, Li L, *et al.* Osteoblast function on electrically conductive electrospun PLA/MWCNTs nanofibers. Biomaterials 2011; 32(11): 2821-33. [http://dx.doi.org/10.1016/j.biomaterials.2011.01.051] [PMID: 21292320]

[55] Vila M, Cicuéndez M, Sánchez-Marcos J, *et al.* Electrical stimuli to increase cell proliferation on carbon nanotubes/mesoporous silica composites for drug delivery. J Biomed Mater Res A 2013; 101(1): 213-21. [http://dx.doi.org/10.1002/jbm.a.34325] [PMID: 22847920]

[56] Abarrategi A, Gutiérrez MC, Moreno-Vicente C, *et al.* Multiwall carbon nanotube scaffolds for tissue

engineering purposes. Biomaterials 2008; 29(1): 94-102.
[http://dx.doi.org/10.1016/j.biomaterials.2007.09.021] [PMID: 17928048]

[57] Sitharaman B, Shi X, Walboomers XF, *et al. In vivo* biocompatibility of ultra-short single-walled carbon nanotube/biodegradable polymer nanocomposites for bone tissue engineering. Bone 2008; 43(2): 362-70.
[http://dx.doi.org/10.1016/j.bone.2008.04.013] [PMID: 18541467]

[58] Lam CW, James JT, McCluskey R, Hunter RL. Pulmonary toxicity of single-wall carbon nanotubes in mice 7 and 90 days after intratracheal instillation. Toxicol Sci 2004; 77(1): 126-34.
[http://dx.doi.org/10.1093/toxsci/kfg243] [PMID: 14514958]

[59] Kang S, Herzberg M, Rodrigues DF, Elimelech M. Antibacterial effects of carbon nanotubes: size does matter! Langmuir 2008; 24(13): 6409-13.
[http://dx.doi.org/10.1021/la800951v] [PMID: 18512881]

[60] Lu FS, Gu LR, Meziani MJ, Wang X, Luo PG, Veca LM, *et al.* Advances in bioapplications of carbon nanotubes. Adv Mater 2009; 21: 139-52.
[http://dx.doi.org/10.1002/adma.200801491]

[61] Feazell RP, Nakayama-Ratchford N, Dai H, Lippard SJ. Soluble single-walled carbon nanotubes as longboat delivery systems for platinum(IV) anticancer drug design. J Am Chem Soc 2007; 129(27): 8438-9.
[http://dx.doi.org/10.1021/ja073231f] [PMID: 17569542]

[62] Liu Z, Sun X, Nakayama-Ratchford N, Dai H. Supramolecular chemistry on water-soluble carbon nanotubes for drug loading and delivery. ACS Nano 2007; 1(1): 50-6.
[http://dx.doi.org/10.1021/nn700040t] [PMID: 19203129]

[63] Bianco A, Kostarelos K, Prato M. Applications of carbon nanotubes in drug delivery. Curr Opin Chem Biol 2005; 9(6): 674-9.
[http://dx.doi.org/10.1016/j.cbpa.2005.10.005] [PMID: 16233988]

[64] Bhavna SP, Sitharaman B. Carbon nanotubes in regenerative medicine.Carbon nanotubes for biomedical applications. Germany: Springer 2012; pp. 27-37.

[65] Liu Z, Chen K, Davis C, *et al.* Drug delivery with carbon nanotubes for *in vivo* cancer treatment. Cancer Res 2008; 68(16): 6652-60.
[http://dx.doi.org/10.1158/0008-5472.CAN-08-1468] [PMID: 18701489]

[66] Veetil JV, Ye K. Tailored carbon nanotubes for tissue engineering applications. Biotechnol Prog 2009; 25(3): 709-21.
[http://dx.doi.org/10.1002/btpr.165] [PMID: 19496152]

[67] Gupta A, Woods MD, Illingworth KD, *et al.* Single walled carbon nanotube composites for bone tissue engineering. J Orthop Res 2013; 31(9): 1374-81.
[http://dx.doi.org/10.1002/jor.22379] [PMID: 23629922]

[68] Klumpp C, Kostarelos K, Prato M, Bianco A. Functionalized carbon nanotubes as emerging nanovectors for the delivery of therapeutics. Biochim Biophys Acta 2006; 1758: 404-12.

[69] Pastorin G. Crucial functionalizations of carbon nanotubes for improved drug delivery: a valuable option? Pharm Res 2009; 26(4): 746-69.
[http://dx.doi.org/10.1007/s11095-008-9811-0] [PMID: 19142717]

[70] Elias KL, Price RL, Webster TJ. Enhanced functions of osteoblasts on nanometer diameter carbon fibers. Biomaterials 2002; 23(15): 3279-87.
[http://dx.doi.org/10.1016/S0142-9612(02)00087-X] [PMID: 12102199]

[71] Chaudhuri O, Mooney DJ. Stem-cell differentiation: Anchoring cell-fate cues. Nat Mater 2012; 11(7): 568-9.
[http://dx.doi.org/10.1038/nmat3366] [PMID: 22717486]

[72] Kilian K A, Bugarija B, Lahn B T, Mrksich M. Geometric cues for 1250 directing the differentiation of mesenchymal stem cells. Proc Natl Acad 1251 Sci USA 2010; 107: 4872-7.

[73] Aoki N, Yokoyama A, Nodasaka Y, *et al.* Journal of Biomedical Materials Research Part B: Applied Biomaterials cells grown on carbon nanotubes. Chem Lett 2006; 35: 508-9. [http://dx.doi.org/10.1246/cl.2006.508]

[74] Aoki N, Akasaka T, Watari F, Yokoyama A. Carbon nanotubes as scaffolds for cell culture and effect on cellular functions. Dent Mater J 2007; 26(2): 178-85. [http://dx.doi.org/10.4012/dmj.26.178] [PMID: 17621932]

[75] Zanello LP, Zhao B, Hu H, Haddon RC. Bone cell proliferation on carbon nanotubes. Nano Lett 2006; 6(3): 562-7. [http://dx.doi.org/10.1021/nl051861e] [PMID: 16522063]

[76] MacDonald RA, Laurenzi BF, Viswanathan G, Ajayan PM, Stegemann JP. Collagen-carbon nanotube composite materials as scaffolds in tissue engineering. J Biomed Mater Res A 2005; 74(3): 489-96. [http://dx.doi.org/10.1002/jbm.a.30386] [PMID: 15973695]

[77] Shi X, Sitharaman B, Pham QP, *et al.* Fabrication of porous ultra-short single-walled carbon nanotube nanocomposite scaffolds for bone tissue engineering. Biomaterials 2007; 28(28): 4078-90. [http://dx.doi.org/10.1016/j.biomaterials.2007.05.033] [PMID: 17576009]

[78] Chen X, Tam UC, Czlapinski JL, *et al.* Interfacing carbon nanotubes with living cells. J Am Chem Soc 2006; 128(19): 6292-3. [http://dx.doi.org/10.1021/ja060276s] [PMID: 16683774]

[79] Mattson MP, Haddon RC, Rao AM. Molecular functionalization of carbon nanotubes and use as substrates for neuronal growth. J Mol Neurosci 2000; 14(3): 175-82. [http://dx.doi.org/10.1385/JMN:14:3:175] [PMID: 10984193]

[80] Hu H, Ni Y, Montana V, Haddon RC, Parpura V. Chemically functionalized carbon nanotubes as substrates for neuronal growth. Nano Lett 2004; 4(3): 507-11. [http://dx.doi.org/10.1021/nl035193d] [PMID: 21394241]

[81] Hu H, Ni Y, Mandal SK, *et al.* Polyethyleneimine functionalized single-walled carbon nanotubes as a substrate for neuronal growth. J Phys Chem B 2005; 109(10): 4285-9. [http://dx.doi.org/10.1021/jp0441137] [PMID: 16851493]

[82] Jia G, Wang H, Yan L, *et al.* Cytotoxicity of carbon nanomaterials: single-wall nanotube, multi-wall nanotube, and fullerene. Environ Sci Technol 2005; 39(5): 1378-83. [http://dx.doi.org/10.1021/es048729l] [PMID: 15787380]

[83] Shvedova AA, Castranova V, Kisin ER, *et al.* Exposure to carbon nanotube material: assessment of nanotube cytotoxicity using human keratinocyte cells. J Toxicol Environ Health A 2003; 66(20): 1909-26. [http://dx.doi.org/10.1080/713853956] [PMID: 14514433]

[84] Cui D, Tian F, Ozkan CS, Wang M, Gao H. Effect of single wall carbon nanotubes on human HEK293 cells. Toxicol Lett 2005; 155(1): 73-85. [http://dx.doi.org/10.1016/j.toxlet.2004.08.015] [PMID: 15585362]

[85] Pantarotto D, Briand JP, Prato M, Bianco A. Translocation of bioactive peptides across cell membranes by carbon nanotubes. Chem Commun (Camb) 2004; 10(1): 16-7. [http://dx.doi.org/10.1039/b311254c] [PMID: 14737310]

[86] Shi Kam NW, Jessop TC, Wender PA, Dai H. Nanotube molecular transporters: internalization of carbon nanotube-protein conjugates into Mammalian cells. J Am Chem Soc 2004; 126(22): 6850-1. [http://dx.doi.org/10.1021/ja0486059] [PMID: 15174838]

[87] Lu Q, Moore JM, Huang G, Mount AS, Rao AM, Larcom LL, *et al.* RNA polymer translocation with single-walled carbon nanotubes. Nano Lett 2004; 4: 2473-7.

[http://dx.doi.org/10.1021/nl048326j]

[88] Cherukuri P, Bachilo SM, Litovsky SH, Weisman RB. Near-infrared fluorescence microscopy of single-walled carbon nanotubes in phagocytic cells. J Am Chem Soc 2004; 126(48): 15638-9. [http://dx.doi.org/10.1021/ja0466311] [PMID: 15571374]

[89] Jan E, Kotov NA. Successful differentiation of mouse neural stem cells on layer-by-layer assembled single-walled carbon nanotube composite. Nano Lett 2007; 7(5): 1123-8. [http://dx.doi.org/10.1021/nl0620132] [PMID: 17451277]

[90] Ni Y, Hu H, Malarkey EB, *et al.* Chemically functionalized water soluble single-walled carbon nanotubes modulate neurite outgrowth. J Nanosci Nanotechnol 2005; 5(10): 1707-12. [http://dx.doi.org/10.1166/jnn.2005.189] [PMID: 16245532]

[91] Sridharan I, Kim T, Wang R. Adapting collagen/CNT matrix in directing hESC differentiation. Biochem Biophys Res Commun 2009; 381(4): 508-12. [http://dx.doi.org/10.1016/j.bbrc.2009.02.072] [PMID: 19233124]

[92] Lovat V, Pantarotto D, Lagostena L, *et al.* Carbon nanotube substrates boost neuronal electrical signaling. Nano Lett 2005; 5(6): 1107-10. [http://dx.doi.org/10.1021/nl050637m] [PMID: 15943451]

[93] Cellot G, Cilia E, Cipollone S, *et al.* Carbon nanotubes might improve neuronal performance by favouring electrical shortcuts. Nat Nanotechnol 2009; 4(2): 126-33. [http://dx.doi.org/10.1038/nnano.2008.374] [PMID: 19197316]

[94] Shi X, Sitharaman B, Pham QP, *et al.* Fabrication of porous ultra-short single-walled carbon nanotube nanocomposite scaffolds for bone tissue engineering. Biomaterials 2007; 28(28): 4078-90. [http://dx.doi.org/10.1016/j.biomaterials.2007.05.033] [PMID: 17576009]

[95] Lin C, Wang Y, Lai Y, *et al.* Incorporation of carboxylation multiwalled carbon nanotubes into biodegradable poly(lactic-co-glycolic acid) for bone tissue engineering. Colloids Surf B Biointerfaces 2011; 83(2): 367-75. [http://dx.doi.org/10.1016/j.colsurfb.2010.12.011] [PMID: 21208787]

[96] Chao TI, Xiang S, Chen CS, *et al.* Carbon nanotubes promote neuron differentiation from human embryonic stem cells. Biochem Biophys Res Commun 2009; 384(4): 426-30. [http://dx.doi.org/10.1016/j.bbrc.2009.04.157] [PMID: 19426708]

[97] Abarrategi A, Gutiérrez MC, Moreno-Vicente C, *et al.* Multiwall carbon nanotube scaffolds for tissue engineering purposes. Biomaterials 2008; 29(1): 94-102. [http://dx.doi.org/10.1016/j.biomaterials.2007.09.021] [PMID: 17928048]

[98] Lee KY, Peters MC, Anderson KW, Mooney DJ. Controlled growth factor release from synthetic extracellular matrices. Nature 2000; 408(6815): 998-1000. [http://dx.doi.org/10.1038/35050141] [PMID: 11140690]

[99] Kim DH, Richardson-Burns SM, Hendricks JL, Sequera C, Martin DC. Effect of immobilized nerve growth factor on conductive polymers: Electrical properties and cellular response. Adv Funct Mater 2007; 17: 79-86. [http://dx.doi.org/10.1002/adfm.200500594]

[100] Aoki N, Yokoyama A, Nodasaka Y, Akasaka T, Uo M, Sato Y, *et al.* Cell culture on a carbon nanotube scaffold. J Biomed Nanotechnol 2005; 1(4): 402-5. [http://dx.doi.org/10.1166/jbn.2005.048]

[101] Khang D, Kim SY, Liu-Snyder P, Palmore GT, Durbin SM, Webster TJ. Enhanced fibronectin adsorption on carbon nanotube/poly(carbonate) urethane: independent role of surface nano-roughness and associated surface energy. Biomaterials 2007; 28(32): 4756-68. [http://dx.doi.org/10.1016/j.biomaterials.2007.07.018] [PMID: 17706277]

[102] Teng YD, Lavik EB, Qu X, *et al.* Functional recovery following traumatic spinal cord injury mediated by a unique polymer scaffold seeded with neural stem cells. Proc Natl Acad Sci USA 2002; 99(5):

3024-9.
[http://dx.doi.org/10.1073/pnas.052678899] [PMID: 11867737]

[103] Wu S, Suzuki Y, Kitada M, *et al.* Migration, integration, and differentiation of hippocampus-derived neurosphere cells after transplantation into injured rat spinal cord. Neurosci Lett 2001; 312(3): 173-6. [http://dx.doi.org/10.1016/S0304-3940(01)02219-4] [PMID: 11602338]

[104] Sato M, Slamovich EB, Webster TJ. Enhanced osteoblast adhesion on hydrothermally treated hydroxyapatite/titania/poly(lactide-co-glycolide) sol-gel titanium coatings. Biomaterials 2005; 26(12): 1349-57. [http://dx.doi.org/10.1016/j.biomaterials.2004.04.044] [PMID: 15482822]

[105] Jin HJ, Chen J, Karageorgiou V, Altman GH, Kaplan DL. Human bone marrow stromal cell responses on electrospun silk fibroin mats. Biomaterials 2004; 25(6): 1039-47. [http://dx.doi.org/10.1016/S0142-9612(03)00609-4] [PMID: 14615169]

[106] Yao C, Slamovich EB, Webster TJ. Enhanced osteoblast functions on anodized titanium with nanotube-like structures. J Biomed Mater Res A 2008; 85(1): 157-66. [http://dx.doi.org/10.1002/jbm.a.31551] [PMID: 17688267]

[107] Khang D, Lu J, Yao C, Haberstroh KM, Webster TJ. The role of nanometer and sub-micron surface features on vascular and bone cell adhesion on titanium. Biomaterials 2008; 29(8): 970-83. [http://dx.doi.org/10.1016/j.biomaterials.2007.11.009] [PMID: 18096222]

[108] Woo KM, Jun JH, Chen VJ, *et al.* Nano-fibrous scaffolding promotes osteoblast differentiation and biomineralization. Biomaterials 2007; 28(2): 335-43. [http://dx.doi.org/10.1016/j.biomaterials.2006.06.013] [PMID: 16854461]

[109] Hu J, Liu X, Ma PX. Induction of osteoblast differentiation phenotype on poly(L-lactic acid) nanofibrous matrix. Biomaterials 2008; 29(28): 3815-21. [http://dx.doi.org/10.1016/j.biomaterials.2008.06.015] [PMID: 18617260]

[110] Wang J, Valmikinathan CM, Liu W, Laurencin CT, Yu X. Spiral-structured, nanofibrous, 3D scaffolds for bone tissue engineering. J Biomed Mater Res A 2010; 93(2): 753-62. [PMID: 19642211]

[111] Tutak W, Chhowalla M, Sesti F. The chemical and physical characteristics of single-walled carbon nanotube film impact on osteoblastic cell response. Nanotechnology 2010; 21(31): 315102-7. [http://dx.doi.org/10.1088/0957-4484/21/31/315102] [PMID: 20622299]

[112] Firkowska I, Godehardt E, Giersig M. Interaction Between Human Osteoblast Cells and Inorganic Two-Dimensional Scaffolds Based on Multiwalled Carbon Nanotubes: A Quantitative AFM Study. Adv Funct Mater 2008; 18: 3765-71. [http://dx.doi.org/10.1002/adfm.200800760]

[113] Tay C Y, Gu H, Shing Leong W, *et al.* Cellular behavior of human mesenchymal stem cells cultured on single-walled carbon nanotube film. Carbon 2010; 48: 1095-104. [http://dx.doi.org/10.1016/j.carbon.2009.11.031]

[114] Kalbacova M, Kalbac M, Dunsch L, Hempel U. Influence of single-walled carbon nanotube films on metabolic activity and adherence of human osteoblasts. Carbon 2007; 45: 2266-72. [http://dx.doi.org/10.1016/j.carbon.2007.06.025]

[115] Li X, Gao H, Uo M, *et al.* Effect of carbon nanotubes on cellular functions *in vitro*. J Biomed Mater Res A 2009; 91(1): 132-9. [http://dx.doi.org/10.1002/jbm.a.32203] [PMID: 18767064]

[116] Shi X, Sitharaman B, Pham QP, *et al.* Fabrication of porous ultra-short single-walled carbon nanotube nanocomposite scaffolds for bone tissue engineering. Biomaterials 2007; 28(28): 4078-90. [http://dx.doi.org/10.1016/j.biomaterials.2007.05.033] [PMID: 17576009]

[117] Vacanti MP, Leonard JL, Dore B, *et al.* Tissue-engineered spinal cord. Transplant Proc 2001; 33(1-2): 592-8.

[http://dx.doi.org/10.1016/S0041-1345(00)02158-8] [PMID: 11266974]

[118] Teng YD, Lavik EB, Qu X, *et al.* Functional recovery following traumatic spinal cord injury mediated by a unique polymer scaffold seeded with neural stem cells. Proc Natl Acad Sci USA 2002; 99(5): 3024-9.
[http://dx.doi.org/10.1073/pnas.052678899] [PMID: 11867737]

[119] Wu S, Suzuki Y, Kitada M, *et al.* Migration, integration, and differentiation of hippocampus-derived neurosphere cells after transplantation into injured rat spinal cord. Neurosci Lett 2001; 312(3): 173-6.
[http://dx.doi.org/10.1016/S0304-3940(01)02219-4] [PMID: 11602338]

[120] Vacanti MP, Leonard JL, Dore B, *et al.* Tissue-engineered spinal cord. Transplant Proc 2001; 33(1-2): 592-8.
[http://dx.doi.org/10.1016/S0041-1345(00)02158-8] [PMID: 11266974]

[121] Li B, Ma Y, Wang S, Moran PM. A technique for preparing protein gradients on polymeric surfaces: effects on PC12 pheochromocytoma cells. Biomaterials 2005; 26(13): 1487-95.
[http://dx.doi.org/10.1016/j.biomaterials.2004.05.004] [PMID: 15522750]

[122] Zhu Y, Gao C, Shen J. Surface modification of polycaprolactone with poly(methacrylic acid) and gelatin covalent immobilization for promoting its cytocompatibility. Biomaterials 2002; 23(24): 4889-95.
[http://dx.doi.org/10.1016/S0142-9612(02)00247-8] [PMID: 12361630]

[123] Ko JY, Lee JY, Park CH, Lee SH. Effect of cell-density on in-vitro dopaminergic differentiation of mesencephalic precursor cells. Neuroreport 2005; 16(5): 499-503.
[http://dx.doi.org/10.1097/00001756-200504040-00016] [PMID: 15770159]

[124] Gannon CJ, Cherukuri P, Yakobson BI, *et al.* Carbon nanotube-enhanced thermal destruction of cancer cells in a noninvasive radiofrequency field. Cancer 2007; 110(12): 2654-65.
[http://dx.doi.org/10.1002/cncr.23155] [PMID: 17960610]

[125] Hirata E, Uo M, Takita H, Akasaka T, Watari F, Yokoyama A. Development of a 3D collagen scaffold coated with multiwalled carbon nanotubes. J Biomed Mater Res B Appl Biomater 2009; 90(2): 629-34.
[http://dx.doi.org/10.1002/jbm.b.31327] [PMID: 19213050]

[126] Atala A. Tissue engineering and regenerative medicine: concepts for clinical application. Rejuvenation Res 2004; 7(1): 15-31.
[http://dx.doi.org/10.1089/154916804323105053] [PMID: 15256042]

[127] Green DE, Longtin JP, Sitharaman B. The effect of nanoparticle-enhanced photoacoustic stimulation on multipotent marrow stromal cells. ACS Nano 2009; 3(8): 2065-72.
[http://dx.doi.org/10.1021/nn900434p] [PMID: 19606849]

[128] Liu Z, Sun X, Nakayama-Ratchford N, Dai H. Supramolecular chemistry on water-soluble carbon nanotubes for drug loading and delivery. ACS Nano 2007; 1(1): 50-6.
[http://dx.doi.org/10.1021/nn700040t] [PMID: 19203129]

[129] Pantarotto D, Singh R, McCarthy D, *et al.* Functionalized carbon nanotubes for plasmid DNA gene delivery. Angew Chem Int Ed Engl 2004; 43(39): 5242-6.
[http://dx.doi.org/10.1002/anie.200460437] [PMID: 15455428]

[130] Balasubramanian K, Burghard M. Chemically functionalized carbon nanotubes. Small 2005; 1(2): 180-92.
[http://dx.doi.org/10.1002/smll.200400118] [PMID: 17193428]

[131] Green DE, Longtin JP, Sitharaman B, *et al.* The effect of nanoparticle-enhanced photoacoustic stimulation on multipotent marrow stromal cells. ACS Nano 2009; 3(8): 2065-72.
[http://dx.doi.org/10.1021/nn900434p] [PMID: 19606849]

[133] Dyke C, Tour J. Covalent functionalization of single-walled carbon nanotubes for materials applications. J Phys Chem A 2004; 108: 11151-9.

[http://dx.doi.org/10.1021/jp046274g]

[134] Hirsch A, Vostrowsky O. Functionalization of carbon nanotubes. Top Curr Chem 2005; 245: 193-238.

[135] Radushkevich LV, Lukyanovich VM. Structure of the carbon produced in the thermal decomposition of carbon monoxide on an iron catalyst. Zurn Fisic Chim 1952; 26: 88-95.

[136] Cai D, Mataraza JM, Qin ZH, *et al.* Highly efficient molecular delivery into mammalian cells using carbon nanotube spearing. Nat Methods 2005; 2(6): 449-54. [http://dx.doi.org/10.1038/nmeth761] [PMID: 15908924]

## CHAPTER 8

# Carbon Nanotubes for Biomedical Imaging

**Swati Jain**[1,*] and **Sruti Chattopadhyay**[2]

[1] *Amity Institute of Nanotechnology, Amity University, Sec. 125, Noida, Uttar Pradesh, India*

[2] *Indian Institute of Technology Delhi (IITD), Haus Khas, New Delhi, India*

**Abstract:** Development of molecular contrast agents for visible determination of biological structural and functional information is an on-going process in relation to ever widening field of imaging techniques. Optical probes highlighting biological subjects in whole or in components serve as tool for disease diagnosis and prognosis deterministic assays for rapid onset of therapeutics. These preserve the specimen and carry wealth of information in a non-destructive and non-invasive way. Various nanotechnological tools are inexplicably used as molecular probes paving the way for smart as well as theragnostic module of action. Carbon nanotubes are an important class of synthetic nanostructures owing to their exceptional properties and are being sought in almost all waking fields of applications. Multiwalled multilayred CNTs are seen as significant cargo encapsulating wide variety of molecules and biological entities. Single walled CNTs on the other hand behave as quasi one dimensional nanomaterials showing unique electronic and optical properties. their high end use in biomedical imaging stems from the fact that they absorb light in NIR having high biological transparency. Though raw SWCNTs are hydrophobic their suitable functionalization yields water dispersible, biocompatible entities while biomolecule attachment lends specificity to the target. The present chapter details approaches for imaging biological subjects *in vitro* and *in vivo* using intrinsic photophysical properties of SWCNTs entailing multitudes of components to meet the current scenario of multiplexed and multimodel imaging. A brief review is presented on optoelectronic properties of SWCNTs, their functionalization to meet the necessitating criteria for biological applications and various optical techniques using them as contrast agents.

**Keywords:** Carbon nanotubes, Imaging, NIR, Optoelectronic properties, Photoacoustic Imaging, raman spectra, Tumour cells.

## 8.1. INTRODUCTION

Carbon nanotubes have garnered substantial interest over the last two decades from scientific community since their discovery in 1991. This enthusiasm in understanding their structure, synthesis, function and applications lies in their

* **Corresponding author Swati Jain**: Amity Institute of Nanotechnology, Amity University, Sec. 125, Noida, Uttar Pradesh, India; Tel: 09818766339; E-mail: sjain@amity.edu

**Jagriti Narang (Ed.)**

exceptional mechanical, electrical and chemical properties. Currently nanotube research is even extending towards theorizing graphitic nanostructures and properties such as phonon dispersion [1]. Owing to this CNTS are in forefront of material science showing great promise in large spectrum of applications. They possess extremely high toughness to weight ratio propelling their use in mechanical reinforcement of industrial polymers and composites. CNTs have reported tensile strength approximately 20 times that of steel owing to hard carbon-carbon bonded network. They display ballistic electron transport in hetroatomic compositions opening up fields of micro and nano electronics.

Advent of molecular tools and improved synthesis, CNTs have been gaining curiosity in biological field. These one dimensional nanomaterials behave distinctly from spherical counterparts in biological medium. This has led to impetus for newer opportunities in biomedical research pertaining to therapeutics and diagnostics. Large surface to volume ratio, abundance of reactive binding sites and possibilities of encapsulating pharmaceutically active ingredient in the hollow core of CNTs as well as creating multiple valencies on their surfaces has been utilized for shuttling Si RNA, DNA, drug, proteins into cells. CNT based drug delivery has shown promising results for chemotherapeutic agents in animal models after suitable chemical modifications [2 - 5].

More recently impressive photophysical properties of CNTs particularly single walled carbon nanotubes SWCNTs have caught the attention of material scientists with a constant surge to explore uncharted direction of biomedical imaging. Quasi one-dimensionality in these nanowires like SWCNTs incur unique optical properties showing greater potential as optical probes as compared with much larger multiply layered multiwalled MWCNTs. Infact MWCNTs cater primarily to drug delivery applications in contrast to single walled nanostructures. SWCNTs are being studies as optical probe for analysing cellular functions, organelle structural, and mechanisms of cellular operations to ultimately seek disease diagnosis and understanding treatment prognosis. Imagining of histopathological, whole or in part biological entities with nanotubes offers unique characterization in near infra-red region which is marked with low background emissions resulting in higher resolution of imaging and better contrasting features.

This chapter focuses on optical properties of SWCNTs illustrating various modes of vibrational spectroscopy. This article also explores the current trend of SWCNTs as imaging probes for various modalities of biomedical imaging techniques including fluorescence, Raman scattering and photoacoustic tomography.

## 8.2. MOLECULAR IMAGING

Molecular imaging is an important technique in biological and medical applications having great potential for rapid and early detection, accurate diagnosis, suitable prognosis and personalized therapeutics. Imaging involving "the visualization, characterization and measurement of biological processes at molecular and cellular levels in humans and other living systems " [6] has become paramount in current biotechnological and pathological assessments.

Introduction of optical probes directly aids in expressing indicative molecular markers at various stages of disease. In addendum, they facilitate readout/signalling of whole intact body with marked accuracy and reliability of evaluation than many of the *in vitro* biochemical tests providing statistically precise analytics in animal models which serves as their internal control also. These probes easily detect any lesion formation in patient and also offer patient stratification helping in possibilities of individualized treatment with dose optimization [7]. Such non-invasive imaging of biological structure and function aids in early disease diagnosis allowing earlier treatment facilitating rapid and enhanced prognosis and monitoring of care. More importantly optical methods are non-destructive, and carry a large amount of information besides the information needed for the assignment itself. Over the last decade this fact has been effectively demonstrated by absorption and emission spectroscopy, time-resolved optical spectroscopy, and two-photon absorption.

Recently single walled carbon nanotubes (SWCNTs) are under thorough investigation as molecular probes, optical tags and biological markers in different modalities of biomedical imaging techniques. Conventional contrast agents (CAs) such as organic dyes are common choice for labelling biological features for understanding structure-function relationships. Many of the reported CAs generally, are unable to cross cell membranes, bleach under laser, blink rapidly, show diminished tagging efficiency, non-specific binding resulting in subsequently erroneous results. Physical and optical limitations of these dyes have propelled the discovery of alternative probes for imaging analytics.

Emerging trends in nano-biotechnology converging physical matter to biological systems through chemical route unearths solutions to some of the limitations associated with common molecular CAs. Semi-conducting quantum dots (QDs), fluorescent nanoparticles including silica and polymeric nano-capsules as well as nano-spheres, gold nano-crystals, coordinated complexes are being explored as *in vivo* and *in vitro* diagnostic markers in bio-sensing and bio-imaging [8, 9]. However, many of them require sophisticated and highly complex synthetic chemistries for obtaining biologically viable nanomaterials. In addition nanomaterials such as QDs are prepared in suitably prepared in organic media needing conversion to water dispersible entities *via* multiple modification steps. Some of the mentioned luminescent nanomaterials suffer from dye leaching, blinking and photobleaching [10, 11]. Table **1** shows important imaging techniques used in biological and medical analysis.

**Table 1. List of important imaging techniques used in biological and medical analysis**

| SI No. | Imaging technique | Resolution | Depth | Signal | Use |
|---|---|---|---|---|---|
| 1 | Fluorescence optical imaging | 0.3 μm | <1 cm | Non-ionizing light radiations | Metabolic, molecular and genetic events |
| 2 | Ultrasound | 50 μm | Mm | Sound waves | Anatomical investigation of metabolic and molecular events |
| 3 | Raman imaging | 0.3 μm | | Non-ionizing light radiations | Metabolic, molecular and genetic events |
| 4 | CT | 50 μm | Deep penetration | Ionizing X rays | Anatomical assessments |
| 5 | SPECT | 0.8-1.4 mm | Deep penetration | Ionizing Gamma (γ) rays | Physiological assessment of metabolic, molecular and genetic systems |
| 6. | PET | 0.8-1.4 mm | Deep penetration | Ionizing Gamma (γ) rays | Physiological assessment of metabolic, molecular and genetic systems |
| 7. | MRI | 10-100 μm | Deep penetration | Radiofrequency non-ionizing waves | All physiological and anatomical evaluations |

Single walled carbon nanotubes offer unparalleled advantage of working in near infrared region (NIR) including classical NIR-I (700-900nm) and the newly defined NIR-II region (1100-1400nm) the most soughtafter spectrum of light for spectroscopic analysis [12]. Relatively high transmission and penetration of NIR light in biological samples allow greater visualization of organism in components or as whole. This region is specifically marked with low background signals of autofluorescent cells, tissues and biomolecules.Also, human blood has a narrow window from 900 to 1400 nm where light can penetrate from 3–5 cm.

In addition recent analysis of SWCNTs has revealed strong resonance enhanced Raman scattering owing to one-dimensional confinement of electronic and photonic states. This has been useful in tracking nanotubes for detection and imaging purpose in cellular organization. Strong absorption of CNTs has motivated their use in other imaging modalities including photoacoustic tomography (PAT), magnetic resonance imaging (MRI), positron emission tomography (PET) and Single Photon Emission Computerized Tomography (SPECT). In most of these techniques CNTs are attached with secondary labels-fluorophores, radionuclides or magnetic contrast agents elucidating enhanced

spectrometers. Photoluminescence, for example, is quenched in nanotube bundles by the presence of metallic tubes, whereas the Raman signal is still present.

The progress in optical characterization of nanotubes was favoured by the post processing method of de-bundlling them by ultrasonication and surfactant coating [14]. This leads to production of large quantities of isolated nanotubes of wide variety of composition as well as chirality. As noted earlier in optical spectroscopy, these isolated nanotubes behave like true 1-D system emitting light in NIR region with sharp peaks and also show resonance Raman scattering. A highly comprehensive study was reported by Dresselhaus and co-workers [19, 20] on vibrational properties of individual NTs. Raman spectroscopic analysis is important these days in looking at detailed and accurate structures of NTs. it also estimates electronic properties of all kinds of NTs thoroughly including multi, single NTs, peapods, large and small NTs.

#### *8.3.2.1. Raman Spectrum of SWCNT Contains Three Characteristic Features*

A Radial Breathing Mode (RBM) is found between 100 and 400 cm-1 for tubes with diameters between ~0.6 and 2 nm. RBM is specific for different chirality of NTs *i.e.* n=m and n≠m, and eigenvector is taken as radial for practical calculations for all types [21]. In RBM all Carbon atoms move in phase in radial direction creating breathing like vibrations in entire tube. With increasing diameter the force needed for this radial deformation of shift decreases. It is extremely significant in Raman analysis as its frequency is inversely dependent on diameter of NTs, thus making it possible for deducing dimensions of unknown nanotubes.

$$\omega\text{RBM} = \frac{C1}{d} + C2$$

where c1 and c2 are coefficients and d is the diameter of tube. C1 is a constant while c2 offset accounts for additional external forces such as interactions with substrate or neighbouring tubes in bundle and is regarded as fitting parameter [22]. After various assumptions and predictions Thomsen 2007 have calculated interdependency of frequency of RBM with tube diameter as:

$$\omega\text{RBM} = \frac{227\ cm - 1.nm}{d}$$

Both are specified completely by ***c***, which is normally given in terms of the graphene lattice vectors ***a***1 and ***a***2

$$C = na1 + na2 \equiv n, m$$

The diameter of CNT is defined as:

$$d = \frac{c}{\pi} = \frac{a_0}{\pi}\sqrt{n^2 + nm + m^2}$$

where $a_0 = 2.460$ is the graphene lattice constant.

The other important parameter for NTs is chiral angle between a1 and chiral vector illustrating the arrangement of graphene hexagons on the wall and this defined as:

$$\theta = \tan^{-1}[\sqrt{3}m/(2n + m)]$$

Chiral angles lying between 0 and 30° encompass all unique SWNT structures, except enantiomers. Variables in the structural parameters ultimately yield variety of electronic and optical properties of SWCNTs.

### 8.3.1. NIR Fluorescence

CNTs are a combination of one-third metallic or semi-metallic component and rest semi-conducting, dictated by chiral index. The one dimensionality or quantum confinement of NTs gives rise to 1D sub bands as opposed to broad electronic energy bands. The order of energy levels and their transitions are defined by the term Density of States (DOS) [13]. DOS indicate the number of allowed electronic states at a particular energy or measure of number of states that can be occupied at each energy level. DOS vary according to dimensionality appearing as continuum function of energy in bulk phase. Whereas in 1-D nanomaterials these are denoted by discontinuous sharp peaks called Van Hove singularities which resemble molecular energy bands. DOS is useful for understanding optical properties and ultimately PL spectroscopy of CNTs. Optical transitions can take place between only mirror sub-bands in 1D DOS. The energy difference between $i_{th}$ conduction and valence bands is denoted as Ei example E11, E22, E33 *etc.* which are unique bands of intraband transition energies. The tight binding model based on DOS states that absorption and emission are dominated by dipole allowed transitions in which electron is promoted from valence to conduction sub-band when light polarized by its electric vector is parallel to the tube axis thus conserving angular momentum projections. When the energy difference between corresponding Van Hove singularities match with that of photon energy, these transitions become very intense (Fig. **2**).

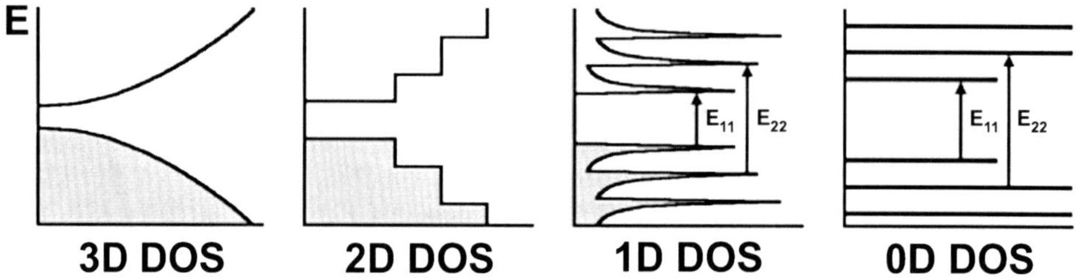


**Fig. (2).** Diagram showing energy *versus* density of states for materials in various dimensions the two transitions - E11 and E22 are labelled for 1-D and 0-D naomaterials. Shaded portion indicates occupied valence band, whereas white portion is for unoccupied conduction band of higher energy.

However initial understanding of CNT optical properties was marked with broad bands which were assigned to un-differentiable absorption features. This was attributed to overlapping of transitions of those species as opposed to sharp resolved spectra. This observation was different from that of theoretical data predicted by mathematical tight binding model. This problem was assigned to the aggregation of NT as bundles resulting in several in-homogenous broadening as a result of mixed energy states. Luminescence was not observed from these bundles as photo-excitation generated in semi-conducting NTs relaxes through non-radiative channels opened by metallic NTs of the bundles. The key solution to this predicament came in 2002 with the breakthrough experiment by O'Connell *et al.* [14] who specifically processed NT bundles countering the Vander wall's forces of attraction which tightly held them together. They prepared disaggregated individual NTs in solution of sodium dodecyl sulphate (SDS) surfactant obtaining micelle encapsulated SWCNTs. This miceller organization prevents re-bundling of separated NTs. They have pioneered encapsulation of NTs in variety of surfactants like micelles, DNA and polymers [15]. The water soluble/dispersible surfactant suspended SWCNTs display resolved optical transitions for distinct NT (n, m) structure as NIR photoluminescence.

Emission peaks are nearly coincident with that of absorption spectrum and very small shift was experimentally observed. Since lowest lying electronic state gives rise to electronic luminescence SWCNT emission is observed for E11 transitions only particularly for individual semi-conducting ones [16] which is consistent with spin-conserving optical transitions, the PL of NTs is therefore classified as fluorescence and not phosphorescence. Spectroscopic studies were used to correlate NT structure with fluorescent band gap transitions or peaks appearing in spectroscopy. Energy of incident photon matched with that of second van-Hove transition E22 generating a hole and electron in second valence and conduction sub-peaks respectively. They relaxed through phonon emission to first sub-bands emitting E11 NIR fluorescence through radiative electron-hole recombination across semi conducting band gap. Therefore this emission wavelength is characteristic of NT (n,m) undergoing E22 excitation (Fig. **3**).

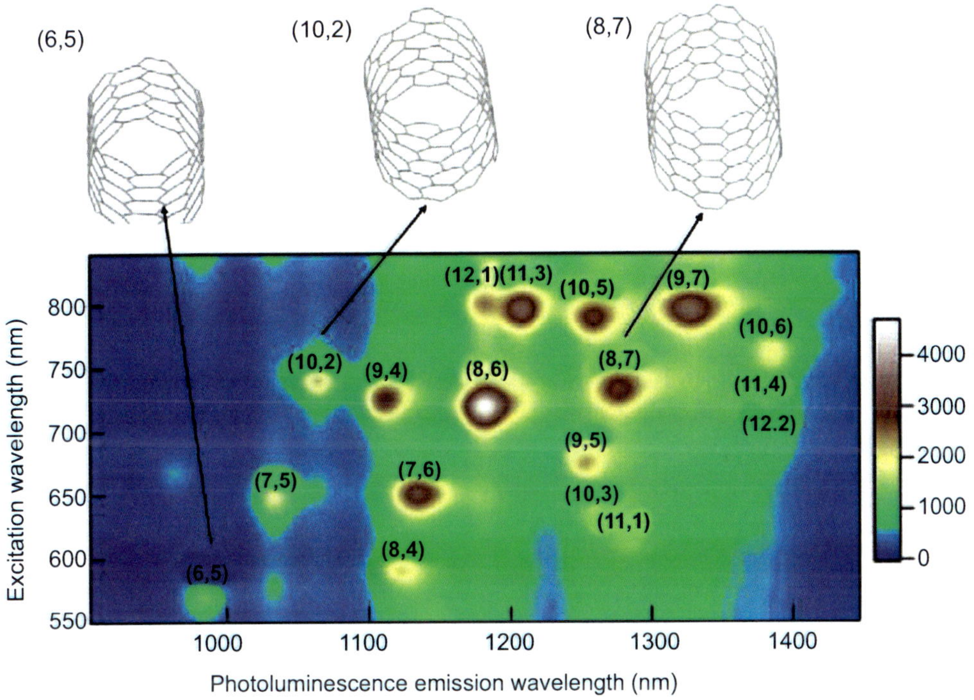


**Fig. (3).** PL excitation E22 and emission E11 spectrum of semiconducting SWCNTs (Source [17]).

Fluorescence is dependent on sample conditions of SWCNTs. An important example of this is the absence of emission from aggregated NTs bundled together by van der Waal forces. Since one-third of NTs are metallic with high chance that several ones would be metallic even in bundles or aggregates. When these bundles of semiconducting SWCNTs absorb light, excitation transfer occur with smaller band gaps owing to electronic coupling. This energy reached metallic NTs in which relaxation now occurs non-radioactively. This is an effective fluorescence quenching process which is helpful in monitoring aggregation of NTs in solution. Spectroscopic analyses have been conducted on SWCNTs for understanding chemical and physical properties. Though the relationships are quite complex but still some predictions have been proposed. Optical properties particularly Fluorescence of NTs is largely dependent on large number of variable including sample conditions, temperature, and chemical environment such as pH, ionic strength & presence of oxidative species. They invariably affect transition energies and emission line widths resulting in complete quenching, drastic shifts and broadening (splaying) of emission peaks. As compared to conventional organic CAs and novel QDs, SWCNT fluorescence is enormously photo-stable while displaying no evidence of blinking or photo-bleaching even after prolonged exposure to excitation at high frequency. With initial experiments, the quantum yield of fluorescence for semiconducting NTs was estimated at 0.1% [13]; however, recent analysis indicates that the actual value for pristine nanotubes is larger and is greatly influenced by length [18]. Emission bands corresponding to unique species of chirality are narrow enough which permit imaging of large number of spectrally separated emitters simultaneously, which is not possible with organic fluoro-phores.

### 8.3.2. Raman Effects

With the advancements of spectroscopic and microscopic techniques, newer imaging modalities are being sought for both chemical and biological analysis. Since photoluminescence of fluoro-phores is drastically affected by their external environment and shows certain unsolicited characteristics such as blinking and photo-bleaching under laser, investigation of entities by Raman technique has displayed several benefits. Since Raman spectra involve phonons, it provides exceptional information on structural and physical properties of carbon materials including SWCNTs. In this phenomenon, energy of incident photons excites electron from low to high energy states emitting or absorbing phonons. Electron relaxes back to ground state *via* emitting a photon or light of lower energy or higher wavelength. Raman plot is measure of phonon frequency of any material calculated by plotting intensity of scattered light against decreased frequency shift. Phonon emission is called Stoke's shift while phonon absorption is termed as Anti-Stoke's shift.

SWCNTS with (n,m) chirality have been characterized with Raman scattering technique since it is simple method operable at room temperature under ambient conditions of pressure. It is non-destructive and non-invasive approach without damaging the inherent structure of nanotubes and the analysis is done quickly.

This technique can easily identify both metallic and semi-conducting NTs in a robust manner as compared with PL signals in bundles which are quenched in presence of metallic NTs. Raman spectroscopy can be done in visible energy range of second optical transition without using infra read detectors and spectrometers. Photoluminescence, for example, is quenched in nanotube bundles by the presence of metallic tubes, whereas the Raman signal is still present.

The progress in optical characterization of nanotubes was favoured by the post processing method of de-bundlling them by ultrasonication and surfactant coating [14]. This leads to production of large quantities of isolated nanotubes of wide variety of composition as well as chirality. As noted earlier in optical spectroscopy, these isolated nanotubes behave like true 1-D system emitting light in NIR region with sharp peaks and also show resonance Raman scattering. A highly comprehensive study was reported by Dresselhaus and co-workers [19, 20] on vibrational properties of individual NTs. Raman spectroscopic analysis is important these days in looking at detailed and accurate structures of NTs. it also estimates electronic properties of all kinds of NTs thoroughly including multi, single NTs, peapods, large and small NTs.

#### *8.3.2.1. Raman Spectrum of SWCNT Contains Three Characteristic Features*

A Radial Breathing Mode (RBM) is found between 100 and 400 cm-1 for tubes with diameters between ~0.6 and 2 nm. RBM is specific for different chirality of NTs *i.e.* n=m and n≠m, and eigenvector is taken as radial for practical calculations for all types [21]. In RBM all Carbon atoms move in phase in radial direction creating breathing like vibrations in entire tube. With increasing diameter the force needed for this radial deformation of shift decreases. It is extremely significant in Raman analysis as its frequency is inversely dependent on diameter of NTs, thus making it possible for deducing dimensions of unknown nanotubes.

$$\omega\mathrm{RBM} = \frac{C1}{d} + C2$$

where c1 and c2 are coefficients and d is the diameter of tube. C1 is a constant while c2 offset accounts for additional external forces such as interactions with substrate or neighbouring tubes in bundle and is regarded as fitting parameter [22]. After various assumptions and predictions Thomsen 2007 have calculated interdependency of frequency of RBM with tube diameter as:

$$\omega\mathrm{RBM} = \frac{227\ cm - 1.nm}{d}$$

which is in good agreement with experimental values. These parameters have also been found empirically using surfactant coated HiPco nanotubes by Bachilo *et al.* [23], Telg *et al.* [24], and Fantini *et al.* [25] for semiconducting and metallic nanotubes.

There are two more Raman active modes in high frequency range originating from optical phonons of graphite structure if nanotubes. These are tangential modes of transitions involving extremely strong sp2 in-plane carbon-carbon bonds with high frequency lying between 1100-1600 cm $^{-1}$. The two groups of Raman active modes are denoted as high energy mode or G band below 1600 cm$^{-1}$ and D mode at around 1350 cm$^{-1}$. G band originally named after graphite bands, consist of multiple peaks in NTs, and is used for (1) characterizing diameter of NTs (2) to differentiate between metallic and semi-conducting SWCNTs (3) to analyse doping effects by measuring charge transfer (4) to study selection rules of various absorption and scattering geometries of SWCNTs.

Both metallic and semiconducting SWNTs demonstrate intense Raman scattering with some of the highest known cross-sections for single molecules (5.7 × 10–21 cm$^2$ sr$^{-1}$ molecule–1 for a 1 lm nanotube excited at 785 nm).Raman spectra explicitly involves phonons is independent of electronic structure of NTs and the laser used for excitation.it was observed that the signal intensity drastically increases when the excitation energy approaches some allowed optical transitions. thus increase in scattering efficiency takes place when Elaser matches the energy difference of optically allowed electronic transitions of the material. This is called resonance Raman scattering effect [26] and is strongly dependent of density of electronic states of NTs particularly for 1-D structures or SWCNTs. In non-resonant Raman scattering excitation of electron from ground state occurs to virtual state followed by either emission (Stokes) or absorption of phonons (Anti-Stokes). However the Raman signal is enhanced when incident of scattered photon energy equates Eii in DOS and one of the excited electrons is in a real state system. This resonance process escalations the signal by a factor of approximately 10$^3$ in comparison to the intensity for a non-resonance Raman process. Thus allowing the measurement of single isolated SWCNTs is possible from a bundle of non-resonating (n, m) nanotubes. This property has been used for analysing SWCNTs is biological tissues for imaging and biosensing. Both nanotube Raman scattering and fluorescence can be measured and tracked over long durations without signal attenuation (Fig. **4**).

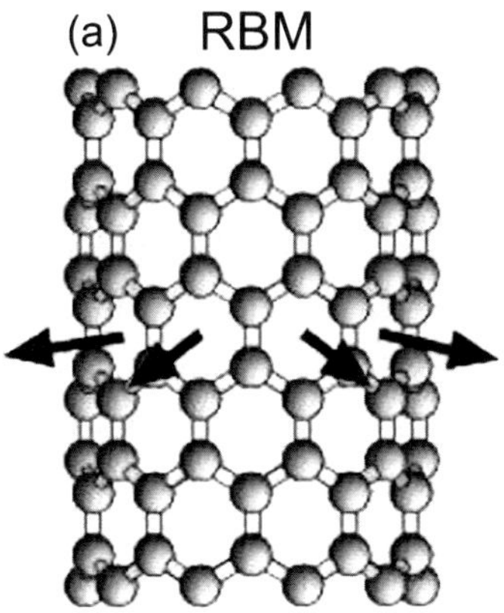


**Fig. (4).** Radial and tangential Movements of carbon atoms depicting RBM and G band of NTs in Raman spectroscopy (source [27]).

### 8.3.3. Photo-Acoustic Imaging

Photo-acoustic effect has also been utilized for developing imaging modality of high resolution where generation of acoustic waves is the prima-facie of the technique. This PA imaging has come a long way from 1880s since the generation of sound waves through light was observed by Bell [28]. non-ionizing ultrasound waves are employed in this technique hence PA is more safe as compared to many other biomedical imaging modalities which utilize ionizing electromagnetic radiations. PA imaging is a combinatorial approach of ultrasound along with high contrast light absorption and is considered as either ultrasound mediated electromagnetic imaging or as ultrasound imaging wit EM contrast enhancers. EM radiations of optical *i.e.* visible to Near-IR and radiofrequency regions are used in PA excitation rendering imaging of biomolecular constituents of biological samples *via* non-ionizing methodology.

The technique takes into account when a short EM pulse is scattered and absorbed into tissues it creates a transient acoustic pressure inside the target tissues subsequently resulting in their thermoelastic expansion. These acoustic waves travel across the tissue area reaching its surface with various time delays. Ultrasound transducers are placed outside the subject of interest which enables detection of these propagated PA waves. Further these receivers map the energy distribution and acoustic source distribution reconstructing a 2D or 3D image after suitable signal amplification and digitization. Maximum imaging depth is scalable with the detected ultrasound bandwidth. This tomographic evaluation of tissue samples demonstrates high spatial resolution and better penetration capacity owing to utilization of sound waves than many other techniques. In addition this thermoelasticity of EM radiation does not damage, change or break the biological subject. PAT has been used for detailing structural and functional information on total haemoglobin concentration, $O_2$ saturation, rate of O2 consumption and blood flow *etc*. [29 - 32]. relying on intrinsic PA contrast agents such as Hb or melanin.

Many limitations have been discovered for intrinsic CAs for PAT as many of them rely on visible spectra delimiting light penetration is hampered by stronger absorption and scattering. Xie *et al*. highlighted that macroscopic PAT is unable to effectively image optically transparent tissues such as lymphatics or bladders [33]. In many diseases particularly in their early development natural PA contrast is not created, thereby necessitating the use of exogenously administered contrast agent to boost up resolution of imaging qualitatively and quantitatively. An ideal probe would encase large optical absorption window maximizing PA signalling while being able to escape uptake by the reticulo-endothelial system (RES). Research focus has been on nano-sized materials and many reported works are highlighting use of QDs, gold nanoparticles and SWCNTs. Strong absorption in NIR, small size and capabilities of attaching other contrast enhancers have motivated tremendous activity in SWCNT front.

## 8.4. FUNCTIONALIZATION OF CNTS

It is difficult to synthesize CNTs with specific characteristics as demanded by different applications. Therefore surface modification and interfacial engineering are good options to render NTs with desirable properties including biocompatibility or coupling ability. Biomedical applications of CNTs require aqueous dispersion of nanostructures for their optimal usage necessitating operation of functionalization chemistries on the surface of nano-sized materials. Since raw SWCNTs are composed of primarily carbon network they are hydrophobic in nature, thus it is imperative to convert these non-polar entities to their hydrophilic analogues. In addition seamless hexagon rings without any dangling bonds are relatively unreactive; however tips of nanotubes are more active than their cylindrical walls. Sidewall having regular graphene framework are more difficult to functionalize. The entire NT shape has been exploited in changing its surface characteristics by carrying out chemical as well as passive changes of the structures as discussed below.

### 8.4.1. Chemical Bond Formation

Primarily covalent bonding has been mediated for the formation of chemical bonds between linker groups and CNTs where outer walls, inner walls and ends have been modified through this approach. The simplest and earliest work reported end group oxidation of NTs using high oxygen containing oxidants including nitric acid, combination of nitric and sulphuric acids, perchloric acid, potassium permanganate and potassium dichromate primarily with extensive ultrasonic treatment. These result in formation of number of functionalities like carboxylic acid, aldehydes and ketonic groups at the open sites and defect positions along with opening of tube caps, production of holes followed by oxidative etching with the release of carbon dioxide. This opens up plethora of chemical coupling steps which can be followed after generation of carboxylic entities though amide and ester bonds. Amino acids Metal nanoparticles, quantum dots, nucleic acids, enzymes, dendrimers, star-bust polymers have been attached on NTs *via* linking them with either a suitable linker or directly onto carboxylic acid terminals [34, 35]. However it was observed that carboxylated CNTs show salting effect and get agglomerated in biological solutions owing to presence of large amounts of salts and ionic species thus cannot be directly used in these applications necessitating their further modifications. Hydrophilic polymer like PEGs are common choice for attaching on oxidized CNTs.

Addition reactions are also popular chemistry for modifying surface of CNTs. These addition reactions primarily on the sidewalls enable direct coupling of π-conjugated carbon network of graphitic structure of NTs with functional groups of linker. 1, 3 dipolar cyclo-addition of azomethineylides, aniline and aryl diazonium salts have been added on NTs. These reactions are paving way to understand changes in thermal and solubility parameters of NTs. Fluorination of NTs has been achieved at elevated temperatures of about 250-400°C on sidewalls as confirmed by FTIT analysis which indicated covalent bonding of Fl. This reaction is also of considerable interest to chemists as fluorinated intermediate can be further functionalized through array of reactions [36]. These nucleo-phillic substitution reactions opened flexible approach for sidewall modification of NTs particularly alkylation method producing Grignard reagent, alkyl lithium compounds, amines di-amines *etc.* with large number of C atoms successfully replaced by other moieties [37 - 39]. However few reports suggested that fluorination significantly modifies conductivity owing to destruction of graphitic structure of nanotubes, thus not much work is reported on biological applications through this chemistry. CNTs are used in many applications pertaining to electrochemistry as electrode owing to their strong electro-catalytic properties and high surface to volume ratio as compared to other carbonaceous materials. This favoured electrochemical functionalization of bundles of CNTs in selective mode. Many organic species react with starting material or reagent present in between cathode and anode of the cell or these chemical species may self-polymerize leading to formation of polymers stacked onto NTs. Electrochemical method is very versatile providing accurate control over the modifications. Similar to electrochemistry photo-irradiation method has also been approached for the functionalization of nanotubes however to somewhat lesser extent (Fig. **5**).

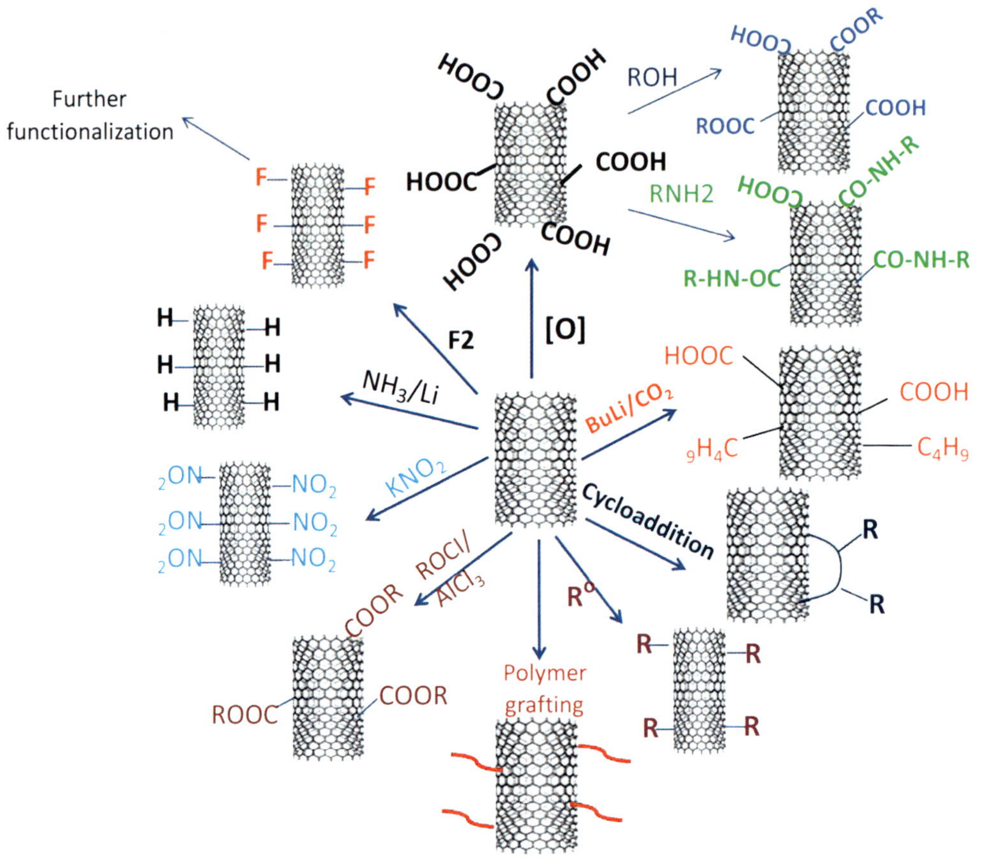


Fig. (5). Covalent modification of nanotubes indicating oxidation, nucleophilic addition by sec-Butyl lithium, cyclo-addition, radical attachment, polymer grafting, electrophilic substitution, nitration, hydrogenation and fluorination depicted in clockwise manner (adapted from 39).

## 8.4.2. Non-Covalent Approaches

In contrast to chemical modifications based on covalent bonding, non-covalent mode has gained popularity as they involve coating or self-assembly of molecules or macromolecules generating thermodynamically stable structures held together by enthalpy and entropy driven interactions such as weak non-covalent forces including van der Walls, hydrogen bonding, electrostatic attractions, polar-no--polar interactions as well as π-π stacking (Fig. **6**). In addendum electronic structure or the π-network of nanotubes is disrupted covalent linkages destroying the Raman features and weakening their NIR photoluminescence intensity. Therefore recently covalent functionalization are not favoured for bio-imaging applications whereas non-covalent reactions preserve the chemical nature of CNTs except shortening of length during sonication process which is done for their solubilization. The physicochemical properties are largely conserved during non-covalent modifications requisite for biomedical imaging applications.

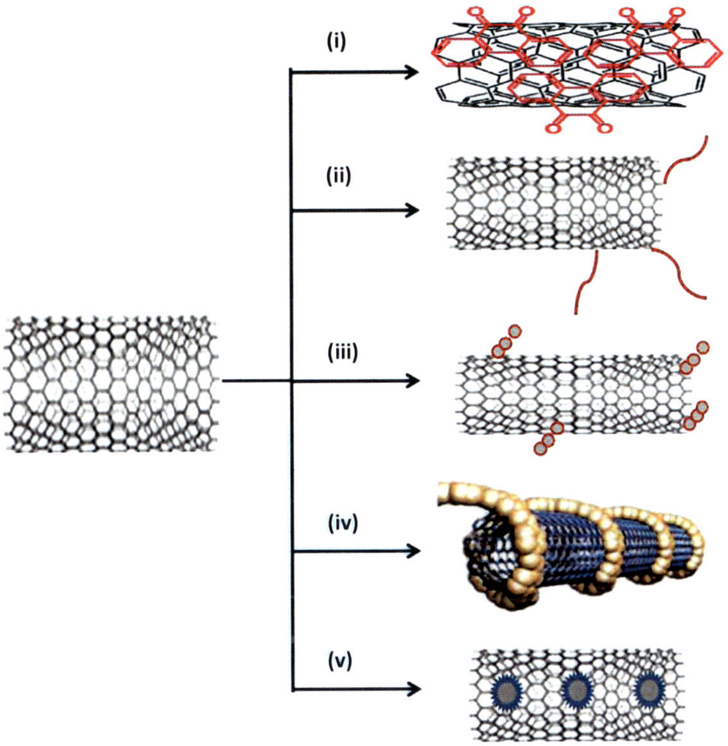


**Fig. (6).** Non-covalent approaches for functionalization of SWCNTs (i) π- π conjugation using pyrene derivatives, (ii) end-tail attachment of polymeric molecule, (iii) exohedral non-covalent functionalization with surfactant molecules, (iv) Wrapping of polymers and (v) endohedral inclusion of external molecules such as dyes, biomolecules *etc* (Fig. **7**).

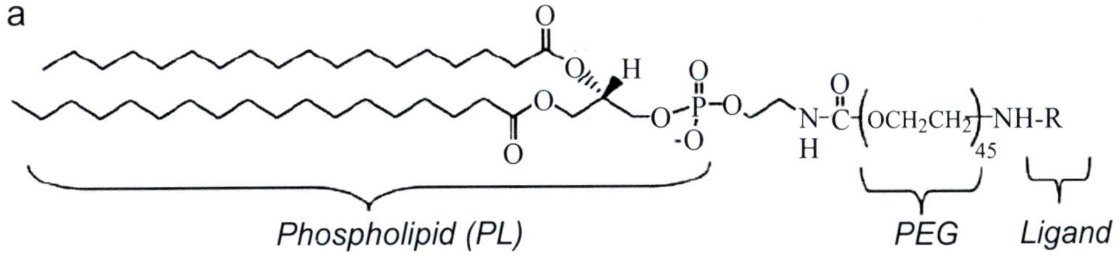


**Fig. (7).** Structure of Pl-PEG attached with ligand through amide bond R = H, Alkyl or aryl group.

Polymers wrapping is a simple way to attain functionalized NTs which show higher stability than surfactant coated CNTs owing to multisite interactions of polymers. Fujigaya and Nakashima [40] have reviewed wrapping by different categories of polymers such as π-conjugated conducting, aromatic, non-aromatic, cationic and block co-polymers [41 - 43].

The thermodynamic driving force for wrapping of polar polymer on nanotube is the need to avoid unfavourable interactions between the non-polar tube walls and the polar solvent (water). Polymer-wrapped CNTs provide improved biocompatibility compared to bare CNTs as indicated by studies on cell attachment. Higher cell viability was reported for the amylose-wrapped SWCNTs than for the non-wrapped SWCNTs [44] which is advantageous for bio-sensing applications. There are many other indicative examples of polymer wrapped NTs having enhanced surface properties essential in biomedical applications of sensing, drug delivery and imaging. O'Connell, *et al.* [45] showed higher water dispersion of NTs by associating them with polyvinyl pyrrolidone (PVP) and polystyrene sulfonate (PSS). Pang *et al.* used a cationic polythiophene as the wrapping polymer, providing the binding sites for glucose oxidase and achieving a highly sensitive amperometric biosensor for glucose [46]. In addition not only classical polymers but also oligonucleotides have been wrapped successfully over CNTs giving them specificity for sensing applications. SWCNTs were wrapped by single stranded DNA in a study conducted by Staii *et al.* [47] A series of molecules including methanol, propionic acid, trimethylamine (TMA), dinitrotoluene (DNT) and dimethyl methylphosphonate (DMMP) have been studied *via* attaching DNA sequences specific for them. Single stranded DNA conjugation on SWCNTs gives improved characteristics for sensing with very high electrical signal as compared with bare nanotubes in the same experimental conditions [48]. Functionalization of CNTs not only given them moieties attachment with other molecules including biological species but also renders them with improved physic-chemical parameters.

Self-assembly is another approach for functionalization of NTs. SA is 'bottom-up' technology for building ordered architectures *via* self-organization of size selected nanoclusters or nano-materials. Self-assembly of SWCNTs is a rapidly evolving area for integrating nano-scale materials into building functional nano-devices. SWCNTs can be ordered into patterned trenches using self-assembly templates with well-defined inter-particle distance. One of the most common and preferred template is using DNA molecule serving as programmable supra-molecule for self-assembly of secondary objects with nano-meter precision [49]. The DNA conjugation be facilitated by covalent binding, electrochemical, physical adsorption and self-assembly. Polyethylene glycols are a class of known biocompatible polymers which do not promote an adaptive immune response finding application as coating material in multitudes of nanomaterials and devices. In-vivo experimentation with SWCNT necessitates their low toxicity, higher bio-distribution and long circulation capacity. Non-covalent modification with PEGs incurs SWNTs with the above mentioned characteristics *via* linking through phospholipids (PL). In addition since both components are biocompatible and are easily removed from body their long term toxicity is low. Sonication of SWNTs in solutions of phospholipid-polyethylene glycol (PL-PEG) is highly preferred method for NT functionalization [50]. The anchoring of this molecule occurs through hydrophobic chains of Pl while PEG renders SWCNTs as aqueous dispersible and biocompatible. PL-PEG modified NTs demonstrate excellent solubility in water medium as well as physiological buffers, whole serum and cell medium. Length of PEG is significant in bio-distribution and predicting circulation time of nanotubes. Extent of PEGylation and length of polymer chain is found to increase circulation time in mice [51].

## 8.5. PHOTOLUMINESCENCE IMAGING

PL imaging of SWCNTs is highly applicable in *in vitro* and *in vivo* biomedical imaging. De-bundled semi-conducting SWCNTs have band gap of approximately 1eV allowing their applicability as NIR probes due to optical transparency of biological tissues in 800-1000 nm where neither haemoglobin nor water exhibits light absorption. Large Stroke-shift between excitation at 550-850 nm of second van Hove absorption transitions and emission at 900-1600nm corresponding to first order van Hove transition drastically reduces auto-fluorescence of biological tissues during imaging.

### 8.5.1. *In Vitro* Evaluations

In a classical work demonstrated by Cherukuri *et al.* [52] in 2004 SWCNTs uptake by cultured mouse peritoneal macrophages was studied under spectro-fluorometer and fluorescence microscope. They estimated an average ingestion rate of approximately 1 NT per second per cell and imaged average of 70000 NTs per cell ingested through phagocytosis process. However significant red shift was observed for CNTS in cells as compared to their suspension in surfactant medium of pluronic. This was attributed to the displacement of nanotube's surfactant coating by proteins in cells. In an extension work to this Jin *et al.* imaged endocytosis and exocytosis of SWCNTS in NIH-3T3 cells in real time on a perfusion microscope stage [53]. Many factors seem to contribute to quantum efficiency such as bundling, sidewall defects and length. Careful preparation of nanotubes is essential for making their suspensions obtain unbundled SWCNTS with bio-inert coatings to realize their full potential in biological imaging.

However extrinsic and intrinsic quenching of SWCNTs PL intensity in terms of environmental factors and metallic NTs resp. is quite deterrent in their use as optical probes. Coupling with plasmonic structures can offer some solution to this predicament as they alter radioactive processes of SWCNTs. Metal-enhanced fluorescence (MEF) by up to 14-fold for surfactant-coated SWNTs using plasmonic gold film was reported by Hong *et al.* [54]. They functionalized nanotubes with arginine–glycine–aspartic acid (RGD) selectively for U87-MG brain cancer cells over MCF-7 breast cancer cells and placed the cells on top of nanostructured gold films synthesized purely in the solution phase (called 'Au/Au films').This PL intensity enhancement by metallic substrate is due to resonance coupling between SWNT emissions and re-radiating plasmonic modes in the Au/Au films offering reduction in radioactive lifetimes of SWNTs, resulting in higher fluorescence QE. They further studied mechanism of internalization of nanotubes using the NIR fluorescence enhancement (NIR-FE) phenomenon on plasmonic Au in the three-dimensional imaging modality with an axial resolution on the order of ~ 10 nm. Individual nanotube-gold conjugates showed strong dependence of fluorescence enhancement to the distance between SWCNT and gold [55]. The work on metal enhanced luminescence is still under investigation and would need a combination of efficient synthetic procedures and improved analytical tools to manifest its potential as an upgrade fluorophore than conventional systems as well as novel ones.

Two photon luminescence and strong plasmonic coupling occurs with conjugation to gold nanoparticles which can operate on combinatorial approach of imaging and therapy. This enhanced intensity will be useful for tracking cellular structures while offering photothermal route for killing specific cancerous cells. This theranostic model has been demonstrated to kill 100% melanoma UACC903 cells with an exposure of 8 min only. This nano-antennaes serve as 2 photon imaging and photothermal source in biological window II *i.e.* NIR II [56]. Newer avenues of thernostics have opened up with tagging of other nanomaterials on SWCNTs.

In an interesting work Ignatova *et al.* have demonstrated study of DNA auto-ionization process using SWCNT based PL contrast agent. Interestingly DNA photo-excitation and SWCNT occurred in UV and visible range of light respectively in two colour spectroscopy. A strong SWCNT photoluminescence quenching was observed by the authors when the UV pump is resonant with the DNA absorption owing to charge transfer from the excited states of the DNA to the SWCNT [57].

### 8.5.2. *In Vivo* PL Imaging

Whole animal models have been investigated using internalization of SWCNTs through PL imaging technique. It is imperative for *in vivo* studies that SWCNTs are functionalized to attain biocompatibility while maintain structural homogeneity as well as integrity, a constraint that ominously limits the available interfaces. In comparison to *in vitro* sensing operations where much optimization has been achieved for selective modulation for target in pool of interfering molecules, *in vivo* adds the complexity of experimental working and designing. It is necessitating maintaining adequate quantum efficiency in order to visualize SWCNTS within highly multifaceted tissue organization.

Dai and co-workers have been working in this field of analysis for long time and have published many results on different aspects of SWCNT based imaging. In another report by the same group semiconducting SWCNTs were used as NIRF tags for selective probing of cell surface receptors. Nanotubes were conjugated with antibodies specific for various cell surface markers and their binding to cells was detected by performing high frame rate fluorescent video imaging of mice [58]. With evolution of non-covalent approaches for modification of nanotubes, Dai and co-workers obtained biocompatible NTs by an interesting coating exchange method. They first solubilised nanotubes in sodium cholate which was later displaced by phospholipid–PEG (PL–PEG) which retained high quantum yield of the nanostructures as the process was mild [59].

Broad classes of dispersants have been identified that individually suspend SWCNTs in water and cell media in addition to reducing nanotube toxicity to cells. New type of PEG amphiphiles are being designed to improve solubilisation and de-bundling of nanotubes. Poly(maleic anhydride-alt-1-octadece-e)-polyethylene glycol has been synthesized by Robinson *et al.* [60] for dispersing SWCNTs and for the first time showed half-life of almost 30h *in vivo* with long blood circulation pattern when injected in murine breast cancer induced mice models. Fluorescent signal were visualized on injecting the suspension of SWCNT in tumour in NIR II window in an advanced imaging system indigenously designed by the group. The signals were attained for 72 h revealing co-localization of tumour vasculatures indicating high penetrating and retention capacity of nanosized SWCNTs in ESR. The study indicates the use of NIR II window for PL imaging along with principle component analysis for acquisition of fluorescent video with high contrast resolution of anatomic data. More recently Hong *et al.* [61] differentiated arterial and venous vessels based on the same technique. In fact NIR II from 1500-1700 nm has been proposed to able cater more effectively in biological imaging owing to lowered auto-fluorescence. It is predicted to be devoid of background noise in tissues including liver where SWCNT home-in preferentially [62].

Apart from utilizing NIR window of light for imaging purpose strong absorbance by SWCNTs in this range can effectively stimulate nanotubes inside targeted living cells for therapeutic effects also. Continuous localized NIR radiation can cause apoptosis impairing cellular functions and organization ultimately leading to cell death. The focussed beam of radiation heats the surrounding environment thereby increasing localized temperature beyond 37°C. These optoelectronic stimulations convert electronic excitations of nanotubes to vibrational energies causing heating, the process known as hyperthermia. This aids in affording cancer cell death *via* internalization of tumour biomarker conjugated SWCNTs specifically in these cells, without affecting receptor-free normal cells. Kam *et al.* [63] showed photothermal therapy through Hipco SWCNTs non-covalently functionalized with PL-PEG and DNA killing He La cells when exposed to radiation at 808 nm laser continuously for 2 minutes. on long exposure to NIR radiations debundling, aggregation and defunctionalisation of nanotubes occurred inside the living cells after releasing the 'cargo' *i.e.* the packaged DNA.

It is also imperative for the NTs to be internalized within cell *via* endocytosis or lysozymes instead being present on the cell surface. Biomolecule-SWCNTs adhered on plasma membrane are not effective in killing cancer cells as compared to being inside the confines of membrane. The ablation mechanism is more pronounced when nanostructures are within cell organization in sub-cellualr organization. Quantitative analysis of NIR mediated thermal killing of tumour cells constructed an effective dose response curve Folate-targeted SWNTs within the vacuolar compartments. A report by Murali *et al.* for the first time correlated with a dose-normalized efficacy of thermal ablation and visualized SWNT subcellular locations Raman imaging of SWNTs merged with fluorescence images of known subcellular markers [64].

In addition of preparing modified NT structures and utilizing NIR II window of absorbance, it is also important to focus on image acquisition to resolve chirally separated SWCNTs. Hyper-spectral microscopy producing continuous emission spectrum at every spatial pixel is in direct contrast to multi-spectral imaging of discrete wavelength bands for multiple fluorophores. Resolution of 17 distinct chiralities of individual NTs on a surface using a single excitation laser was done by constructing NIR hyper-spectral microscope incorporating Bragg grating between the emission port of a standard inserted fluorescence microscope and the NIR camera. Light passing from sample is reflected by grating to detector encounters two times the grating system which reduces bandwidth. Sodium deoxycholate suspended Hipco NTs adsorbed on poly-Lysine and incubated in human cervical cancer cells (HeLa CCL-2) for analysis in air and biological medium showed chirality dependent hyper-spectral resolution of 8-12 NTs. This is also a first ever report on analysing spectral resolution in animal models (Fig. **8**).

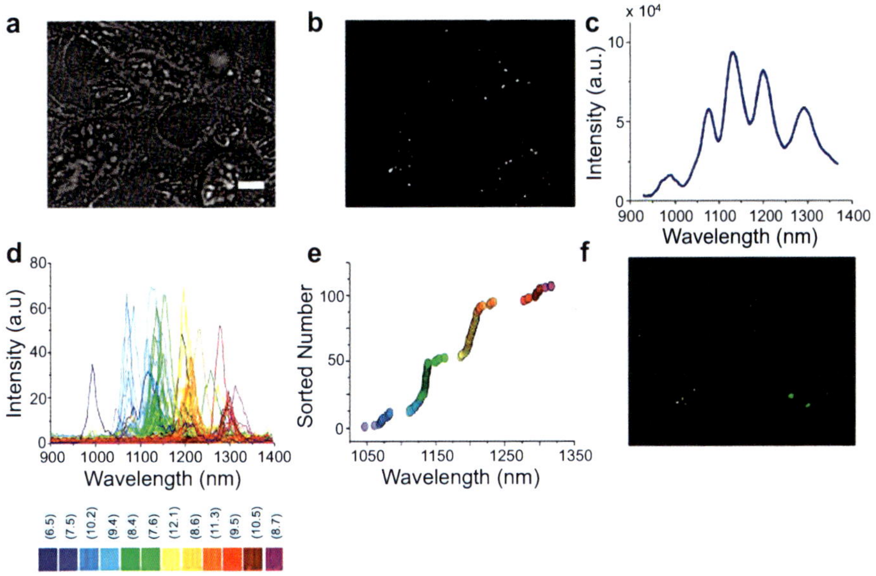


**Fig. (8).** Hyperspectral microscopy of individual carbon nanotubes in live mammalian cells. (a) Transmitted light image of live HeLa cells incubated with SDC-dispersed nanotubes for 30 minutes. Scale bar, 10 μ m. (**b**) Broadband nIR image (900–1600 nm) of the same region. (**c**) Near-infrared spectrum of carbon nanotubes from within HeLa cells as acquired from a conventional spectrometer/nIR detector. (**d**) Spectra from within the live HeLa cells, acquired from 10 cubes of hyperspectral data. (**e**) Fitted emission peak wavelength values sorted in ascending order by emission wavelength. (**f**) False-colored image of the same region as shown in (**b**), colored by nanotube chirality (Source 65).

NIR PL imaging of sodium cholate suspended NTs injected subcutaneously into a hairless SKH1 mouse afforded the multiplexed detection of 8 nanotube chiralities in murine tissues *ex vivo*. Introduction of DNA-encapsulated nanotube complexes in anesthetized zebra-fish embryos *via* cardinal vein injection was carried out for *in vivo* experiments. Hyperspectral imaging showed nanotubes within cells of the zebra-fish vessel wall [65].

## 8.6. RAMAN IMAGING

### 8.6.1. Raman from SWCNTs

Quasi 1-D nanostructure of SWCNT incurs strong and sharp Raman signals as RBM, G and D bands. These peaks are distinguishable from background with narrow spectral features with full width at half maximum (FWHM) in few nm offering multiplicity for detection, an advantageous feature for analyzing many samples at a time. Many diseases can lead to changes in molecular composition of the affected tissues. Raman spectroscopy can reflect these changes in many cases, thereby providing the physician with invaluable information for diagnosing certain diseases in real time. Also the simple NIR Raman scattering spectra of SWNTs permits facile spectral de-convolution of images without using any sophisticated algorithmic and statistical complexities for ruling out auto-fluorescence as well as background data variable in biological samples. Raman spectroscopy, in conjunction with SERS nanoparticles, offers pico-molar sensitivity *in vivo* (at limited depths), as opposed to the nanomolar sensitivity achievable using conventional fluorescence imaging in conjunction with quantum dots *in vivo*.

Intrinsic properties of SWCNT showing Raman characteristics was first utilized by Heller *et al.* [18] imaging SWCNTs in liver cells and tissue slices using oligonucleotide wrapped nanotubes. Robust-Raman signals were acquired from 3T3 cells with excitation at 785 nm wavelength for RBM mode without showing limitations of organic dyes such as photo-bleaching and gradual quenching. They observed higher photo-stability of SWCNTs as compared to QDs and dyes. This was further supported by work by Liu *et al.* [66] with ex-vivo Raman spectroscopy by tracking SWCNTs in mice over three months. They used branched PEG modified SWCNTs which showed highest blood circulation time of approximately 1 day relatively low uptake in RES. Since then many reports have surfaced regarding biological analysis using varied Raman modes of SWCNTs. Weisman group in 2007 [67] utilized Raman spectroscopic analysis of the first time in imaging nanotubes inside animal system. *Drosophila melanogaster* was fed with 10ppm of individual SWCNTs and dissected tissue specimens were imaged at Near IR. Their biodistribution was monitored by counting the nanotubes. In another work by Gambhir group [68, 69] RGD conjugated PEGylated SWCNTs noninvasively localized them in U87MG tumour model in mice over several days. Their Raman spectral analysis publicized nanotube accumulation in the excised liver and spleen of both experimental and control animals owing to localization of these nanotubes in these tissues. Strong G bands at 1593 $cm^{-1}$ appeared in tumours with RGD nanotube clusters for 72 h as compared to complete absence of this band in non-targeted control. These analyses proved that non-invasive Raman mapping would prove to be an important and sensitive modality for pre-clinical investigation of pharmacokinetic evaluation of targeted nano-system.

### 8.6.2. Multiplicity

In addendum Raman offers multiplicity for noting different analytes while revealing geometrical and functional relations between different targets in the same experimental procedure owing to appearance of many significant narrow Raman G-band peaks with FWHM <2 nm. Multiplexing with QDs and dyes shows several limitations in terms of wide fluorescence spectra with broad NIR window and confocal imaging cannot be done for longer duration of time since quenching greatly affects the intensity of emission. Carbon isotopes have different Raman G-band peaks, in consequence variation in composition of SWCNT with $^{13}C$ changes the G band frequency by 62 $cm^{-1}$. Phospholipid-polyethylene glycol-amine (PL-PEG5000-$NH_2$) functionalized SWCNTs having water affinity in debundled form have been conjugated with different targeting ligands for specifically labelling and imaging of biological samples. Five different targeting antibodies and peptide such as Erbitux (anti-EGFR/Her1) for targeting epidermal growth factor receptor (EGFR/Her1), cyclic arginine-glycine-aspartic acid (RGD, cyclo-RGDyK) peptide for integrins like $\alpha v\beta 3$, goat anti-human CEA antibody targeting carcino embryonic antigen (CEA), Rituxan (anti-CD20) to recognize CD20 protein and finally Herceptin (anti-Her2) to target Her2/neu Protein were designated as five different Raman colors. Under single laser excitation five variables used in the study were measured [70]. It is appreciable that five colour coding is not achievable with PL imaging technique using NIR excitation and emission analysis owing to wide fluorescence emission spectra (Fig. **9**).

Multimodality can also be achieved by utilizing chirality/diameter separated SWCNTs beside isotope dependent G-peak shift [18] and diameter dependent RBM peak [71]. Dyes can also be incorporated inside CNTs giving them colour coding and giant Raman scattering effect. Rod like dyes encapsulated with the confines of nanotubes resemble highly polarizable J-aggregates with a resonant Raman cross-section of $(3 \pm 2) \times 10^{-21}$ $cm^2$ $sr^{-1}$, measurable at the highest optical resolution required for detecting individual aggregates. This effect allows multispectral anlaysis with the realization of a library of functionalized nanoprobe Raman labels [72].

### 8.6.3. SERS Labels

The time required for image acquisition is rather large shifting the focus on using Raman tags along with utilizing inherent properties of nanomaterials. Surface enhanced Raman scattering uses active noble metallic species on the surface of nanomaterials increasing signal intensity manifolds while reducing data time interval. SERS is a plasmonic effect where molecules on nano-roughened surfaces increasing Raman signals by increasing incident electric field. Many articles have been published on SERS based analysis [73, 74] and we are currently exploiting this technique the sensitive detection of bacteria in an integrated approach utilizing multiple facets of Au-Iron oxide nanoparticles in our lab. SERS enhances detection sensitivity up to 10–14 orders of magnitude over conventional Raman spectroscopy resulting in ultra-sensitive detection comparable with fluorescence

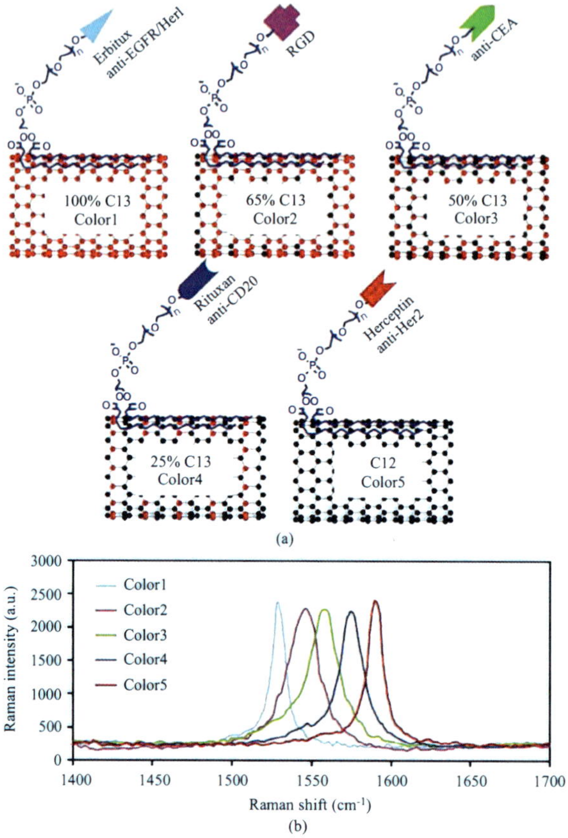


**Fig. (9).** Different Raman "colors". (a) Scheme of isotopically modified SWNTs, grown from FeRu catalysts, conjugated with different targeting ligands. Color1, 2, 3, 4, and 5 represent FeRu-grown SWNTs with C13 percentages (determined by the C13-methane/C12-methane growth gas ratios) of 100%, 65%, 50%, 25%, and 0%, respectively. (b) Raman spectra of the five different SWNT samples in aqueous solutions. The shift of SWNT Raman G-band peak is clearly dependent on the C13/C12 ratio in the SWNTs. The average shift between two adjacent colours is about 15 $cm^{-1}$.

measurements. Electro-deposition by applying certain voltage to electrochemical cell containing SWCNTs in contact with gold salt and requisite reagents like surfactants can initiate nucleation of nanoparticle formation at the defect sites of sidewall of nanotubes. SERS is wavelength dependent phenomenon where emission considerably increases when excited at particular excitation wavelength. This is the famous resonant Raman scattering RRS effect when absorbed or scattered photons are in resonance with energy between the van Hove singularities of nanotubes. Scolari observed additional enhancement of intensity upon gold nanoparticle decoration stemming from the excited surface plasmons resulting in SERS effect [75]. Both the resonant and surface enhancement combine giving rise to surface enhanced resonance Raman scattering (SERS) effect when excited at a particular wavelength of 568nm for large sized nanoparticle of 70 nm as compared with 35 nm gold nanoparticles.

Many methods have been developed for attaching gold nanoparticles on the surface of covalent functionalized SWCNTs *via* their terminal entities [76, 77]. However intrinsic Raman signals of SWCNTs are rather low owing to structural damage caused by general covalent attachment of linkers on the surface of nanotubes. Non-covalent attachment facilitates not only binding of external nanomaterials but also preserves intensity of Raman signals of the pristine nanotubes.

### 8.6.4. Theragnostic Approaches

Imaging cancerous growth is particularly vital in understanding location, malignancy and other pathological conditions initiating requisite treatment modality. Similar to gold-SWCNT hybrid demonstrating combinatorial mode of imaging and therapy in NIR fluorescence imaging system mentioned by Hong *et al.* [30], Raman spectral features can be coupled with photo-thermal therapy. Gold nanoparticles here would not only act as hyper-thermic source but also enhance Raman spectral characteristics *via* plasmonic effect indicative of SERS. These serve as an interesting optical theranostic probe for cancer imaging as well as therapeutic action. In addition these nano-hybrids can be easily functionalized with biocompatible layer of PEG derivatives and be made targeted entities by linking them with necessary biomolecules. Such nano-hybrids have been developed by Wang *et al.* [78] where seed mediated growth of gold nanoparticles was carried out on CVD synthesized SWCNTs. Liu and co-workers [79] coated SWCNTs with poly(allylamine hydrochloride) (PAH) which rendered nanotubes positively charged. These then acted as anchoring sites for the attachment of negatively charged gold nano-seeds which in turn mediated growth of gold or silver nano-shell on their surface. Water solubilisation and debundling was done by coating them with PEG and further conjugated with folic acid attaining SWNT-Au-PEG-FA composition. The group showed noble metal coated maximal enhancement factor of over 20 times than unmodified nanotubes in solution phase (Fig. **10**).

Very recently SERS has been employed for the detection of circulating tumor cells (CTC) and circulating tumor DNA (ctDNA) in the peripheral blood to evaluate malignancy and tumour burden. SWCNTs suitably modified with plasmonic nanostructures such as combination of silver and gold nanoparticles or nano-rods. They can be further functionalized with biomolecules including antibodies and oligonucleotides specific for circulating tumour species for imaging cancerous cells by imaging Raman spectra. Zeid [80] in his dissertation has shown multicolour spectroscopic nanomaterial based on silver nanoparticle decorated with gold nanorods with four different Raman active labels. These were further attached with breast cancer targeting molecules for the detection in peripheral blood samples within 10 sec of acquisition time.

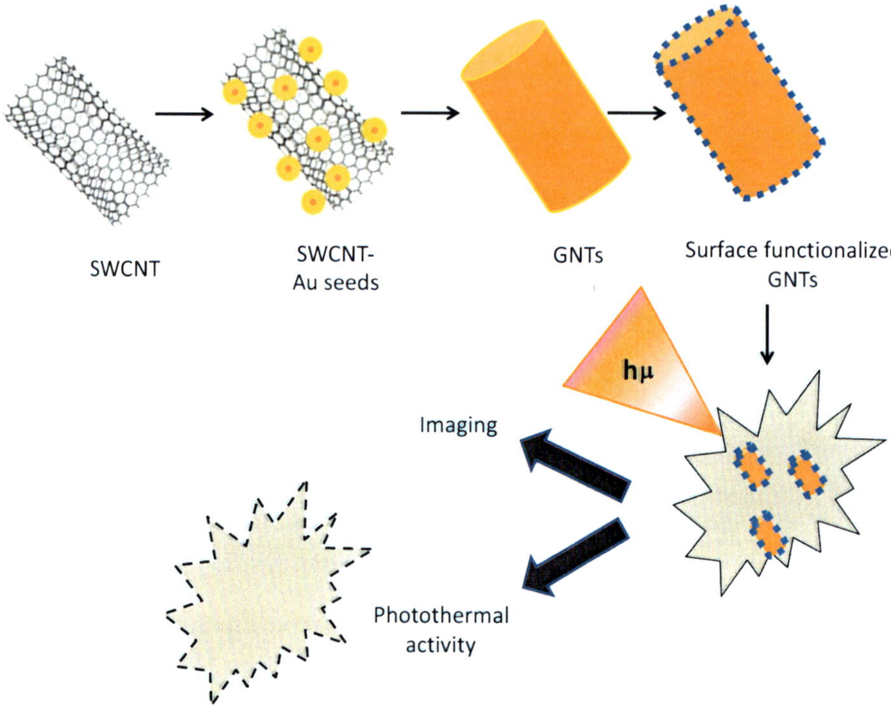


**Fig. (10).** Schematic representation of synthesis of gold SWCNTs or GNTs *via* seed mediated growth mechanism. Surface functionalized enable receptor targeting capabilities to nanotubes which can internalize in tumour cells only. The cells can be imaged using radiations in NIR window or using Raman spectra. In addition they demonstrate ablation properties *via* increasing the temperature of cells locally.

### 8.6.5. Recent Progress

In another work SERS based SWCNTs were employed for the detection circulating tumour DNA (ctDNA) in peripheral blood for the first time for marking malignancy progression. DNA mediated SERS using SWCNTs created a triple helical structure as a molecular switch which was able detect 0.3 fM levels of target DNA as content of KRAS G12DM from sample volume of 5.0 μL [81].

Spectral fingerprints of up to 10 different types of SERS nanoparticles utilizing Picomolar sensitivity of Raman technique were carried out exploiting multiplexing capabilities in living mice model after sub-cutaneous injection. In consequence, intravenous injection of SERS based biomolecule conjugated SWCNTs was followed to image their accumulation. This targeting was due to natural amassing of RGD-SWCNTS in liver as compared to other tissues. Reineshaw Raman imaging system was able to resolve five most intense and spectrally unique signatures of SERS nanomaterials [82].

Very recently SWCNT delivery in nucleus is shown with unambiguous control of localization and distribution within cells. Non-covalent attachment of tail domain of the nuclear protein lamin B1 (LB1) was done on nanotube's surface giving stable dispersion in water and cell media displaying sharp peaks in NIR fluorescence spectra indicating individual dispersion of nanotubes by LB1. Plethora of these conjugates internalizes into cellular complex of HeLa cells localizing in nucleus and interact with the DNA as measured by confocal Raman spectroscopy and microscopy, NIR fluorescence and fluorescence lifetime imaging microscopy. This shows possibility of nuclear targeting by SWCNTs as gene delivery carrier which can be imaged using PL and Raman modes of spectroscopy [83].

## 8.7. PHOTOACOUSTIC IMAGING/TOMOGRAPHY

Photoacoustic imaging (PAI) yield precise depth information, sub-millimetre resolution with nanomolar level sensitivity. Recent work on SWCNTs in PAI indicates their potential use as contrast agents in this imaging modality. Avti etal [84] adopted PA microscopy to map and quantify nano-gram amount of SWCNT in various histo-pathological samples consisting of benign and cancer tissue biopsies with scalable spatial resolution and sufficient depth penetration. This indicates that very small amounts of nanotubes can be detected as very strong PA signal s are generated by NTs providing significant contrast to separate background tissue from the target region of interest. Statistical analysis can further quantitate and map SWCNTS in the specimens.

SWCNTs were conjugated with cyclic RGD peptide without affecting their photo-acoustic properties, significantly the signal was in linear relation with the concentration of nanotubes [85]. SWCNT-RGD complex detected tumour in targeted fashion as compared to non-targeted SWCNTs in U87MG *in vivo* brain xenograft (~100 mm$^3$) model *via* the tail-vein (IV). The validation of PA signals was done by Raman microscope demonstrated the similar results of target based imaging of tumour cells *via* SWCNT-RGD complex with high signal acquisition (Fig. **11**).

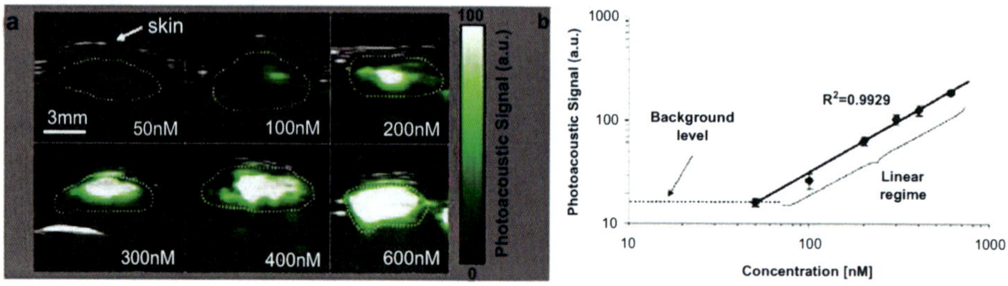


**Fig. (11).** Photoacoustic detection of SWCNT in living mice (a) one vertical slice in 3D PA image (green) was overlaid on the corresponding slice (grey), ultrasound images depict skin while PA images show SWCNTs. (b) the photoacoustic signal was calculated and background level represents endogenous signals of the tissues (Source: de la Zerda A [86]).

Further modifications are being explored for acquiring higher intensity PA signals by many researchers [87, 88]. Molecular CA especially Indocyanine green has been the top choice of tagging molecule on SWCNT for better visualization under the microscope. Gambhir *et al.* reported attachment of ICG on PEGylated SWCNT's surface *via* Π-Π stacking which resulted in almost 20 fold increase in optical density exhibiting sub-nanomolar detection sensitivity [86]. ICG can also be attached on NTs *via* covalent linkages. SWCNT-ICG conjugate primarily marks tumour boundary and will pose highly valuable for surgeons during surgeries to differentiate tumour lining from normal cells *in vivo* and also has the possibility of enhancing surgical re-sectioning of tumours.

Other small optical dyes can also enhance PA contrast and signal on conjugation with PEGylated nanotubes. This improvement is postulated owing to higher optical absorption of the particle as well as their emission is red-shifted than unmodified NTs where the background tissue photo-acoustic signal is greatly reduced. This work by Gambhir *et al.* [89] is significant in illustrating multiplexed detection of analytes where peptide-SWCNTs tagged with different dyes can be imaged simultaneously. Dye enhanced PA imaging of SWCNTs can be done on many types of samples including histo-pathological specimens, as well as living organism.

Though ICG significantly improves signal intensity providing 128-196% enhancement however the total time of data acquisition is very large. PAT is though comparable with other imaging techniques but prolonged data acquisition time is a serious consideration. Apart from dyes, other kinds of materials are being studies to enhance intensity of photo-acoustic signals while decreasing the time of analysis. A layer of gold and other species with NIR absorbance were coupled with SWCNTs to improve their characteristics. Golden nanotubes'(GNTs) have higher intrinsic PA signal owing to coating of gold over

NTs. In addition, Aurate lining induces thermal activity in SWCNTs providing therapeutic action in terms of photo-thermal activity in targeted fashion. Zharov and co-workers [90] used these GNTs for targeting lymphatic vessels in mice with approximately $10^2$ fold improvements in NIR PA imaging contrast than plain NTs. After antibody conjugation GNTs were utilized for mapping lymphatic endothelia receptor. Both PA and photo-thermal signals in the antibody conjugated GNTs exceeded that of endogenous background preferentially in the wall of vessels. The same group extended this work in 2009 showing detection of CTCs under PA imaging module [91]. Very recently a new dispersion agent Evans blue (EB) has been formulated for long circulation. This complex was endowed with multiple characteristics for fluorescent imaging and photodynamic therapy *via* albumin coated fluorescent photosensitizer Chlorin e6 though self-assembly and reaction of albumin with EB. The SWCNT based complex facilitated image guided synergistic PDT and PTT achieving efficient tumour ablation effect in optimized therapeutic window. The dual modality of tumour damage was shown to diminish cancerous growth without recurrence [92].

Though a plethora of research work is going on in directions of understanding depth penetration capabilities of PA imaging systems, signal enhancers, discovery of new materials and molecules for coating over SWCNTs as dispersing agents however the focus is also required to improve photoacoustic instrumentation. The series of work demonstrated by Gambhir and Galzana *et al.* were carried out on custom built PA instruments with lasers, fibre optics and signal transducers. Further improvement in background reduction, hardware and image reconstruction softwares will pave the way for improving signal acquisition enhancing clinical and research use of PA modality of imaging. Several companies are actively commercializing PAT instruments with more patented technology in this direction.

## 8.8. NUCLEAR IMAGING WITH RADIONUCLIDES

Intrinsic properties of SWCNTs have widely been appreciated for optical imaging techniques as mentioned above. Apart from using these characteristics exogenous labels have also find uses with nanotubes for gathering visual information using radio-isotopes. These radiolabel based imaging techniques such as single photon emission computed tomography (SPECT) and positron emission tomography (PET) have gained clinical relevance.

In SPECT, radionuclides are given in dosage form to patient for analysis. Gamma γ-rays emitted by these decaying radionuclides are detected by different rotating detectors and converted to electrical signals and processed into image format. In PET positrons are emitted by radionuclides which are bombarded into electrons annihilating one another while emitting two γ-rays which are perpendicular to one another. These rays are detected by an array of detectors and processing the signals to image form. These both techniques give 2-dimensional cross-section of the study materials from various angles owing to rotating detectors and reconstructing to create a stacked 3-D image. Nuclear imaging is used for analysing functional information. Pet offers high specificity in terms of molecular targeting while investigating biological active substance in its unchanged format.

Radioiodine, $Tc^{m99}$ and $^{111}In$ are common isotopes for labelling in SPECT and PET. Low toxicity and water dispersibility make SWCNT interesting candidates specially PEG derivatives, for attaching these radioisotopes and tracing them by electron microscope. PET and SPECT based studies with radionuclides as the imaging component of the CNT [30, 93, 94] have been used for analysing pharmacokinetic behaviour and distribution pattern of nanotubes in general as well as drug loaded NTs *in vivo* and *ex vivo*.

### 8.8.1. PET

PET is an extremely sensitive tomographic imaging modality and diverse assortments of radionuclides are available to optimize the radiochemical half-life and chemistry. The most informative PK data were obtained using dynamic PET imaging. Radio-active tracers permit measurement of the activity balance and account for the entire administered dose. Wang *et al.* reported for the first time the use of $^{125}I$ ($t_{1/2}$ = 59 days) for tracking bio-distribution of SWCNTS in animal model [95]. Since then many reports have surfaces on attaching radiolabels on functionalized and biomolecule conjugated NTs empowering them with targeting capabilities. Mc Devitt and coworkers used 86Y as radionuclide, attached on the surface of SWCNTs for PET imaging [96]. They conjugated SWCNT with both tumor neovascular-targeting antibody E4G10 and metal chelate of $^{225}Ac^{+3}$ an alpha emitter for radiotherapy chelated by 1,4,7,10-tetraazacyclododecane-1,4,7,-0-tetraacetic acid (DOTA), as well as a β+ emitter $^{89}Zr^{4+}$, for PET imaging (t1/2 = 78 h) chelated by desferrioxamine B. they showed approximately five fold increase in the specific activity using these modified radiolabels with high signal to noise ratio. RGD peptide conjugated PEGylated-SWCNT were used for bindinG with 64Cu for *in vivo* tumor-targetting PET imaging of U87MG tumors. This was facilitated by specific binding between the RGD peptide and integrin avβ3 overexpressed on U87MG tumor cells as well as tumor vasculatures [97].

### 8.8.2. SPECT

In addition to chemical linkages, radiolabels can be filled inside hollow nanotube structure while the outer surface can be tailored according to desired requirement. Na125 sealed inside SWCNT hollow nanostructure for SPECT/CT imaging. The developed high-density radio-emitting crystals were covalently linked with biantennary carbohydrates, for improving dispersion and biocompatibility. The authors observed lung specific accumulation of these nano-materials which conserved their payload without leaking in stomach or excretion. Nano-encapsulation of iodide enabled its bio-distribution to be redirected from tissue with to lungs which was affected by glycosylation [98]. Histological analysis of lungs, liver, spleen, and kidneys corroborated the SPECT imaging data and showed no sign of tissue necrosis or other acute (24 h) or long-term (30 d) abnormalities.

Though there are several advantages of using SWCNT based radionuclide imaging systems such as high sensitivity and being able to gather information of superfacial tissue organization, still the progress in this direction has been slow. New developments in nuclear imaging focus on luminous, fast scintillation detectors, compact light sensors and reconstruction algorithms taking into account the precise system response model.

Currently nuclear images are often overlaid with computed tomography CT image to generate a composite compound image providing anatomical information of the subject. Multimodality instruments use the combination of functional (PET) and anatomical information (CT) in one device (Fig. **12**). Faraj *et al.* showed preferential inclusion of CD44 antibody-conjugated SWCNTs in tumour site and they checked their bio-distribution by three techniques of MRI, SPECT and near-infrared fluorescence imaging. These non-invasive imaging modalities revealed uptake of modified NTs in CD44 rich areas only [99].

## 8.9. CONCLUSION

In this chapter we have comprehensively reviewed SWCNTs as molecular probes for non-invasive imaging of histopathological, cellular or whole animal. Since they are carbonaceous materials having an upper hand than quantum dots containing toxic heavy metals. Different modalities of SWCNTs based biomedical imaging have their benefits and limitations. These nanotubes provide an interesting platform for tracking, detection and imaging with strong absorption in NIR window, photoluminescence and Raman scattering generated owing to their unique photo-physical properties. Their biomedical application is largely dependent on synthetic procedures and surface functionalization protocols. Covalent and non-covalent approaches have been investigated over the years however the thrust primarily is to create functional moieties on NTs without damaging their inherent characteristics. PEGylation is the most sought after derivatization methodology adopted for created long and branched network on the surface of nanotubes imparting water suspension and functional moieties. This

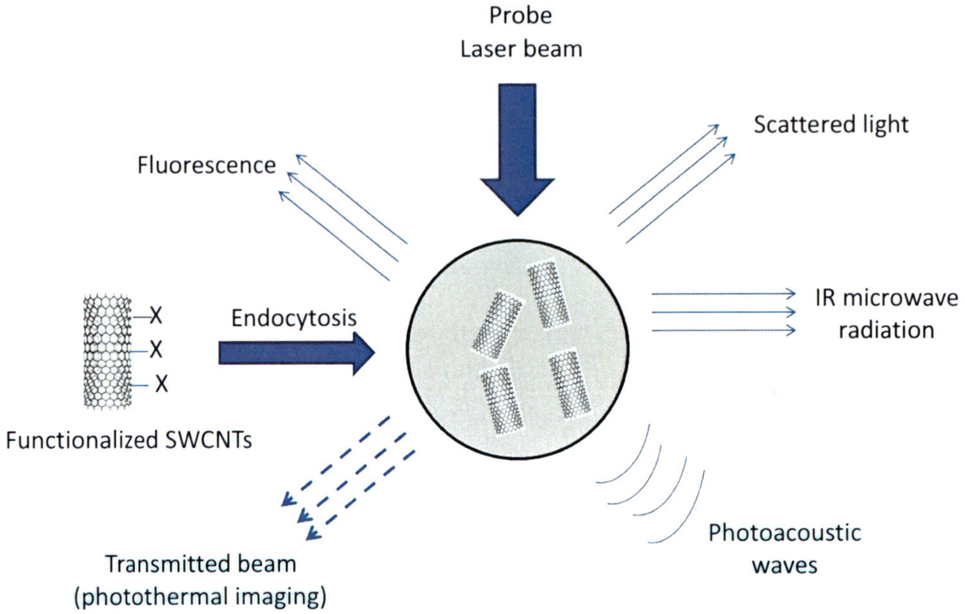


**Fig. (12).** Internalization of SWCNTs in cells and their interaction with light paving the way for various modes of imaging techniques.

also ensures minimum toxicity of CNTs reducing non-specific binding to biological species such as proteins of serum and cell surface. Many of the examples documented in the chapter indicate that appropriately functionalized SWCNTS are versatile material for wide range of imaging with an overlay of therapeutic action in nanomedicine.

SWCNTS absorb maximally in NIR I (700-1000 nm) and newly identified NIR-II of biological transparent window enabling deep tissue penetration and spatial resolution with minimum background absorption. They have large scattering cross-section and strong resonance in Raman scattering, thus they have very strong potential as probe in Raman imaging. Unlike conventional organic dyes and novel fluorophores, SWCNTS are not photo-bleached or easily quenched and show multiplicity in Raman peaks. In addition coating of more biocompatible metal such as gold or silver imparts plasmonic characteristics to this new hybrid termed GNTs. This allows facile functionalization through thiolation and with controllable by recently developed aqueous -phase solid-phase mono-functionalization methods. They show resonance *via* plasmonic interaction paving the way for SERS mode of imaging with very signal to noise ratio in quick response time of imaging. Highly enriched optical and magnetic properties provide multimodality to SWCNTs without the necessity to create complicated multifunctional nanostructures. They provide data of multiple region of interests (ROIs) simultaneously under the laser and Raman mode of imaging. Moreover they can be tuned to provide various payloads at specific site of action while tracking and monitoring their response through imaging systems in advanced mode of theragnostic approach of disease diagnosis and treatment. Combinatorial imaging systems are gaining tremendous popularity in clinical setting proving not only high resolution of imaging but also 3-D reconstruction of images is possible for marking structural as well as functional features. SWCNTs can serve as optical probe for both techniques simultaneously specially with production of 'Gadonanotubes' containing Gadolinium based contrast agents for MRI in combination with either SPECT or PET.

The current limitations of nanotubes are the future challenges to be worked on for researchers across the panel. Since the quantum yield of SWCNTs is still low, it is mandatory to develop fabrication procedures which would not yield heterogeneous mixtures of NTs with different diameter, chirality and lengths. Purification procedures and development of single diameter and chirality nanotubes may bring revolution in semiconductors and imaging systems also. QY increase will play a major role in photoluminescence imaging. Chirality-purified NTs together in resonance with laser would significantly enhance Raman signals. Besides modulating these intrinsic characteristics the major challenge is to produce non-toxic entities for long term *in vivo* applications. Since NTS are non-biodegradable their debris, exudates or any leaching component produced owing to interactions with lasers or biological surrounding should also be non-toxic. Efforts are required to optimize and tailor surface chemistry of SWNTs to enhance biocompatibility. It is foremost concern for biomedical scientist to convert hydrophobic nanotubes into water dispersible biocompatible entities *via* surface modification enabling long circulation time in blood with low RES uptake.

Certain solutions to these problems have been identified with work related to density column-based length separation method [100], selective synthesis protocols [101], removal of SWNTs with undesired diameters by etching [102], and chirality separation based on chromatography [103] It has been learned that CNTs can be enzymatically degraded by peroxidases [104] in cellular organelles including macrophages, eosinophils, neutrophyls, microglia and extracellular space [105 - 108]. This discovery directly implies their toxicity can be modulated easily as compared to many other nanomaterials. However more work is required in this direction before the real clinical applications of SWCNTs in imaging can be set forth.

## CONFLICT OF INTEREST

The authors declare no conflict of interest, financial or otherwise.

## ACKNOWLEDGEMENTS

Authors are thankful to Dr. Apoorva Sau, Associate Professor, National Institute of Immunology (NII), New Delhi, India for proof reading this article.

## REFERENCES

[1] Maultzsch J, Reich S, Thomsen C, Requardt H, Ordejón P. Phonon dispersion in graphite. Phys Rev Lett 2004; 92(7): 075501.
[http://dx.doi.org/10.1103/PhysRevLett.92.075501] [PMID: 14995866]

[2] Liu Z, Chen K, Davis C, *et al.* Drug delivery with carbon nanotubes for *in vivo* cancer treatment. Cancer Res 2008; 68(16): 6652-60.
[http://dx.doi.org/10.1158/0008-5472.CAN-08-1468] [PMID: 18701489]

[3] Liu Z, Sun X, Nakayama-Ratchford N, Dai H. Supramolecular chemistry on water-soluble carbon nanotubes for drug loading and delivery. ACS Nano 2007; 1(1): 50-6.
[http://dx.doi.org/10.1021/nn700040t] [PMID: 19203129]

[4] Bhirde AA, Patel V, Gavard J, *et al.* Targeted killing of cancer cells *in vivo* and *in vitro* with EGF-directed carbon nanotube-based drug delivery. ACS Nano 2009; 3(2): 307-16.
[http://dx.doi.org/10.1021/nn800551s] [PMID: 19236065]

[5] Chen J, Chen S, Zhao X, Kuznetsova LV, Wong SS, Ojima I. Functionalized single-walled carbon nanotubes as rationally designed vehicles for tumor-targeted drug delivery. J Am Chem Soc 2008; 130(49): 16778-85.
[http://dx.doi.org/10.1021/ja805570f] [PMID: 19554734]

[6] Mankoff DA. A definition of molecular imaging. J Nucl Med 2007; 48(6): 18N-21N, 21N.
[PMID: 17536102]

[7] Rudin M, Weissleder R. Molecular imaging in drug discovery and development. Nat Rev Drug Discov 2003; 2(2): 123-31.
[http://dx.doi.org/10.1038/nrd1007] [PMID: 12563303]

[8] Jackeray R, Abid CK, Singh G, *et al.* Selective capturing and detection of Salmonella typhi on polycarbonate membrane using bioconjugated quantum dots. Talanta 2011; 84(3): 952-62.
[http://dx.doi.org/10.1016/j.talanta.2011.02.052] [PMID: 21482309]

[9] Goldys EM, Tomsia K, Zhu G, *et al.* Fluorescence labeling. Optica Applitica 2006; 36: 217-25.

[10] Hammond VJ, Aylott JW, Greenway GM, Watts P, Webster A, Wiles C. An optical sensor for reactive oxygen species: encapsulation of functionalised silica nanoparticles into silicate nanoprobes to reduce fluorophore leaching. Analyst (Lond) 2008; 133(1): 71-5.
[http://dx.doi.org/10.1039/B711995J] [PMID: 18087616]

[11] Sperling RA, Parak WJ. Surface modification, functionalization and bioconjugation of colloidal inorganic nanoparticles. Philos Trans A Math Phys Eng Sci 2010; 368(1915): 1333-83.
[http://dx.doi.org/10.1098/rsta.2009.0273] [PMID: 20156828]

[12] Kosaka N, Ogawa M, Choyke PL, Kobayashi H. Clinical implications of near-infrared fluorescence imaging in cancer. Future Oncol 2009; 5(9): 1501-11.
[http://dx.doi.org/10.2217/fon.09.109] [PMID: 19903075]

[13] Carlson LJ, Krauss TD. Photophysics of individual single-walled carbon nanotubes. Acc Chem Res 2008; 41(2): 235-43.

[http://dx.doi.org/10.1021/ar700136v] [PMID: 18281946]

[14] O'Connell MJ, Bachilo SM, Huffman CB, *et al.* Band gap fluorescence from individual single-walled carbon nanotubes. Science 2002; 297(5581): 593-6. [http://dx.doi.org/10.1126/science.1072631] [PMID: 12142535]

[15] Zheng M, Jagota A, Strano MS, *et al.* Structure-based carbon nanotube sorting by sequence-dependent DNA assembly. Science 2003; 302(5650): 1545-8. [http://dx.doi.org/10.1126/science.1091911] [PMID: 14645843]

[16] Kasha M. Characterization of electronic transitions in complex molecules. Discuss Faraday Soc 1950; 9: 14-9. [http://dx.doi.org/10.1039/df9500900014]

[17] Liu Z, Tabakman S, Welsher K, Dai H. Carbon Nanotubes in Biology and Medicine: *In vitro* and *in vivo* Detection, Imaging and Drug Delivery. Nano Res 2009; 2(2): 85-120. [http://dx.doi.org/10.1007/s12274-009-9009-8] [PMID: 20174481]

[18] Heller DA, Baik S, Eurell TE, Strano MS. Single walled carbon nanotube spectroscopy in live cells: Towards long-term labels and optical sensors. Adv Mater 2005; 17: 2793-9. [http://dx.doi.org/10.1002/adma.200500477]

[19] Dresselhaus MS, Dresselhaus G, Jorio A, Souza Filho AG, Saito R. Raman spectroscopy on isolated single wall carbon nanotubes. Carbon 2002; 40: 2043-61. [http://dx.doi.org/10.1016/S0008-6223(02)00066-0]

[20] Filho A G, Chou S G, Samsonidze G G, *et al.* Stokes and anti-Stokes Raman spectra of small-diameter isolated carbon nanotubes. Phys Rev B 2004; 69(11): 115428.1-.

[21] Maultzsch J, Telg H, Reich S, Thomsen C. Radial breathing mode of single-walled carbon nanotubes: Optical transition energies and chiral-index assignment. Phys Rev B 2005; 72: 205438-16. [http://dx.doi.org/10.1103/PhysRevB.72.205438]

[22] Thomsen C, Reich S, Rafailov PM, Jantoljak H. Phys Status Solidi, B Basic Res 1999; 214: R15. [http://dx.doi.org/10.1002/(SICI)1521-3951(199908)214:2<R15::AID-PSSB999915>3.0.CO;2-I]

[23] Bachilo SM, Strano MS, Kittrell C, Hauge RH, Smalley RE, Weisman RB. Structure-assigned optical spectra of single-walled carbon nanotubes. Science 2002; 298(5602): 2361-6. [http://dx.doi.org/10.1126/science.1078727] [PMID: 12459549]

[24] Telg H, Maultzsch J, Reich S, Hennrich F, Thomsen C. Chirality distribution and transition energies of carbon nanotubes. Phys Rev Lett 2004; 93: 177401.1-.

[25] Fantini C, Jorio A, Souza M, Strano M S, Dresselhaus M S, Pimenta M A. Optical transition energies for carbon nanotubes from resonant raman spectroscopy: environment and temperature effects. Phys Rev Lett 2004; 93: 147406.1-.

[26] Martin RM, Falicov LM. Raw Single-Wall Carbon Nanotubes Induce Oxidative Stress and Activate MAPKs, AP-1, NF-[kappa] B, and Akt in Normal and Malignant Human Mesothelial Cells. Environmental health perspectives 1983; 116(9): 1211.

[27] Jorio A, Pimenta M A, Souza Filho A G, Saito R, Dresselhaus G, Dresselhaus M S. Characterizing carbon nanotube samples with resonance Raman scattering. Toxicology letters 2003; 181(3): 182-9. [http://dx.doi.org/10.1088/1367-2630/5/1/139]

[28] Bell AG. On the production and reproduction of sound by light. Am J Sci 1880; 20: 305-24. [http://dx.doi.org/10.2475/ajs.s3-20.118.305]

[29] Li ML, Oh JT, Xie X, *et al.* Simultaneous molecular and hypoxia imaging of brain tumors *in vivo* using spectroscopic photoacoustic tomography. Proc IEEE 2008; 96: 481-9. [http://dx.doi.org/10.1109/JPROC.2007.913515]

[30] Liu Z, Cai W, He L, *et al.* *In vivo* biodistribution and highly efficient tumour targeting of carbon nanotubes in mice. Nat Nanotechnol 2007; 2(1): 47-52.

[http://dx.doi.org/10.1038/nnano.2006.170] [PMID: 18654207]

[31] Kim C, Cho EC, Chen J, *et al. In vivo* molecular photoacoustic tomography of melanomas targeted by bioconjugated gold nanocages. ACS Nano 2010; 4(8): 4559-64. [http://dx.doi.org/10.1021/nn100736c] [PMID: 20731439]

[32] Fang H, Maslov K, Wang LV. Photoacoustic Doppler effect from flowing small light-absorbing particles. Phys Rev Lett 2007; 99(18): 184501. [http://dx.doi.org/10.1103/PhysRevLett.99.184501] [PMID: 17995411]

[33] Xie Z, Roberts W, Carson P, Liu X, Tao C, Wang X. Evaluation of bladder microvasculature with high-resolution photoacoustic imaging. Opt Lett 2011; 36(24): 4815-7. [http://dx.doi.org/10.1364/OL.36.004815] [PMID: 22179893]

[34] Zeng L, Alemany LB, Edwards CL, Barron AR. Demonstration of covalent sidewall functionalization of single wall carbon nanotubes by NMR spectroscopy: Side chain length dependence on the observation of the sidewall sp3 carbons. Nano Res 2008; 1: 72-88. [http://dx.doi.org/10.1007/s12274-008-8004-9]

[35] Katz E, Willner I. Biomolecule-functionalized carbon nanotubes: applications in nanobioelectronics. ChemPhysChem 2004; 5(8): 1084-104. [http://dx.doi.org/10.1002/cphc.200400193] [PMID: 15446731]

[36] Boul PJ, Liu J, Mickelson ET, *et al.* Reversible sidewall functionalization of buckytubes. Chem Phys Lett 1999; 310: 367. [http://dx.doi.org/10.1016/S0009-2614(99)00713-7]

[37] Khabashesku VN, Billups WE, Margrave JL. Fluorination of single-wall carbon nanotubes and subsequent derivatization reactions. Acc Chem Res 2002; 35(12): 1087-95. [http://dx.doi.org/10.1021/ar020146y] [PMID: 12484797]

[38] Stevens JL, Huang AY, Peng HQ, Chiang LW, Khabashesku VN, Margrave JL. Sidewall Amino-Functionalization of Single-Walled Carbon Nanotubes through Fluorination and Subsequent Reactions with Terminal Diamines. Nano Lett 2003; 3: 331. [http://dx.doi.org/10.1021/nl025944w]

[39] Wu HC, Chang X, Liu L, Zhaoa F, Zhao Y. Chemistry of carbon nanotubes in biomedical applications. J Mater Chem 2010; 20: 1036-52. [http://dx.doi.org/10.1039/B911099M]

[40] Fujigaya T, Nakashima N. Non-covalent polymer wrapping of carbon nanotubes and the role of wrapped polymers as functional dispersants. Sci Technol Adv Mater 2015; 16(2): 024802. [http://dx.doi.org/10.1088/1468-6996/16/2/024802] [PMID: 27877763]

[41] Star A, Stoddart JF, Steuerman D, *et al.* Heath Preparation and Properties of Polymer-Wrapped Single-Walled Carbon Nanotubes. Angew Chem 2001; 40: 1721-5. [http://dx.doi.org/10.1002/1521-3773(20010504)40:9<1721::AID-ANIE17210>3.0.CO;2-F]

[42] Gomulya W, Gao J, Loi M. Conjugated polymer-wrapped carbon nanotubes: physical properties and device applications. Eur Phys J B 2013; 86: 404. [http://dx.doi.org/10.1140/epjb/e2013-40707-9]

[43] Gerstel P, Klumpp S, Hennrich F, *et al.* Highly Selective Dispersion of Single-Walled Carbon Nanotubes *via* Polymer Wrapping: A Combinatorial Study *via* Modular Conjugation. ACS Macro Lett 2014; 3: 10-5. [http://dx.doi.org/10.1021/mz400472q]

[44] Zhang X, Meng L, Lu Q. Cell behaviors on polysaccharide-wrapped single-wall carbon nanotubes: a quantitative study of the surface properties of biomimetic nanofibrous scaffolds. ACS Nano 2009; 3(10): 3200-6. [http://dx.doi.org/10.1021/nn9006362] [PMID: 19719149]

[45] O'Connell MJ, Boul P, Ericson LM, Huffman C, Wang Y, Haroz E. Reversible water-solubilization of

single-walled carbon nanotubes by polymer wrapping. Chem Phys Lett 2001; 342: 265-71. [http://dx.doi.org/10.1016/S0009-2614(01)00490-0]

[46] Pang X, Imin P. Zhitomir sky I, and Adronov A. Amperometric Detection of Glucose Using a Conjugated Polyelectrolyte Complex with Single-Walled Carbon Nanotubes Macromolecules 2010; 43: 10376-81.

[47] Staii C, Johnson AT Jr, Chen M, Gelperin A. DNA-decorated carbon nanotubes for chemical sensing. Nano Lett 2005; 5(9): 1774-8. [http://dx.doi.org/10.1021/nl051261f] [PMID: 16159222]

[48] Johnson AT, Staii C, Chen M, *et al.* DNA-decorated carbon nanotubes for chemical sensing. Semicond Sci Technol 2006; 21: S17. [http://dx.doi.org/10.1088/0268-1242/21/11/S03]

[49] Maune HT, Han SP, Barish RD, *et al.* Self-assembly of carbon nanotubes into two-dimensional geometries using DNA origami templates. Nat Nanotechnol 2010; 5(1): 61-6. [http://dx.doi.org/10.1038/nnano.2009.311] [PMID: 19898497]

[50] Kam NW, Liu Z, Dai H. Functionalization of carbon nanotubes *via* cleavable disulfide bonds for efficient intracellular delivery of siRNA and potent gene silencing. J Am Chem Soc 2005; 127(36): 12492-3. [http://dx.doi.org/10.1021/ja053962k] [PMID: 16144388]

[51] Hadidi N, Kobarfard F, Nafissi-Varcheh N, Aboofazeli R. Optimization of single-walled carbon nanotube solubility by non-covalent PEGylation using experimental design methods. Int J Nanomedicine 2011; 6: 737-46. [PMID: 21556348]

[52] Cherukuri P, Bachilo SM, Litovsky SH, Weisman RB. Near-infrared fluorescence microscopy of single-walled carbon nanotubes in phagocytic cells. J Am Chem Soc 2004; 126(48): 15638-9. [http://dx.doi.org/10.1021/ja0466311] [PMID: 15571374]

[53] Jin H, Heller DA, Strano MS. Single-particle tracking of endocytosis and exocytosis of single-walled carbon nanotubes in NIH-3T3 cells. Nano Lett 2008; 8(6): 1577-85. [http://dx.doi.org/10.1021/nl072969s] [PMID: 18491944]

[54] Hong G, Tabakman SM, Welsher K, *et al.* Near-infrared-fluorescence-enhanced molecular imaging of live cells on gold substrates. Angew Chem Int Ed Engl 2011; 50(20): 4644-8. [http://dx.doi.org/10.1002/anie.201100934] [PMID: 21506225]

[55] Hong G, Wu JZ, Robinson JT, Wang H, Zhang B, Dai H. Three-dimensional imaging of single nanotube molecule endocytosis on plasmonic substrates. Nat Commun 2012; 3: 700. [http://dx.doi.org/10.1038/ncomms1698] [PMID: 22426221]

[56] Tchounwou C, Sinha SS, Viraka Nellore BP, *et al.* Hybrid Theranostic Platform for Second Near-IR Window Light Triggered Selective Two-Photon Imaging and Photothermal Killing of Targeted Melanoma Cells. ACS Appl Mater Interfaces 2015; 7(37): 20649-56. [http://dx.doi.org/10.1021/acsami.5b05225] [PMID: 26327304]

[57] TIgnatova A. Balaeff M, Blades M, Zheng P, Stoeckl S V, Rotkin. Two-color spectroscopy of UV excited ssDNA complex with a single-wall nanotube photoluminescence probe Fast relaxation by nucleobase autoionization mechanism. Nano Res 2016; 9: 571-83. [http://dx.doi.org/10.1007/s12274-015-0938-0]

[58] Welsher K, Liu Z, Daranciang D, Dai H. Selective probing and imaging of cells with single walled carbon nanotubes as near-infrared fluorescent molecules. Nano Lett 2008; 8(2): 586-90. [http://dx.doi.org/10.1021/nl072949q] [PMID: 18197719]

[59] Welsher K, Liu Z, Sherlock SP, *et al.* A route to brightly fluorescent carbon nanotubes for near-infrared imaging in mice. Nat Nanotechnol 2009; 4(11): 773-80. [http://dx.doi.org/10.1038/nnano.2009.294] [PMID: 19893526]

[60] Robinson JT, Welsher K, Tabakman SM, *et al.* High Performance *In vivo* Near-IR (>1 μm) Imaging and Photothermal Cancer Therapy with Carbon Nanotubes. Nano Res 2010; 3(11): 779-93. [http://dx.doi.org/10.1007/s12274-010-0045-1] [PMID: 21804931]

[61] Hong G, Lee JC, Robinson JT, *et al.* Multifunctional *in vivo* vascular imaging using near-infrared II fluorescence. Nat Med 2012; 18(12): 1841-6. [http://dx.doi.org/10.1038/nm.2995] [PMID: 23160236]

[62] Diao S, Hong G, Antaris AL. Biological imaging without autofluorescence in the second near-infrared region. Nano Res 2015; 8: 3027-34. [http://dx.doi.org/10.1007/s12274-015-0808-9]

[63] Kam NW, O'Connell M, Wisdom JA, Dai H. Carbon nanotubes as multifunctional biological transporters and near-infrared agents for selective cancer cell destruction. Proc Natl Acad Sci USA 2005; 102(33): 11600-5. [http://dx.doi.org/10.1073/pnas.0502680102] [PMID: 16087878]

[64] Murali VS, Wang R, Mikoryak CA, Pantano P, Draper RK. The impact of subcellular location on the near infrared-mediated thermal ablation of cells by targeted carbon nanotubes. Nanotechnology 2016; 27(42): 425102. [http://dx.doi.org/10.1088/0957-4484/27/42/425102] [PMID: 27632056]

[65] Roxbury D, Jena P V, Williams R M, *et al.* Cytotoxicity of single-wall carbon nanotubes on human Wbroblasts. Toxicol In Vitro 2015; 20: 1202-12.

[66] Liu Z, Davis C, Cai W, He L, Chen X, Dai H. Circulation and long-term fate of functionalized, biocompatible single-walled carbon nanotubes in mice probed by Raman spectroscopy. Proc Natl Acad Sci USA 2008; 105(5): 1410-5. [http://dx.doi.org/10.1073/pnas.0707654105] [PMID: 18230737]

[67] Leeuw T K, Reith R M, Simonette R A, *et al.* Cytotoxicity of carbon nanomaterials: single-wall nanotube, multi-wall nanotube, and fullerene. Environmental Science and Technology 2007; 39(5): 1378-83.

[68] Keren S, Zavaleta C, Cheng Z, de la Zerda A, Gheysens O, Gambhir SS. Noninvasive molecular imaging of small living subjects using Raman spectroscopy. Proc Natl Acad Sci USA 2008; 105(15): 5844-9. [http://dx.doi.org/10.1073/pnas.0710575105] [PMID: 18378895]

[69] Zavaleta C, de la Zerda A, Liu Z, *et al.* Noninvasive Raman spectroscopy in living mice for evaluation of tumor targeting with carbon nanotubes. Nano Lett 2008; 8(9): 2800-5. [http://dx.doi.org/10.1021/nl801362a] [PMID: 18683988]

[70] Liu Z, Tabakman S, Sherlock S, *et al.* Multiplexed five-color molecular imaging of cancer cells and tumor tissues with carbon nanotube Raman tags in the near-infrared. Nano Res 2010; 3(3): 222-33. [http://dx.doi.org/10.1007/s12274-010-1025-1] [PMID: 21442006]

[71] Rao AM, Richter E, Bandow S, *et al.* Diameter-Selective Raman Scattering from Vibrational Modes in Carbon Nanotubes. Science 1997; 275(5297): 187-91. [http://dx.doi.org/10.1126/science.275.5297.187] [PMID: 8985007]

[72] Gaufrès E. N. Y.-Wa Tang, F. Lapointe, J. Cabana, M.-A. Nadon, N. Cottenye, F. Raymond, T. Szkopek & R. Martel Giant Raman scattering from J-aggregated dyes inside carbon nanotubes for multispectral imaging. Nat Photonics 2014; 8: 72-8. [http://dx.doi.org/10.1038/nphoton.2013.309]

[73] Qian XM, Nie SM. Single-molecule and single-nanoparticle SERS: from fundamental mechanisms to biomedical applications. Chem Soc Rev 2008; 37(5): 912-20. [http://dx.doi.org/10.1039/b708839f] [PMID: 18443676]

[74] Schlücker S. SERS microscopy: nanoparticle probes and biomedical applications. ChemPhysChem 2009; 10(9-10): 1344-54.

[http://dx.doi.org/10.1002/cphc.200900119] [PMID: 19565576]

[75] Scolari M, Mews A, Fu N, *et al.* Surface Enhanced Raman Scattering of Carbon Nanotubes Decorated by Individual Fluorescent Gold Particles. J Phys Chem C 2008; 112: 391-6. [http://dx.doi.org/10.1021/jp076190i]

[76] Chen YC, Young RJ, Macpherson JV, Wilson NR. Silver-decorated carbon nanotube networks as SERS substrates. J Raman Spectrosc 2011; 42: 1255-62. [http://dx.doi.org/10.1002/jrs.2862]

[77] Chu H, Wang J, Ding L, *et al.* Decoration of gold nanoparticles on surface-grown single-walled carbon nanotubes for detection of every nanotube by surface-enhanced Raman spectroscopy. J Am Chem Soc 2009; 131(40): 14310-6. [http://dx.doi.org/10.1021/ja9035972] [PMID: 19764748]

[78] Wang X, Wang C, Cheng L, Lee ST, Liu Z. Noble metal coated single-walled carbon nanotubes for applications in surface enhanced Raman scattering imaging and photothermal therapy. J Am Chem Soc 2012; 134(17): 7414-22. [http://dx.doi.org/10.1021/ja300140c] [PMID: 22486413]

[79] Liu L, Grunlan JC. Tailoring of Carbon Nanotube Microstructure Using Poly(acrylic acid) and Poly(allylamine hydrochloride). MRS Proceedings. 922.

[80] Nima ZA. Raman spectroscopy as a tool for single circulating tumor cell (CTC) detection and imaging. PhD, University of Arkansas at Little Rock 2014; 160: 3685745.

[81] Zhou Q, Zheng J, Qing Z, *et al.* Detection of Circulating Tumor DNA in Human Blood *via* DNA-Mediated Surface-Enhanced Raman Spectroscopy of Single-Walled Carbon Nanotubes. Anal Chem 2016; 88(9): 4759-65. [http://dx.doi.org/10.1021/acs.analchem.6b00108] [PMID: 27028517]

[82] Zavaleta CL, Smith BR, Walton I, *et al.* Multiplexed imaging of surface enhanced Raman scattering nanotags in living mice using noninvasive Raman spectroscopy. Proc Natl Acad Sci USA 2009; 106(32): 13511-6. [http://dx.doi.org/10.1073/pnas.0813327106] [PMID: 19666578]

[83] Boyer PD, Ganesh S, Qin Z, *et al.* Delivering Single-Walled Carbon Nanotubes to the Nucleus Using Engineered Nuclear Protein Domains. ACS Appl Mater Interfaces 2016; 8(5): 3524-34. [http://dx.doi.org/10.1021/acsami.5b12602] [PMID: 26783632]

[84] Avti PK, Hu S, Favazza C, *et al.* Detection, mapping, and quantification of single walled carbon nanotubes in histological specimens with photoacoustic microscopy. PLoS One 2012; 7(4): e35064. [http://dx.doi.org/10.1371/journal.pone.0035064] [PMID: 22496892]

[85] De la Zerda A, Zavaleta C, Keren S, *et al.* Carbon nanotubes as photoacoustic molecular imaging agents in living mice. Nat Nanotechnol 2008; 3(9): 557-62. [http://dx.doi.org/10.1038/nnano.2008.231] [PMID: 18772918]

[86] de la Zerda A, Liu Z, Bodapati S, *et al.* Ultrahigh sensitivity carbon nanotube agents for photoacoustic molecular imaging in living mice. Nano Lett 2010; 10(6): 2168-72. [http://dx.doi.org/10.1021/nl100890d] [PMID: 20499887]

[87] Zanganeh S, Li H, Kumavor PD, *et al.* Photoacoustic imaging enhanced by indocyanine green-conjugated single-wall carbon nanotubes. J Biomed Opt 2013; 18(9): 096006. [http://dx.doi.org/10.1117/1.JBO.18.9.096006] [PMID: 24002193]

[88] Nguyen V P, Oh Y, Ha K, Oh J, Kang H W. Enhancement of high-resolution photoacoustic imaging with indocyanine green-conjugated carbon nanotubes. Jap J Appl Phys 2015; 54(7S1)

[89] de la Zerda A, Bodapati S, Teed R, *et al.* Family of enhanced photoacoustic imaging agents for high-sensitivity and multiplexing studies in living mice. ACS Nano 2012; 6(6): 4694-701. [http://dx.doi.org/10.1021/nn204352r] [PMID: 22607191]

[90] Kim J-W, Galanzha EI, Shashkov EV, Moon H-M, Zharov VP. Golden carbon nanotubes as multimodal photoacoustic and photothermal high-contrast molecular agents. Nat Nanotechnol 2009; 4(10): 688-94.
[http://dx.doi.org/10.1038/nnano.2009.231] [PMID: 19809462]

[91] Galanzha EI, Shashkov EV, Kelly T, Kim J-W, Yang L, Zharov VP. *In vivo* magnetic enrichment and multiplex photoacoustic detection of circulating tumour cells. Nat Nanotechnol 2009; 4(12): 855-60.
[http://dx.doi.org/10.1038/nnano.2009.333] [PMID: 19915570]

[92] Xie L, Wang G, Zhou H, *et al.* Functional long circulating single walled carbon nanotubes for fluorescent/photoacoustic imaging-guided enhanced phototherapy. Biomaterials 2016; 103: 219-28.
[http://dx.doi.org/10.1016/j.biomaterials.2016.06.058] [PMID: 27392290]

[93] Lacerda L, Soundararajan A, Singh R. Dynamic imaging of functionalized multi-walled carbon nanotube systemic circulation and urinary excretion. Adv Mater 2008; 20: 225-30.
[http://dx.doi.org/10.1002/adma.200702334]

[94] Al Faraj A, Shaik AS, Al Sayed B, Halwani R, Al Jammaz I. Specific targeting and noninvasive imaging of breast cancer stem cells using single-walled carbon nanotubes as novel multimodality nanoprobes. Nanomedicine (Lond) 2016; 11(1): 31-46.
[http://dx.doi.org/10.2217/nnm.15.182] [PMID: 26673059]

[95] Wang H, Wang J, Deng X, *et al.* Biodistribution of carbon single-wall carbon nanotubes in mice. J Nanosci Nanotechnol 2004; 4(8): 1019-24.
[http://dx.doi.org/10.1166/jnn.2004.146] [PMID: 15656196]

[96] McDevitt MR, Chattopadhyay D, Jaggi JS, *et al.* PET imaging of soluble yttrium-86-labeled carbon nanotubes in mice. PLoS One 2007; 2(9): e907.
[http://dx.doi.org/10.1371/journal.pone.0000907] [PMID: 17878942]

[97] Liu Z, Cai W, He L, *et al. In vivo* biodistribution and highly efficient tumour targeting of carbon nanotubes in mice. Nat Nanotechnol 2007; 2(1): 47-52.
[http://dx.doi.org/10.1038/nnano.2006.170] [PMID: 18654207]

[98] Hong SY, Tobias G, Al-Jamal KT, *et al.* Filled and glycosylated carbon nanotubes for *in vivo* radioemitter localization and imaging. Nat Mater 2010; 9(6): 485-90.
[http://dx.doi.org/10.1038/nmat2766] [PMID: 20473287]

[99] Al Faraj A, Shaik AS, Al Sayed B, Halwani R, Al Jammaz I. Specific targeting and noninvasive imaging of breast cancer stem cells using single-walled carbon nanotubes as novel multimodality nanoprobes. Nanomedicine (Lond) 2016; 11(1): 31-46.
[http://dx.doi.org/10.2217/nnm.15.182] [PMID: 26673059]

[100] Sun X, Zaric S, Daranciang D, *et al.* Optical properties of ultrashort semiconducting single-walled carbon nanotube capsules down to sub-10 nm. J Am Chem Soc 2008; 130(20): 6551-5.
[http://dx.doi.org/10.1021/ja8006929] [PMID: 18426207]

[101] Li X, Tu X, Zaric S, *et al.* Selective synthesis combined with chemical separation of single-walled carbon nanotubes for chirality selection. J Am Chem Soc 2007; 129(51): 15770-1.
[http://dx.doi.org/10.1021/ja077886s] [PMID: 18052285]

[102] Zhang G, Qi P, Wang X, *et al.* Selective etching of metallic carbon nanotubes by gas-phase reaction. Science 2006; 314(5801): 974-7.
[http://dx.doi.org/10.1126/science.1133781] [PMID: 17095698]

[103] Tu X, Zheng M. A DNA-based approach to the carbon nanotube sorting problem. Nano Res 2008; 1: 185-94.
[http://dx.doi.org/10.1007/s12274-008-8022-7]

[104] Kotchey GP, Zhao Y, Kagan VE, Star A. Peroxidase-mediated biodegradation of carbon nanotubes *in vitro* and *in vivo*. Adv Drug Deliv Rev 2013; 65(15): 1921-32.
[http://dx.doi.org/10.1016/j.addr.2013.07.007] [PMID: 23856412]

[105] Kagan VE, Kapralov AA, St Croix CM, *et al.* Lung macrophages "digest" carbon nanotubes using a superoxide/peroxynitrite oxidative pathway. ACS Nano 2014; 8(6): 5610-21. [http://dx.doi.org/10.1021/nn406484b] [PMID: 24871084]

[106] Bhattacharya K, Sacchetti C, El-Sayed R, *et al.* Enzymatic 'stripping' and degradation of PEGylated carbon nanotubes. Nanoscale 2014; 6(24): 14686-90. [http://dx.doi.org/10.1039/C4NR03604B] [PMID: 25377797]

[107] Farrera C, Bhattacharya K, Lazzaretto B, *et al.* Extracellular entrapment and degradation of single-walled carbon nanotubes. Nanoscale 2014; 6(12): 6974-83. [http://dx.doi.org/10.1039/c3nr06047k] [PMID: 24835568]

[108] Bussy C, Al-Jamal KT, Boczkowski J, *et al.* Microglia determine brain region-specific neurotoxic responses to chemically functionalized carbon nanotubes. ACS Nano 2015; 9(8): 7815-30. [http://dx.doi.org/10.1021/acsnano.5b02358] [PMID: 26043308]

**CHAPTER 9**

# Carbon Nanotubes and its Toxicity

**Suman Lata**[1,*], **Bhawana Sharma**[1], **C.S. Pundir**[1] and **Rajesh Dabur**[1]

[1] *Department of Biochemistry, Maharshi Dayanand University University in Rohtak, 124001; Haryana, India*

**Abstract:** Many researchers are fascinated by carbon nanotubes (CNTs) as it is one of the most exploited and unique engineered nano-materials in biomedical application. Inspite fascinating properties, CNTs was also questioned for their biocompatibility. CNTs have remarkable properties such as small size, conductivity, mechanical strength and surface chemistry. Due to these properties, CNTs are exploited in biomedical applications therefore there is a need for the assessment of toxicity of CNTs. Researchers monitor the toxicity level of CNTS both *in vitro* and *in vivo*. Various tests are available to determine the cytotoxicity level of CNTs. This chapter describes the various aspects of toxicity related to CNTs.

**Keywords:** Cellular toxicity analysis, CNTs, *In vivo* and *in vitro* toxicity of CNTs, Mechanism of CNT toxicity, Toxicity.

## 9.1. INTRODUCTION

The demand of engineered nano-materials is swiftly growing up due to their application not only in material sciences but also into pharmaceutical, chemical and bioelectronics fields, because of the high specific surface area of these nano-materials, their conductivity, magnetic susceptibility and catalytic activity [1, 2]. Out of these nano-materials, carbon nanotubes (CNTs) are the most widely used that form the foremost structural block of this latest technology. CNTs have distinctive thermal, mechanical, and electrical properties, with potential wide application in the, computer, aerospace and other electronics industries. CNTs are present in two major form, single walled (SWCNTs) and multi walled (MWCNTs), which can be further modified further based on various kinds of functionalisation. Though these materials are widely used, there is a big concern because of their occupational inhalation exposure hazard [3 - 5]. These are very light in weight so, these can be suspended as a particulate matter (PM) of respirable size and can produce inflammation, epithelioid granulomas

* **Corresponding author Suman Lata**: Bionanotechnology Department; Guru Jambheshwar University of Science and Technology, Hisar, 125001; Haryana, India; E-mail: dr.sumanrajesh@gmail.com

**Jagriti Narang & Chandra Shekhar Pundir (Eds.)**

(microscopic nodules), fibrosis, and biochemical/toxicological changes in different organs. From various studies, the ability of SWCNTs was examined that distress cell viability, proliferation, reactive oxygen species (ROS) producing ability, genotoxicity, extra cellular matrix remodelling/tissue remodelling and mutagenicity [6 - 9]. It has been found out that the manufacturing and combustion sources of MWCNTs are the main cause of air pollution in the environment that could have adverse effect on human health which led to various cardiopulmonary diseases. The National Institute for Occupational Safety and Health (NIOSH) has recommended a coverage perimeter of 1$\mu g/m^3$ with other safety guidelines [10 - 15]. The CNTs were found to enhance pathological changes in the lungs that produce respiratory function impairments, atherosclerotic lesions in the brachiocephalic artery of the heart, damage to the mitochondrial DNA in aorta and it can increase the percent of aortic plaque formation [16 - 21].

Hence, this chapter will provide the valuable brief description for theCNTs nano-toxicity studies with an emphasis on cellular toxicity analysis, their mechanism and its risk assessment with reference to human health. The description of the current *in vitro* and *in vivo* toxicity analysis with dosage-dependant analysis of CNTs on different human cell lines has also been discussed [22].

## 9.2. CELLULAR TOXICITY ANALYSIS

The CNTs, carbon nanofibers (CNFs), and carbon nanoparticles (CNPs) were studied for its cellular toxicity *in vitro* and *in vivo* on various cancerous cells [23]. The toxic effects of CNTs depend on the size, shape, various modifications and functionalization (Fig. **1**). The carbon nano based materials were studied for their cellular toxicity as a function of their aspect ratio and surface chemistry.

SWCNTs have been highly studied and considered more toxic as compared to MWCNTs owing to their smaller size, but recent studies show comparable toxicity varying with size and functionalization [24 - 26]. It has been reported that the hazardous effects were dependent on their size and the toxicity level gets enhanced when the surface of the particles is functionalized after an acid treatment [27]. Fiber-shape particles of CNTs are similar to asbestos and their activities are thus supposed to be similar.

The size and the fibrous form of asbestos produce a strong influence on their toxicity. The carbon nanotubes have been employed for the modification of various biomolecules such as DNA and proteins, and used for the drug delivery system that could be a potential health hazard [28 - 32]. A number of studies evaluate that the carbon based nano-particles (CBNPs) showed cytotoxic and cell

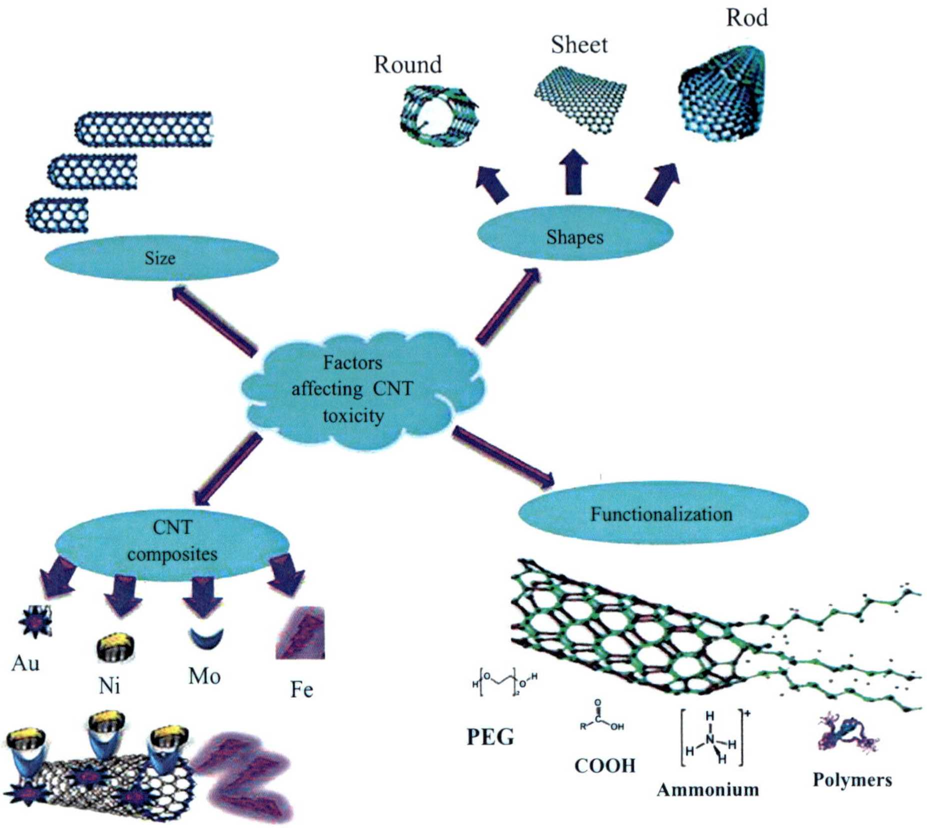


**Fig. (1).** Schematic representation of various factors affecting CNTs.

proliferation on different human cell lines (*e.g.* H596, H446, and Calu-1). The possibility for this toxicity could be the presence of dangling bonds, which are highly reactive sites. These bonds have high density in carbon black, while in CNTs these are present at the lattice end caps. From the experimental findings it has been reported that the CNTs are comparatively less hazardous than that of carbonyl (CdO), carboxyl (COOH), and/ or hydroxyl (OH) groups functionalized carbon fibers and carbon nanoparticles [33 - 40].

Therefore, toxicity of CNT and their derivatives becomes a major issue in the cases where it has been injected into human body for drug delivery [41 - 43].

### 9.2.1. *In vivo* Toxicity of CNTs

There is great variation in the physicochemical properties of engineered nanoparticles from particles with equivalent composition but larger size. A number of studies have assessed the *in vivo* toxicity of CNTs regarding their possible effects on human health. According to Oberdorster *et al.*, 2005 exposure to some engineered nanoparticles especially single-walled carbon nanotubes (SWCNT) leads to unsympathetic health effects, predominantly in the elderly or people with compromised cardiovascular problems [44 - 49]. It has been revealed that MWCNTs can persuade genotoxic effects in lung epithelial cells and have the possibility to induce a strong inflammatory and fibrotic reaction.

Several *in vivo* observations confirmed that both SWCNTs and MWCNTs both were capable to persuade cytotoxicity and apoptosis in diverse cell types and they produce toxic responses in multiple organ systems [50 - 53]. From the studies of Koyama *et al.*, 2009 and Murray *et al.*, 2009 it has been found that the toxicity of CNTs increases, especially when it combine with metal (mainly iron) after interacting with skin that initiate the oxidative stress which causes depletion of glutathione and an increase of dermal cell number with localized alopecia and skin thickening. A number of organs can be affected by the exposure of carbon nano-materials. One of the finding by Mercer *et al.*, (2010) showed that when CNTs gets aggregate, their mobility and bioavailability increase the general cytotoxicity or temporary organ injury. Similarly the asbestos deposited in the alveolar layer and reaches the subpleura after injected through lungs. Therefore, lungs are the main route of exposure and inhalation as well as the most relevant method for determining the toxicity of carbon nano-materials [54 - 57].

A number of studies of carbon nano-materials cytotoxicity demonstrated that the high dose and their site of exposure leads to dose-dependent granulomas, fibrosis or animal death. The impact of lethality was found to be associated with agglomeration of these nano-materials in the airways of major experimental rats while not the inherent toxicity of CNTs or metal catalyst [58, 59]. Davoren *et al.*, 2007 tested different % concentrations of SWCNT (400 or 800 lg/ml) and concluded that the higher concentration of metals (10% of iron) in human lung epithelial cell line was found to be cytotoxic.

A few investigators have examined the grievous effects of nano-materials on cultured mammalian cells for their cell viability by using MTT assay [60 - 63]. The possible reason for the cytotoxic effect was found to be increased in intracellular ROS production by the catalytic activity of iron that decrease the adhesive ability and apoptosis in human embryo kidney cells. It has been found that these indications of apoptosis increased with increase in exposure time as well as concentration of nano-materials.The mechanism of action of CNTs has been depicted in Fig. (**2**).

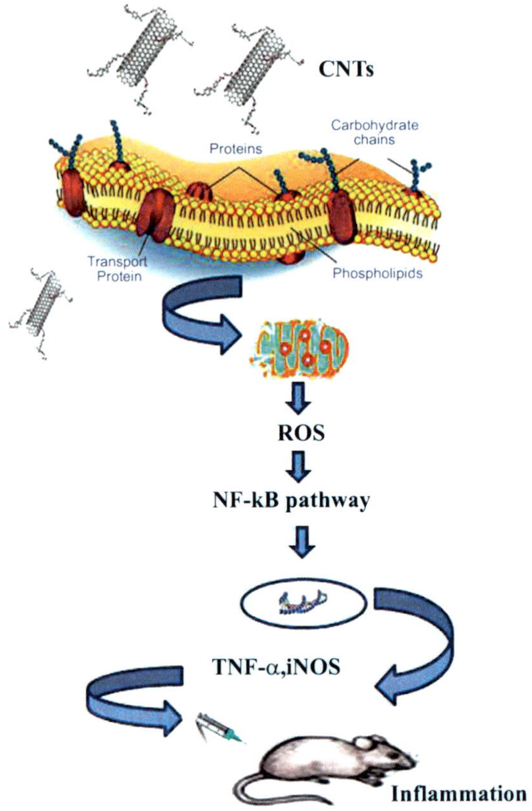


**Fig. (2).** Possible mechanism of action of CNTs.

### 9.2.2. Evaluation of Carbon Nanotubes Toxicity *In Vitro*

Persuade or realistic toxic effects of CNTs and their derivatives on different cells are crucial focus in evaluating and understanding CNT biocompatibility *versus* cytotoxicity.These smart nano-materials are most widely used for the diagnosis and therapeutic purpose for which focus has to be taken for their interaction and uptake by different routes, effects on different biological processes such as cytokines production, cell signaling, perturbations of membrane, chemokines, reactive oxygen species (ROS), and apoptosis [64]. Several types of cell lines (human macrophage cells, A549 cells, a human lung cell, human embryo kidney cells (HEK293, human epithelial-like lung cell line) have been used for the testing of CNTs toxicity. It has been reported from various studies that carbon nanotubes and their derivatives seem to be more toxic than that of other metal oxide nanoparticles. CNTs especially when bound with Nickel it has been found that they influence more cytotoxic effect due to the change in physic chemical behaviour of CNTs and those nanotubes called "metal compounded CNTs".

The studies showed that CNTs are more cytotoxic than quartz or fullerenes and the sign of cytotoxicity of SWCNTs seems to be higher than that of MWCNTs. It has been concluded by the various studies performed and evaluated on SWCNTs by varying percentages of catalysts and other impurities [7, 43, 61].

### 9.2.3. Toxicity of CNTs on Other Species

Leaching of hazardous materials or release of nano-materials (CNTs,CNFs or CNPs) may be accidentally or incidentally from a variety of products have been considered to have a potential mobility and therefore impacts in and across water, soil, air and biota released to the environment at different stages of their life cycle that not only to the humans but also to the ecosystem health.The risk of exposure to nano-materials (CNTs) is particularly to the, as it acts as a sink for most environmental contaminants.For example, the 10-day exposure of rainbow trout to SWCNTs was found to cause respiratory toxicity, including histopathological changes (hyperplasia, altered mucocytes and oedema) in the gills [65].

It has been reported that various environmental factors like temperature, pH, and interactions with natural substances may alter the surface chemistry of CNTs that are released into the environment may be more complicated than lab studies, because of the complexity of ecosystems.

### 9.2.4. CNT Toxicity on Biomacromolecules

In a biological fluid the biomolecules (proteins) are the first to be come in contact with CNTs.

The interaction of protein with carbon nanomaterials (CNTs,CNFs,CNPs) would be enormous, but few protein (*e.g.* serum albumin, enzymes) could be damaged by CNTs and lose or alter their activity and influence the performance of the organism. It has been found that bovine serum albumin non-covalently interact to the surface of cMWCNTs at the cost of partly losing its secondary and tertiary structure through electrostatic interaction.

Mu and Colleagues have reported that the electrostatic and stereo-chemical properties of both nanotubes and proteins govern nanotubes -protein binding and Oberdoster *et al.*, 2004, observed that CNTs can enter the body *via* some other route like translocation of inhaled nanoparticles to the brain through the olfactory neuronal pathway and may affect the brain tissues. Gua *et al.*, 2008, reported that the presence of additional, non-essential amino acids (phenylalanine) and vitamins like folate may passively adsorb onto CNTs and leads to artifactual cell death [52, 62 - 65]. It has been reported that the penetration of nano-particles is an energy- independent, passive process, where CNTs diffuses across the cellular membrane.On the other hand, fewer efforts have been made in determining the exact mechanism of action of CNTs at molecular level. However, in this context, nano-toxicogenomics may prove beneficial upcoming field correlating the gene expression of cells or tissues, which are exposed to CNTs by using c-DNA microarray techniques. This becomes much more important as micro RNA regulation and geno-toxicity are found to be factors affecting CNTs action. 2D electrophoresis (proteomics) and mass spectrometry (metabolomics) techniques may provide the new insight for complete understanding of biological consequences prompted by CNTs. Therefore, there is an emergent need of complete understanding of CNTs mechanistic approach, which can further be helpful in resolving the safety and risk measures of CNTs.

CNTs have been reported to be least biodegradable artificially synthesized material. They tend to get accumulate with their increased toxic concentrations in the food chain because of their ability to remain inside the organism and lipophilic nature [66].

## 9.3. MECHANISM OF CNT TOXICITY

It has been reported that the CNTs cause toxic interaction that produce impairment to the cells by activating various pathways at once, that generally involving DNA damage. According to Jacoban *et al.*, 2008, the exposed concentration of ~25 μg/cm$^2$ of SWCNT to the mesothelial cells could activate DNA recovery along with changes in the cell cycle and generation of apoptotic signals. From the studies of Ulrich *et al.*, 1999 it has been observed that CNTs (96 μg SWCNT/cm$^2$)/DNA interaction was the preferred route of toxicity in a 3- hour incubation which induced DNA damage (through micronucleus generation) in lung fibroblasts [66 - 68]. Lacerda *et al.*, 2008, in their study observed that the cytotoxic effect of pristine SWCNTs in the spleen, liver, and lungs was the indicator for oxidative stress due to dose- dependent manner and it was reported that the stress remained comparatively invariable in the spleen as the nanotubes dosage increased.

## 9.4. RISK ASSESSMENT TO HUMAN HEALTH

Recently, more and more, the attention of the community of nano-scientists has been focusing on technological applications. The CNTs have been used very widely now a days in various fields, mainly for their capability to be functionalized, both covalently as well as non-covalently, with various moieties, and their aspect ratio and geometry, which enables them to penetrate into the cell. Therefore, it is essential to assess its toxicity risk, especially to human health, which is complex and intricate process. Fig. (**3**) shows the possible hazardous impacts of CNTs on human health. The CNTs are highly light weight and could be potentially exposed to the lungs through route of inhalation. By the emergence of toxicological data sets, the health hazardous risk assessment of these materials could be identified [69, 70].

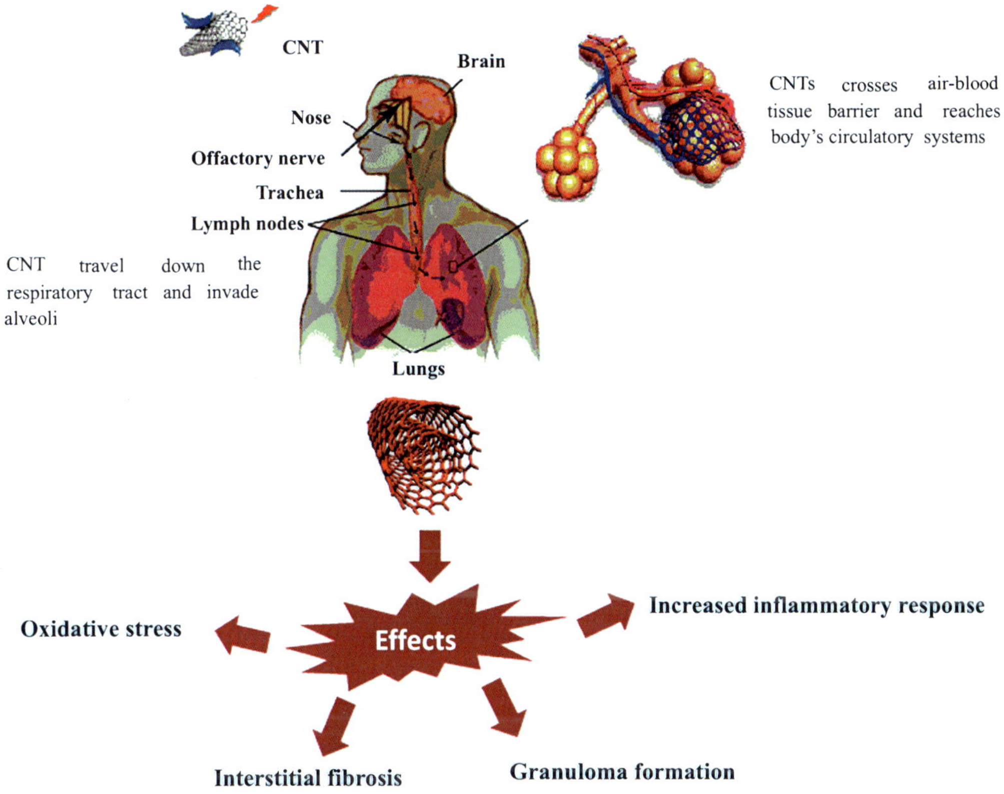


**Fig. (3).** Possible *in vivo* effects of CNTs on human health.

### 9.4.1. Identification of Hazard

Various researches on different animal groups indicates that the CNTs showed pulmonary toxicity and produce numerous biochemical cytotoxic effects like epithelioid granulomas, inflammation, fibrosis and change in the lungs that might impair pulmonary functions. The experimental studies of cell culture demonstrated that CNTs particulates can reach to the lungs in sufficient quantity after exposure directly or indirectly to work place and produce a cytotoxic damaging response [51 - 55].

### 9.4.2. Response of Dose

The quantitative and morphometric studies of Shvedova *et al.*, 2005, on mice revealed that the total mass of granulomas increase with increase in treatment doses of CNTs for 60 days. Similarly, the data result of Muller *et al.*, 2005 also reported a dose dependent increase in hydroxyproline and collagen contents in the lungs of rats treated with increase in CNTs content. Lam *et al.*, 2004, and Shvedova *et al.*, 2005 both have demonstrated that the doses and duration of time of CNTs played a very important role on the cytotoxic effect. It has been found that the extent of fibrosis and the biomarkers of toxicity in mice had focally severe foreign-body reactions that have been characterized by extensively major foci of particle-laden macrophages and giant cells, and prominent peri-bronchial and perivascular lymphocytic infiltrates and the lesions in mouse lungs became worse with time. It was found that the high doses (0.0029 mg of CNTs) and long duration exposure (8 hours) killed some mice immediately after the instillation of CNT dust that could leads to the mechanical blockage of the airways in mice lungs [44 - 49].

### 9.4.3. CNTs Exposure Risk Assessment

The observations of Baron *et al.*, 2003 showed that CNTs are more toxic than quartz, which are considered to be more hazardous in chronic inhalation exposure. Because CNTs are neither soluble nor biodegradable nanoparticles, these get accumulated in the interstitium for prolonged period. The CNTs are present in the clumping form and it has been found that the exposed risk depend upon the respirable size of CNTs that has been dissociated from the bulk amount of clumping CNTs [36, 45, 46]. CNTs being non-biodegradable remains in the lungs for prolonged periods could be removed by the macrophage-mucociliary clearance mechanism [53 - 55].

## 9.5. CONCLUSION

In the present chapter, we describe the over view of cytotoxic effect of CNTs being a most promising and unique engineered nano-materials. There is a need for the assessment of toxicity of CNTs at various biological levels such as cells, organ, or on an entire organism, so that nano-toxicity mechanisms can be explained and the safe and sound use of CNTs can be achieved. Though a lot of efforts have been made to cautiously explore the *in vitro* and *in vivo* toxicity of CNTs, researchers still fail to attain consensus on the toxicity of CNTs and their mechanism of action. Among the various uses from Carbon nano-materials developed for cancer treatment showed a good quality application in overcoming the issues of deleterious side effects confronted in traditional cancer therapy. There are some safeties issues which must be considered before using these nano-materials. There should be some set of standard reference nano-materials, so that the potential risk of any given nanomaterial could be compared with the known nano-materials concentration and standard experimental protocols should be there to compare the experimental studies so that the nano-materials could be used safely for different biological applications.

## CONFLICT OF INTEREST

The authors declare no conflict of interest, financial or otherwise.

## ACKNOWLEDGEMENTS

Thanks to all scientists referenced throughout the chapter whose valuable work has guided the way through to this research work.

## REFERENCES

[1] Bianco A, Kostarelos K, Partidos CD, Prato M. Biomedical applications of functionalised carbon nanotubes. Chem Commun (Camb) 2005; (5): 571-7.
[http://dx.doi.org/10.1039/b410943k] [PMID: 15672140]

[2] Sayes CM, Liang F, Hudson JL, *et al.* Functionalization density dependence of single-walled carbon nanotubes cytotoxicity in vitro. Toxicology letters 2006; 161(2): 135-42.

[3] Muller J, Huaux F, Moreau N, *et al.* Respiratory toxicity of multi-wall carbon nanotubes. Toxicol Appl Pharmacol 2005; 207(3): 221-31.
[http://dx.doi.org/10.1016/j.taap.2005.01.008] [PMID: 16129115]

[4] Nel A, Xia T, Mädler L, Li N. Toxic potential of materials at the nanolevel. Science 2006; 311(5761): 622-7.
[http://dx.doi.org/10.1126/science.1114397] [PMID: 16456071]

[5] Kipen HM, Laskin DL. Smaller is not always better: nanotechnology yields nanotoxicology. Am J Physiol Lung Cell Mol Physiol 2005; 289(5): L696-7.
[http://dx.doi.org/10.1152/ajplung.00277.2005] [PMID: 16214820]

[6] Lam CW, James JT, McCluskey R, Arepalli S, Hunter RL. A review of carbon nanotube toxicity and assessment of potential occupational and environmental health risks. Crit Rev Toxicol 2006; 36(3): 189-217.
[http://dx.doi.org/10.1080/10408440600570233] [PMID: 16686422]

[7] Lam CW, James JT, McCluskey R, Hunter RL. Pulmonary toxicity of single-wall carbon nanotubes in mice 7 and 90 days after intratracheal instillation. Toxicol Sci 2004; 77(1): 126-34.
[http://dx.doi.org/10.1093/toxsci/kfg243] [PMID: 14514958]

[8] Li N, Kim S, Wang M, Froines J, Sioutas C, Nel A. Use of a stratified oxidative stress model to study the biological effects of ambient concentrated and diesel exhaust particulate matter. Inhal Toxicol 2002; 14(5): 459-86.
[http://dx.doi.org/10.1080/089583701753678571] [PMID: 12028803]

[9] Magrez A, Kasas S, Salicio V, *et al.* Cellular toxicity of carbon-based nanomaterials. Nano Lett 2006; 6(6): 1121-5.
[http://dx.doi.org/10.1021/nl060162e] [PMID: 16771565]

[10] Ullrich SE, Lyons HJ. Mechanisms involved in the immunotoxicity induced by dermal application of JP-8 jet fuel. Toxicol Sci 2000; 58(2): 290-8.
[http://dx.doi.org/10.1093/toxsci/58.2.290] [PMID: 11099641]

[11] Warheit DB, Laurence BR, Reed KL, Roach DH, Reynolds GA, Webb TR. Comparative pulmonary toxicity assessment of single-wall carbon nanotubes in rats. Toxicol Sci 2004; 77(1): 117-25.
[http://dx.doi.org/10.1093/toxsci/kfg228] [PMID: 14514968]

[12] Lam C, James JT, McCluskey R, Hunter RL. Pulmonary toxicity of carbon nanotubes in mice 7 and 90 days after intratracheal instillation. 2003. [http://dx.doi.org/10.1093/toxsci/kfg243]

[13] Shvedova AA, Castranova V, Kisin ER, *et al.* Exposure to carbon nanotube material: assessment of nanotube cytotoxicity using human keratinocyte cells. J Toxicol Environ Health A 2003; 66(20): 1909-26. [http://dx.doi.org/10.1080/713853956] [PMID: 14514433]

[14] Warheit DB, Hansen JF, Yuen IS, Kelly DP, Snajdr SI, Hartsky MA. Inhalation of high concentrations of low toxicity dusts in rats results in impaired pulmonary clearance mechanisms and persistent inflammation. Toxicol Appl Pharmacol 1997; 145(1): 10-22. [http://dx.doi.org/10.1006/taap.1997.8102] [PMID: 9221819]

[15] Driscoll KE, Costa DL, Hatch G, *et al.* Intratracheal instillation as an exposure technique for the evaluation of respiratory tract toxicity: uses and limitations. Toxicol Sci 2000; 55(1): 24-35. [http://dx.doi.org/10.1093/toxsci/55.1.24] [PMID: 10788556]

[16] Handy RD, Sims DW, Giles A, Campbell HA, Musonda MM. Metabolic trade-off between locomotion and detoxification for maintenance of blood chemistry and growth parameters by rainbow trout (Oncorhynchus mykiss) during chronic dietary exposure to copper. Aquat Toxicol 1999; 47(1): 23-41. [http://dx.doi.org/10.1016/S0166-445X(99)00004-1]

[17] Kammer F, Baborowski M, Friese K. Field-flow fractionation coupled to multi-angle laser light scattering detectors: Applicability and analytical benefits for the analysis of environmental colloids. Anal Chim Acta 2005; 552(1): 166-74. [http://dx.doi.org/10.1016/j.aca.2005.07.049]

[18] Oberdörster E. Manufactured nanomaterials (fullerenes, C60) induce oxidative stress in the brain of juvenile largemouth bass. Environ Health Perspect 2004; 112(10): 1058-62. [http://dx.doi.org/10.1289/ehp.7021] [PMID: 15238277]

[19] Oberdorster E, Zhu S, Blickley TM, McClellan-Green P, Haasch ML. Ecotoxicology of carbon-based engineered nanoparticles: effects of fullerene (C 60) on aquatic organisms. Carbon 2006; 44(6): 1112-20. [http://dx.doi.org/10.1016/j.carbon.2005.11.008]

[20] Shvedova AA, Kisin ER, Mercer R, *et al.* Unusual inflammatory and fibrogenic pulmonary responses to single-walled carbon nanotubes in mice. Am J Physiol Lung Cell Mol Physiol 2005; 289(5): L698-708. [http://dx.doi.org/10.1152/ajplung.00084.2005] [PMID: 15951334]

[21] Warheit DB, Laurence BR, Reed KL, Roach DH, Reynolds GA, Webb TR. Comparative pulmonary toxicity assessment of single-wall carbon nanotubes in rats. Toxicol Sci 2004; 77(1): 117-25. [http://dx.doi.org/10.1093/toxsci/kfg228] [PMID: 14514968]

[22] Roberts AP, Mount AS, Seda B, *et al. In vivo* biomodification of lipid-coated carbon nanotubes by *Daphnia magna.* Environ Sci Technol 2007; 41(8): 3025-9. [http://dx.doi.org/10.1021/es062572a] [PMID: 17533874]

[23] Zhang H, Li HX, Cheng HM. Water-soluble multiwalled carbon nanotubes functionalized with sulfonated polyaniline. J Phys Chem B 2006; 110(18): 9095-9. [http://dx.doi.org/10.1021/jp060193y] [PMID: 16671720]

[24] Shi Kam NW, Jessop TC, Wender PA, Dai H. Nanotube molecular transporters: internalization of carbon nanotube-protein conjugates into Mammalian cells. J Am Chem Soc 2004; 126(22): 6850-1. [http://dx.doi.org/10.1021/ja0486059] [PMID: 15174838]

[25] Cai D, Mataraza JM, Qin ZH, *et al.* Highly efficient molecular delivery into mammalian cells using carbon nanotube spearing. Nat Methods 2005; 2(6): 449-54.

[http://dx.doi.org/10.1038/nmeth761] [PMID: 15908924]

[26] Pacurari M, Yin XJ, Zhao J, *et al.* Raw single-wall carbon nanotubes induce oxidative stress and activate MAPKs, AP-1, NF-kappaB, and Akt in normal and malignant human mesothelial cells. Environ Health Perspect 2008; 116(9): 1211-7.
[http://dx.doi.org/10.1289/ehp.10924] [PMID: 18795165]

[27] Yang ST, Wang X, Jia G, *et al.* Long-term accumulation and low toxicity of single-walled carbon nanotubes in intravenously exposed mice. Toxicol Lett 2008; 181(3): 182-9.
[http://dx.doi.org/10.1016/j.toxlet.2008.07.020] [PMID: 18760340]

[28] Yang H, Liu C, Yang D, Zhang H, Xi Z. Comparative study of cytotoxicity, oxidative stress and genotoxicity induced by four typical nanomaterials: the role of particle size, shape and composition. J Appl Toxicol 2009; 29(1): 69-78.
[http://dx.doi.org/10.1002/jat.1385] [PMID: 18756589]

[29] Raja PM, Connolley J, Ganesan GP, *et al.* Impact of carbon nanotube exposure, dosage and aggregation on smooth muscle cells. Toxicol Lett 2007; 169(1): 51-63.
[http://dx.doi.org/10.1016/j.toxlet.2006.12.003] [PMID: 17275220]

[30] Liu Z, Davis C, Cai W, He L, Chen X, Dai H. Circulation and long-term fate of functionalized, biocompatible single-walled carbon nanotubes in mice probed by Raman spectroscopy. Proc Natl Acad Sci USA 2008; 105(5): 1410-5.
[http://dx.doi.org/10.1073/pnas.0707654105] [PMID: 18230737]

[31] Guo J, Zhang X, Li Q, Li W. Biodistribution of functionalized multiwall carbon nanotubes in mice. Nucl Med Biol 2007; 34(5): 579-83.
[http://dx.doi.org/10.1016/j.nucmedbio.2007.03.003] [PMID: 17591558]

[32] Dobson J. Gene therapy progress and prospects: magnetic nanoparticle-based gene delivery. Gene Ther 2006; 13(4): 283-7.
[http://dx.doi.org/10.1038/sj.gt.3302720] [PMID: 16462855]

[33] Murthy N, Thng YX, Schuck S, Xu MC, Fréchet JM. A novel strategy for encapsulation and release of proteins: hydrogels and microgels with acid-labile acetal cross-linkers. J Am Chem Soc 2002; 124(42): 12398-9.
[http://dx.doi.org/10.1021/ja026925r] [PMID: 12381166]

[34] Mitchell DT, Lee SB, Trofin L, *et al.* Smart nanotubes for bioseparations and biocatalysis. J Am Chem Soc 2002; 124(40): 11864-5.
[http://dx.doi.org/10.1021/ja027247b] [PMID: 12358526]

[35] Wang H, Wang J, Deng X, *et al.* Biodistribution of carbon single-wall carbon nanotubes in mice. J Nanosci Nanotechnol 2004; 4(8): 1019-24.
[http://dx.doi.org/10.1166/jnn.2004.146] [PMID: 15656196]

[36] Cui D, Tian F, Ozkan CS, Wang M, Gao H. Effect of single wall carbon nanotubes on human HEK293 cells. Toxicol Lett 2005; 155(1): 73-85.
[http://dx.doi.org/10.1016/j.toxlet.2004.08.015] [PMID: 15585362]

[37] Tkac J, Whittaker JW, Ruzgas T. The use of single walled carbon nanotubes dispersed in a chitosan matrix for preparation of a galactose biosensor. Biosens Bioelectron 2007; 22(8): 1820-4.
[http://dx.doi.org/10.1016/j.bios.2006.08.014] [PMID: 16973345]

[38] Guldi DM, Rahman GM, Jux N, *et al.* Functional single-wall carbon nanotube nanohybrids--associating SWNTs with water-soluble enzyme model systems. J Am Chem Soc 2005; 127(27): 9830-8.
[http://dx.doi.org/10.1021/ja050930o] [PMID: 15998087]

[39] Zhao B, Hu H, Yu A, Perea D, Haddon RC. Synthesis and characterization of water soluble single-walled carbon nanotube graft copolymers. J Am Chem Soc 2005; 127(22): 8197-203.
[http://dx.doi.org/10.1021/ja042924i] [PMID: 15926849]

[40] Asuri P, Bale SS, Pangule RC, Shah DA, Kane RS, Dordick JS. Structure, function, and stability of enzymes covalently attached to single-walled carbon nanotubes. Langmuir 2007; 23(24): 12318-21. [http://dx.doi.org/10.1021/la702091c] [PMID: 17944500]

[41] Wang X, Deng XY, Wang HF, *et al.* [Bio-effects of water soluble taurine multi-wall carbon nanotubes on lungs of mice]. Zhonghua Yu Fang Yi Xue Za Zhi 2007; 41(2): 85-90. [PMID: 17605231]

[42] Sayes CM, Fortner JD, Guo W, *et al.* The differential cytotoxicity of water-soluble fullerenes. Nano Lett 2004; 4(10): 1881-7. [http://dx.doi.org/10.1021/nl0489586]

[43] Tian F, Cui D, Schwarz H, Estrada GG, Kobayashi H. Cytotoxicity of single-wall carbon nanotubes on human fibroblasts. Toxicol In Vitro 2006; 20(7): 1202-12. [http://dx.doi.org/10.1016/j.tiv.2006.03.008] [PMID: 16697548]

[44] Warheit DB, Laurence BR, Reed KL, Roach DH, Reynolds GA, Webb TR. Comparative pulmonary toxicity assessment of single-wall carbon nanotubes in rats. Toxicol Sci 2004; 77(1): 117-25. [http://dx.doi.org/10.1093/toxsci/kfg228] [PMID: 14514968]

[45] Jia G, Wang H, Yan L, *et al.* Cytotoxicity of carbon nanomaterials: single-wall nanotube, multi-wall nanotube, and fullerene. Environ Sci Technol 2005; 39(5): 1378-83. [http://dx.doi.org/10.1021/es048729l] [PMID: 15787380]

[46] Monteiro-Riviere NA, Inman AO. Challenges for assessing carbon nanomaterial toxicity to the skin. Carbon 2006; 44(6): 1070-8. [http://dx.doi.org/10.1016/j.carbon.2005.11.004]

[47] Smart SK, Cassady AI, Lu GQ, Martin DJ. The biocompatibility of carbon nanotubes. Carbon 2006; 44(6): 1034-47. [http://dx.doi.org/10.1016/j.carbon.2005.10.011]

[48] Kong J, Cassell AM, Dai H. Chemical vapor deposition of methane for single-walled carbon nanotubes. Chem Phys Lett 1998; 292(4): 567-74. [http://dx.doi.org/10.1016/S0009-2614(98)00745-3]

[49] Jia G, Wang H, Yan L, *et al.* Cytotoxicity of carbon nanomaterials: single-wall nanotube, multi-wall nanotube, and fullerene. Environ Sci Technol 2005; 39(5): 1378-83. [http://dx.doi.org/10.1021/es048729l] [PMID: 15787380]

[50] McClellan RO. Human health risk assessment: a historical overview and alternative paths forward. Inhal Toxicol 1999; 11(6-7): 477-518. [http://dx.doi.org/10.1080/089583799196880] [PMID: 10380179]

[51] Pope CA III, Burnett RT, Thun MJ, *et al.* Lung cancer, cardiopulmonary mortality, and long-term exposure to fine particulate air pollution. JAMA 2002; 287(9): 1132-41. [http://dx.doi.org/10.1001/jama.287.9.1132] [PMID: 11879110]

[52] Salvetat JP, Bonard JM, Thomson NH, *et al.* Mechanical properties of carbon nanotubes. Appl Phys, A Mater Sci Process 1999; 69(3): 255-60. [http://dx.doi.org/10.1007/s003390050999]

[53] Schleh C, Hohlfeld JM. Interaction of nanoparticles with the pulmonary surfactant system 2009. [http://dx.doi.org/10.1080/08958370903005744]

[54] Morrow PE. Possible mechanisms to explain dust overloading of the lungs. Fundam Appl Toxicol 1988; 10(3): 369-84. [http://dx.doi.org/10.1016/0272-0590(88)90284-9] [PMID: 3286345]

[55] Baron PA, Deye GJ, Chen BT, Schwegler-Berry DE, Shvedova AA, Castranova V. Aerosolization of single-walled carbon nanotubes for an inhalation study. Inhal Toxicol 2008; 20(8): 751-60. [http://dx.doi.org/10.1080/08958370801975303] [PMID: 18569097]

[56] Bottini M, Bruckner S, Nika K, *et al.* Multi-walled carbon nanotubes induce T lymphocyte apoptosis. Toxicol Lett 2006; 160(2): 121-6. [http://dx.doi.org/10.1016/j.toxlet.2005.06.020] [PMID: 16125885]

[57] Bruckner S, Rhamouni S, Tautz L, *et al.* Yersinia phosphatase induces mitochondrially dependent apoptosis of T cells. J Biol Chem 2005; 280(11): 10388-94. [http://dx.doi.org/10.1074/jbc.M408829200] [PMID: 15632192]

[58] Iijima S. Helical microtubules of graphitic carbon. Nature 1991; 354(6348): 56-8. [http://dx.doi.org/10.1038/354056a0]

[59] Shi Kam NW, Jessop TC, Wender PA, Dai H. Nanotube molecular transporters: internalization of carbon nanotube-protein conjugates into Mammalian cells. J Am Chem Soc 2004; 126(22): 6850-1. [http://dx.doi.org/10.1021/ja0486059] [PMID: 15174838]

[60] Baughman RH, Zakhidov AA, de Heer WA. Carbon nanotubes--the route toward applications. Science 2002; 297(5582): 787-92. [http://dx.doi.org/10.1126/science.1060928] [PMID: 12161643]

[61] Dybdahl M, Risom L, Bornholdt J, Autrup H, Loft S, Wallin H. Inflammatory and genotoxic effects of diesel particles *in vitro* and *in vivo*. Mutat Res 2004; 562(1-2): 119-31. [http://dx.doi.org/10.1016/j.mrgentox.2004.05.010] [PMID: 15279835]

[62] Fiorito S, Serafino A, Andreola F, Togna A, Togna G. Toxicity and biocompatibility of carbon nanoparticles. J Nanosci Nanotechnol 2006; 6(3): 591-9. [http://dx.doi.org/10.1166/jnn.2006.125] [PMID: 16573109]

[63] Warheit DB, Laurence BR, Reed KL, Roach DH, Reynolds GA, Webb TR. Comparative pulmonary toxicity assessment of single-wall carbon nanotubes in rats. Toxicol Sci 2004; 77(1): 117-25. [http://dx.doi.org/10.1093/toxsci/kfg228] [PMID: 14514968]

[64] Shvedova AA, Kisin ER, Mercer R, *et al.* Unusual inflammatory and fibrogenic pulmonary responses to single-walled carbon nanotubes in mice. Am J Physiol Lung Cell Mol Physiol 2005; 289(5): L698-708. [http://dx.doi.org/10.1152/ajplung.00084.2005] [PMID: 15951334]

[65] Tian F, Cui D, Schwarz H, Estrada GG, Kobayashi H. Cytotoxicity of single-wall carbon nanotubes on human fibroblasts. Toxicol In Vitro 2006; 20(7): 1202-12. [http://dx.doi.org/10.1016/j.tiv.2006.03.008] [PMID: 16697548]

[66] Bahr JL, Tour JM. Covalent chemistry of single-wall carbon nanotubes. J Mater Chem 2002; 12(7): 1952-8. [http://dx.doi.org/10.1039/b201013p]

[67] Jia G, Wang H, Yan L, *et al.* Cytotoxicity of carbon nanomaterials: single-wall nanotube, multi-wall nanotube, and fullerene. Environ Sci Technol 2005; 39(5): 1378-83. [http://dx.doi.org/10.1021/es048729l] [PMID: 15787380]

[68] Warheit DB, Laurence BR, Reed KL, Roach DH, Reynolds GA, Webb TR. Comparative pulmonary toxicity assessment of single-wall carbon nanotubes in rats. Toxicol Sci 2004; 77(1): 117-25. [http://dx.doi.org/10.1093/toxsci/kfg228] [PMID: 14514968]

[69] Mutlu GM, Budinger GR, Green AA, *et al.* Biocompatible nanoscale dispersion of single-walled carbon nanotubes minimizes *in vivo* pulmonary toxicity. Nano Lett 2010; 10(5): 1664-70. [http://dx.doi.org/10.1021/nl9042483] [PMID: 20377197]

[70] Baughman RH, Zakhidov AA, de Heer WA. Carbon nanotubes--the route toward applications. Science 2002; 297(5582): 787-92. [http://dx.doi.org/10.1126/science.1060928] [PMID: 12161643]

# SUBJECT INDEX

## A



## B



## C



Jagriti Narang & Chandra Shekhar Pundir (Eds.)

## D



## E

## T



## U



## W

Made in United States
North Haven, CT
04 November 2026

21727482R00175